T0321715

NEUROMETHODS ☐ 14

Neurophysiological Techniques

NEUROMETHODS

Program Editors: Alan A. Boulton and Glen B. Baker

1. **General Neurochemical Techniques**
   Edited by *Alan A. Boulton and Glen B. Baker*, 1985
2. **Amines and Their Metabolites**
   Edited by *Alan A. Boulton, Glen B. Baker, and Judith M. Baker*, 1985
3. **Amino Acids**
   Edited by *Alan A. Boulton, Glen B. Baker, and James D. Wood*, 1985
4. **Receptor Binding Techniques**
   Edited by *Alan A. Boulton, Glen B. Baker, and Pavel D. Hrdina*, 1986
5. **Neurotransmitter Enzymes**
   Edited by *Alan A. Boulton, Glen B. Baker, and Peter H. Yu*, 1986
6. **Peptides**
   Edited by *Alan A. Boulton, Glen B. Baker, and Quentin Pittman*, 1987
7. **Lipids and Related Compounds**
   Edited by *Alan A. Boulton, Glen B. Baker, and Lloyd A. Horrocks*, 1988
8. **Imaging and Correlative Physicochemical Techniques**
   Edited by *Alan A. Boulton, Glen B. Baker, and Donald P. Boisvert*, 1988
9. **The Neuronal Microenvironment**
   Edited by *Alan A. Boulton, Glen B. Baker, and Wolfgang Walz*, 1988
10. **Analysis of Psychiatric Drugs**
    Edited by *Alan A. Boulton, Glen B. Baker, and Ronald T. Coutts*, 1988
11. **Carbohydrates and Energy Metabolism**
    Edited by *Alan A. Boulton, Glen B. Baker, and Roger F. Butterworth*, 1989
12. **Drugs as Tools in Neurotransmitter Research**
    Edited by *Alan A. Boulton, Glen B. Baker, and Augusto V. Juorio*, 1989
13. **Psychopharmacology**
    Edited by *Alan A. Boulton, Glen B. Baker, and Andrew J. Greenshaw*, 1989
14. **Neurophysiological Techniques: Basic Methods and Concepts**
    Edited by *Alan A. Boulton, Glen B. Baker, and Case H. Vanderwolf*, 1990
15. **Neurophysiological Techniques: Applications to Neural Systems**
    Edited by *Alan A. Boulton, Glen B. Baker, and Case H. Vanderwolf*, 1990
16. **Molecular Neurobiological Techniques**
    Edited by *Alan A. Boulton, Glen B. Baker, and Anthony T. Campagnoni*, 1990
17. **Neuropsychology**
    Edited by *Alan A. Boulton, Glen B. Baker, and Merrill Hiscock*, 1990

NEUROMETHODS

Program Editors: Alan A. Boulton and Glen B. Baker

NEUROMETHODS ☐ 14

Neurophysiological Techniques

Basic Methods and Concepts

Edited by

Alan A. Boulton

University of Saskatchewan, Saskatoon, Canada

Glen B. Baker

University of Alberta, Edmonton, Canada

and

Case H. Vanderwolf

University of Western Ontario, London, Canada

Humana Press • Clifton, New Jersey

Library of Congress Cataloging–in–Publication Data

Neurophysiological techniques. Basic methods and concepts / edited by
Alan A. Boulton, Glen B. Baker, and Case H. Vanderwolf.
p. cm. —(Neuromethods : 14)
Includes bibliographical references and idex.
ISBN 0-89603-160-8
1. Neurophysiology—Research—Methodology. 2. Electrophysiology—Research—Methodology. I. Boulton, A. A. (Alan A.) II. Baker, Glen B., 1947– . III. Vanderwolf, C. H. IV. Series.
[DNLM: 1. Electrophysiology. 2. Neurophysiology—methods. W1 NE337G v. 14 / WL 102 N4958]
QP356.N48284 1990
591.1'88–dc20
DNLM/DLC
for Library of Congress 89–26855
CIP

Crescent Manor
PO Box 2148
Clifton, NJ 07015

Printed in the United States of America

Preface to the Series

When the President of Humana Press first suggested that a series on methods in the neurosciences might be useful, one of us (AAB) was quite skeptical; only after discussions with GBB and some searching both of memory and library shelves did it seem that perhaps the publisher was right. Although some excellent methods books have recently appeared, notably in neuroanatomy, it is a fact that there is a dearth in this particular field, a fact attested to by the alacrity and enthusiasm with which most of the contributors to this series accepted our invitations and suggested additional topics and areas. After a somewhat hesitant start, essentially in the neurochemistry section, the series has grown and will encompass neurochemistry, neuropsychiatry, neurology, neuropathology, neurogenetics, neuroethology, molecular neurobiology, animal models of nervous disease, and no doubt many more "neuros." Although we have tried to include adequate methodological detail and in many cases detailed protocols, we have also tried to include wherever possible a short introductory review of the methods and/or related substances, comparisons with other methods, and the relationship of the substances being analyzed to neurological and psychiatric disorders. Recognizing our own limitations, we have invited a guest editor to join with us on most volumes in order to ensure complete coverage of the field. These editors will add their specialized knowledge and competencies. We anticipate that this series will fill a gap; we can only hope that it will be filled appropriately and with the right amount of expertise with respect to each method, substance or group of substances, and area treated.

Alan A. Boulton
Glen B. Baker

Preface

The development of neurophysiology, the study of the activity of living nervous tissue, has relied heavily on the techniques of electrophysiology. This emphasis is revealed in volumes 14 and 15 of this series, which show how electrophysiological techniques can be applied to research topics ranging from ion channels to human behavior. Kitai and Park show how cellular neurophysiology can be related to classical neuroanatomy, an important basis for any type of functional analysis. Wonderlin, French, Arispe, and Jones describe new (single channel) and more traditional (whole cell) techniques for studying the role of ion channels in cellular processes, a field that is currently developing very rapidly. An exciting nontraditional approach to the study of cellular electrophysiology is discussed by Hopp, Wu, Xiao, Rioult, London, Zecevic, and Cohen in their paper on optic measurement of membrane potentials. Humphrey and Schmidt offer a thoughtful review of the uses and limitations of the technique of recording extracellular unit potentials in the brain. Hoffer presents an introduction to a field that is of great interest but is technically very difficult—the recording from cells and axons in the spinal cord and peripheral nervous system in freely moving animals. An electrophysiological approach to the analysis of the neural mechanisms of normal behavior is presented by Halgren in a wide-ranging review of the field of evoked potentials in humans. The papers by Carlini and Ransom (ion-selective electrodes) and Maidment, Martin, Ford, and Marsden (in vivo voltammetry) describe techniques that could be said to lie at the border between neurophysiology and neurochemistry, and that promise new insights into the dynamics of neurochemically defined classes of neuronal activity. Finally, Leung discusses the increasing use and sophistication of computational techniques in the analysis of neurophysiological data.

Not all readers will agree with my selection of topics. However, I think that all will agree that the contributors have presented excellent discussions of their respective fields that will be of wide and enduring interest. For this, I thank the contributors most heartily.

C. H. Vanderwolf

Contents

RECORDING AND ANALYSIS OF CURRENTS FROM SINGLE ION CHANNELS

William F. Wonderlin, Robert J. French, and Nelson J. Arispe

WHOLE-CELL AND MICROELECTRODE VOLTAGE CLAMP

Stephen W. Jones

MULTISITE OPTICAL MEASUREMENT OF MEMBRANE POTENTIAL

Hans-Peter Höpp, Jian-Young Wu, Chun X. Falk, Marc G. Rioult, Jill A. London, Dejan Zecevic, and Lawrence B. Cohen

FABRICATION AND IMPLEMENTATION OF ION-SELECTIVE MICROELECTRODES

Walter G. Carlini and Bruce R. Ransom

Contributors

NELSON J. ARISPE • *Department of Medical Physiology, University of Calgary, Calgary, Alberta, Canada*

GLEN B. BAKER • *Neurochemical Research Unit, Department of Psychiatry, University of Alberta, Edmonton, Alberta, Canada*

ALAN A. BOULTON • *Neuropsychiatric Research Unit, University of Saskatchewan, Saskatoon, Saskatchewan, Canada*

WALTER G. CARLINI • *Department of Neurology, Yale University School of Medicine, New Haven, Connecticut*

LAWRENCE B. COHEN • *Department of Physiology, Yale University School of Medicine, New Haven, CT*

CHUN X. FALK • *Department of Physiology, Yale University School of Medicine, New Haven, CT*

ANTHONY P. D. W. FORD • *Department of Physiology and Pharmacology, Queen's Medical Center, Nottingham, UK*

ROBERT J. FRENCH • *Department of Medical Physiology, University of Calgary, Calgary, Alberta, Canada*

HANS-PETER HÖPP • *Department of Physiology, Yale University School of Medicine, New Haven, CT*

STEPHEN W. JONES • *Department of Physiology and Biophysics, Case Western Reserve University, Cleveland, Ohio*

STEPHEN T. KITAI • *Department of Anatomy and Neurobiology, University of Tennessee, Memphis, Tennessee*

JILL A. LONDON • *Department of Biostructure and Function, University of Connecticut Health Science Center, Farmington, CT*

NIGEL T. MAIDMENT • *Department of Physiology and Pharmacology, Queen's Medical Centre, Nottingham, UK*

CHARLES A. MARSDEN • *Department of Physiology and Pharmacology, Queen's Medical Centre, Nottingham, UK*

KEITH F. MARTIN • *Department of Physiology and Pharmacology, Queen's Medical Centre, Nottingham, UK*
MELBURN R. PARK • *Department of Anatomy and Neurobiology, University of Tennessee, Memphis, Tennessee*
BRUCE R. RANSOM • *Department of Neurology, Yale University School of Medicine, New Haven, Connecticut*
MARC G. RIOULT • *Physiological Institut, University of Bern, Bern, Switzerland*
CASE H. VANDERWOLF • *University of Western Ontario, London, Canada*
WILLIAM F. WONDERLIN • *Department of Medical Physiology, University of Calgary, Calgary, Alberta, Canada*
JIAN-YOUNG WU • *Department of Physiology, Yale University School of Medicine, New Haven, CT*
DEJAN ZECEVIC • *Institute of Biological Research, Belgrade, Yugoslavia*

Intracellular Electrophysiological Techniques

Stephen T. Kitai and Melburn R. Park

1. Introduction

The technique of intracellular recording is useful in studies seeking information about the functioning of individual neurons, their place in neuronal circuitry, and their membrane properties. For some of the just named applications, intracellular recording is the only method for obtaining the desired information. The functional significance of neurons is examined in the whole animal in intracellular experiments studying the responses of neurons during behavioral tasks or following sensory stimulation. The role of neurons in networks is studied by observing the synaptic responses generated by stimulation of connected structures. There are several standard tests of connectivity that can be made using intracellular recording techniques:

1. Verification of a projection through antidromic activation can be made (e.g., Andersen et al., 1964; Kitai et al., 1972). This is usually done as a screening procedure to select projection neurons during an experiment.
2. The nature of synaptic inputs can be examined. Here the sign (inhibitory or excitatory) of an input, its mechanism (i.e., ionic specie, conductance increase, disfacillitation, and so on, and, in many cases, the neurotransmitter can be learned (e.g., Coombs et al., 1955, Kita et al., 1983;).
3. The conduction velocity of the afferent pathway can be obtained from latency measurements (Kita et al., 1983), and often it is possible to determine whether the pathway being activated is monosynaptic or not (Kitai et al., 1976; Park, 1987b).

4. Divergence of axonal pathways can be tested via collision experiments involving time-locked activation at two antidromic stimulation sites. A neuron having a branched axon will exhibit a critical window of stimulus timings in which the independently activated antidromic action potentials triggered by stimulation of each branch will extinguish themselves and fail to reach the soma (Kita et al., 1983).

Membrane properties that can be determined include input resistance, dendritic dominance, current voltage characteristic, and synaptic and action potential generating mechanisms. Particularly in slice preparations where the constituents of the extracellular milieu can be experimentally manipulated, it is possible to study membrane ionic conductance mechanisms (Kita et al., 1984).

These standard physiological tests can now be combined with morphological identification through intracellular staining with horseradish peroxidase (HRP) or other substances, such as lucifer yellow (Stewart, 1978). The method of intracellular injection of HRP, introduced in 1976 (Cullheim and Kellerth, 1976; Jankowska et al., 1976; Kitai et al., 1976; Light and Durkovic, 1976; Snow et al., 1976), has established itself as an enormously productive tool for neurobiology (Kitai and Bishop, 1981; Kitai and Wilson, 1982; Kitai and Kita, 1984). The fundamental power of the technique is that it allows direct observation of the cellular physiology and the morphology of a single neuron. First, as a physiological tool, use of an HRP-filled microelectrode in no way compromises its suitability for conventional intracellular recording. Properly fashioned, HRP-filled microelectrodes can be used to measure spontaneous activity, evoked responses, and membrane characteristics. Second, as a morphological tool, intracellular iontophoresis of HRP leads to labeling of the entire soma-dendritic extent of a neuron, including such specializations as dendritic and somatic spines, as well as much of the axon, axon collaterals, and terminals as survival time permits. The morphological rendition of the HRP-filled neuron revealed by enzyme histochemistry is equal to or better than the results of the best Golgi stains, and offers several additional advantages over the Golgi technique. The intracellularly stained neuron is the only stained element in a region of tissue and can be easily reconstructed over many sections, a task that is often difficult or impossible in Golgi-impregnated material. Moreover, the method can be applied to a variety of preparations including the whole animal (Park 1987a), the in vitro brain slice (Kitai and Kita, 1984), and tissue culture (Neale et al., 1978). Advanced techniques in which intracellular recordings can be made from intact organ preparations (Bourque and Renaud, 1983) or from isolated perfused brain (Llinas et al., 1981; Richerson and Getting, 1987) are compatible with intracellular HRP labeling.

The HRP reaction product is robust, so that sections processed for HRP histochemistry cannot only be subsequently exposed to a variety of rigorous treatments, but can also be stored permanently, allowing the building up of a morphological library. With a few minor modifications to the method, the intracellularly HRP-labeled material can be prepared for electron microscopy (Wilson and Groves, 1979; Kitai and Wilson, 1982). Thus, electron microscopic examination can be conducted on a neuron that has previously been physiologically characterized, and then examined, drawn, reconstructed, and photographed using the light microscope. Finally, the intracellularly HRP-labeled procedure is compatible with subsequent processing for immunocytochemistry (Kitai and Penny, 1987).

In this chapter, we will describe the procedures that we have adopted in our laboratories for in vivo and in vitro preparations. This begins with an outline of the electronic principles and components used for intracellular recording.

2. The Recording Apparatus

The apparatus (Fig. 1) for intracellular recording and HRP injection includes a recording amplifier with the capability for capacitance compensation and intracellular current injection, one or more cathode ray oscilloscopes, an audio amplifier, and a stimulator. In addition to these, a micromanipulator, and a mechanism for storing the electrophysiological traces, including resting potentials, are necessary.

2.1. Recording Amplifiers

A single element of the intracellular recording preparation, the recording micropipet, sets the requirements of the electronic circuitry that is used in intracellular recording experiments. It is not

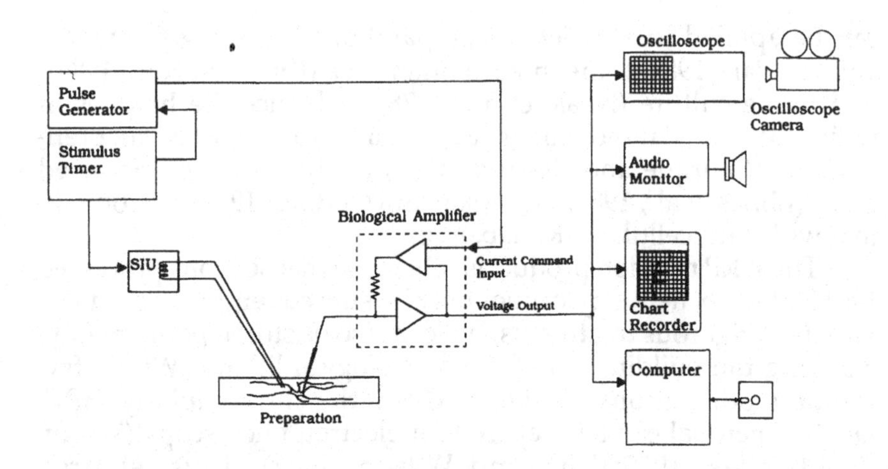

Fig. 1. Block diagram of a conventional intracellular recording setup. Shown are the individual instruments and the flow of data among them. The Chart Recorder and Oscilloscope Camera provide permanent records of slow and fast events, respectively. The Computer may also perform this role, but is primarily used to sum traces. The Audio Monitor is an aid to the experimenter in penetrating and monitoring cells. Commercial stimulation units often combine the functions of Stimulus Timer, Pulse Generator, and SIU (Stimulus Isolation Unit) in a single instrument. The Biological Amplifier used for recording has the capability of injecting current through an active bridge circuit.

the neuron itself whose signals are in the range of millivolts and tens of millivolts and are, therefore, large enough to be visualized with conventional oscilloscopes and voltmeters. Indeed, if we were able to place an amplifier directly within a neuron, it would be, because of the size of these signals, a very simple one. Instead, we must content ourselves with using a micropipet that is small enough to penetrate a neuron, without excessively damaging it, to act as a salt bridge between the metallic contacts of the amplifier and the interior of the cell and then devise an amplifier to, in effect, make that pipet disappear from the circuit. Put another way, the path from the first electrical node that we can reliably control, the input lead of the biological amplifier, to the node that interests us, the interior of the cell, is a long and tenuous one. The path consists of a very narrow, and thus high-resistance, salt bridge to the interior of the neuron and at least one (more often several) metal and metallic compound contacts between the conducting salt solution and the amplifier input. For a significant part of its length, the recording electrode is immersed in a volume conductor that further degrades the signal through capacitance coupling present along its axis. Through all of this, we require that the voltage at the amplifier input be the same as that at the tip of the microelectrode. The way that this is accomplished places the first major constraint on the design of the biological amplifier: that it itself not disturb the system by drawing any current across the recording electrode and neuronal membrane.

The electrical properties of the micropipet that constrain our amplifier are its high resistance, typically between 10–100 MΩ, and the significant capacitance that develops between its central conducting core and the preparation's ground when the electrode is immersed even a few millimeters into the brain or bathing medium. The high resistance of the pipet means that, if any current at all is drawn across it, a large and artifactual voltage signal will develop. Fortunately for the pioneers in this field, a type of amplifier, the electrometer, already existed that could record a voltage signal while drawing virtually no current from its source. Capacitance is a separate problem, and a solution did not come right away, although now circuitry for capacitance compensation is a part of every biological amplifier. Its adjustment is a critical part of the intracellular recording technique and will be discussed under a separate heading.

2.1.1. The Electrometer Amplifier

The biological amplifier must present as close to no load on the recording microelectrode as is possible. Were it to do this, then current would be drawn across the high resistance of the recording electrode and across the cell membrane. The latter produces a change in membrane potential, and the former results in an error in the measurement of that potential.

The means of accomplishing high-input impedance amplifiers has gone through an evolution that has left several historical terms in current usage. The accepted solution, in the days before the transistor, was the use of a special vacuum tube, termed an "electrometer tube" that, in itself, had the necessary very high-input impedance, on the order of 10^{12} Ω. There is no electrical

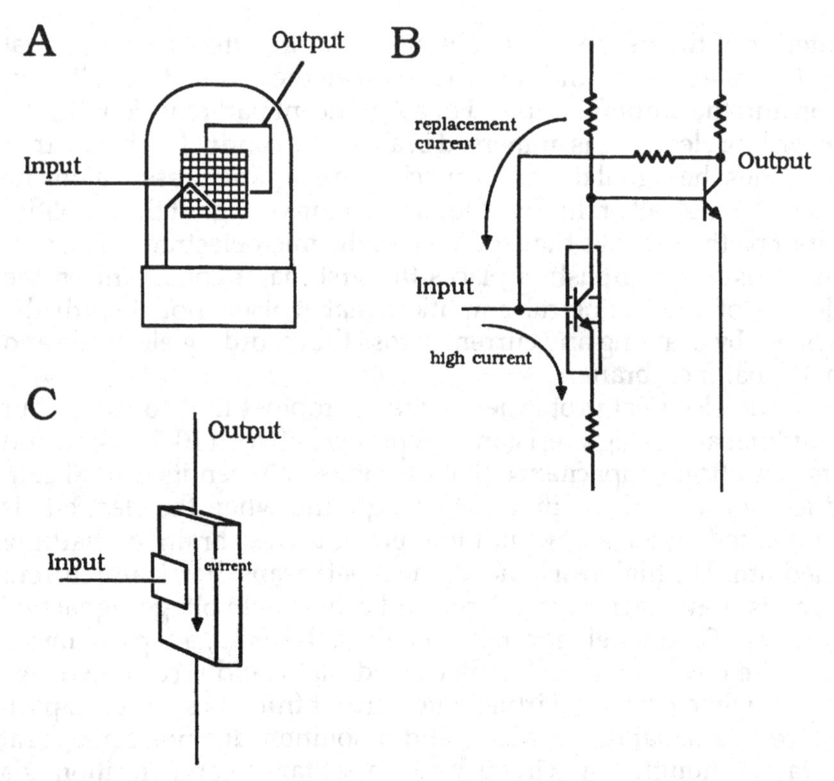

Fig. 2. (A) In the electometer tube, the input is separated by a vacuum from all other electrical nodes, and the stream of electrons flowing in this vacuum is kept low so that the electrometer tube has the essential property of high-input impedance. (B) Relatively high current flows through the bipolar transistor and must be replaced by a carefully matched source of feedback current. This is bootstrapping. (C) Like the vacuum tube, the input of the MOSFET transistor is electrically insulated from the current path that it controls. Current flow in the semiconductor part of the MOSFET is exquisitely sensitive to voltage applied at the input.

contact in the electrometer tube between input and output elements. Instead, the electrical charge that appears at the input is able to control the flow of electrons between the two output nodes (Fig. 2 A). The input is not degraded in this process. Electrical engineers have come to assign the name electrometer to a class of amplifiers having very high input impedance, good stability, and modest voltage gain, and this term is still correct for the modern transistorized biological amplifier. Electrometer amplifiers were originally developed as part of a physics apparatus to measure the electrical potential developed between two charged electrodes. Here the requirements are identical to our own. A substantial voltage exists, requiring little or no multiplication to be within the range of conventional display instruments, but any current drawn from the system during measurement would perturb the signal too much to make it valid. The vacuum tube solution to this problem was the electrometer tube incorporated into a cathode-follower circuit. This latter is a term that every electrophysiologist has more than likely heard, since many still refer to our modern field effect transistor (FET) amplifiers as cathode followers. Cathode followers were particularly well suited for intracellular recording.

For a long time, single transistors could not, by themselves, duplicate the properties of an electrometer vacuum tube, or indeed of any vacuum tube. Their principles of operation are different and, in particular, they inherently place a large load on the device driving them. Nonetheless, useful—even superior—circuits were devised that differed from vacuum tube implementations and that accounted for the particular electrical properties of transistors. The biological amplifiers of this era used positive feedback to disguise the low input impedance of the transistor (Fig. 2 B). This technique is referred to as "bootstrapping," a term that is still used; although, for the most part, it is inappropriate for the newest generation of biological amplifiers. The present technology makes use of another kind of transistor, the field effect transistor or FET (Fig. 2 C), that is close to the vacuum tube in its principle of operation. In particular, metal-oxide FETs (MOSFETs), like electrometer tubes, are inherently high-input impedance devices, and require no feedback or other circuitry to operate as the first stage of a biological amplifier. The input of a MOSFET device is actually a small conducting plate that is separated from a semiconductor junction by a thin film of insulating material. Potential on the conducting plate, called the gate, is able to affect the flow of current on the underlying semiconductor without there being any current flow from the gate and thus yields input impedances on the order of 10^{13} Ω. The FET may exist as a discrete transistor in the head stage (also called the probe) of an amplifier, or it may be integrated into an FET-input operational amplifier, such as the RCA CA3130, likewise built into the head stage. For the purposes of intracellular recording, amplifier head stages that use MOSFET operational amplifiers are close to ideal. They are reliable, stable, require no maintenance or adjustment, and produce no offset or error in measurement. The investigator can forget about them, except for one disagreeable property—they are easily destroyed by the buildup of static charge. Thus, the investigator must always ground himself or herself and any instruments being used before touching the input to a biological amplifier.

A basic intracellular recording amplifier could consist of just the input amplifier. A single MOSFET integrated circuit operational amplifier wired as a voltage follower with two 9 V batteries for power meets this requirement admirably and at virtually no cost. However, three remaining elements are generally included in practical circuits: a circuit for adjusting the DC level of its output, a circuit for current injection, and a circuit to compensate for the excessive capacitance at the input node. The first two are desirable options. The last is a necessity.

2.1.2. Capacitance Neutralization

Having found a suitable means of not loading the electrode with a low resistance, a means must be devised for unloading it of the parasitic capacitive coupling to the conducting fluid, be it the bath of an in vitro preparation or the surrounding tissue in an in vivo preparation. This capacitive shunting, if uncorrected, would cause an intolerable loss of the higher frequency components that are present at the intracellular node. For example, just 2 pF of coupling along the shank of a 50 MΩ electrode would cause 50% attenuation of the signal at 1600 Hz and a 3 dB/octave roll-off above that point. The formula for that determination is:

$$f_{3\ dB} = (2\ \pi\ R\ C)^{-1} \quad (1)$$

For these same values of *R* and C, the loss would be 75% at 3200 Hz and 87% at 6400 Hz. We recognize that 50 MΩ is a reasonable, even a low, value for an intracellular electrode, whereas 2 pF of shunting capacitance can be attained with just 2 mm of immersion in bath or brain (Nastuk and Hodgkin, 1950; Freygang, 1958). Moreover, many biological amplifiers intended for intracellular recording already have more than 2 pF of capacitive shunting associated with just their input circuitry and their physical construction. Five pico-

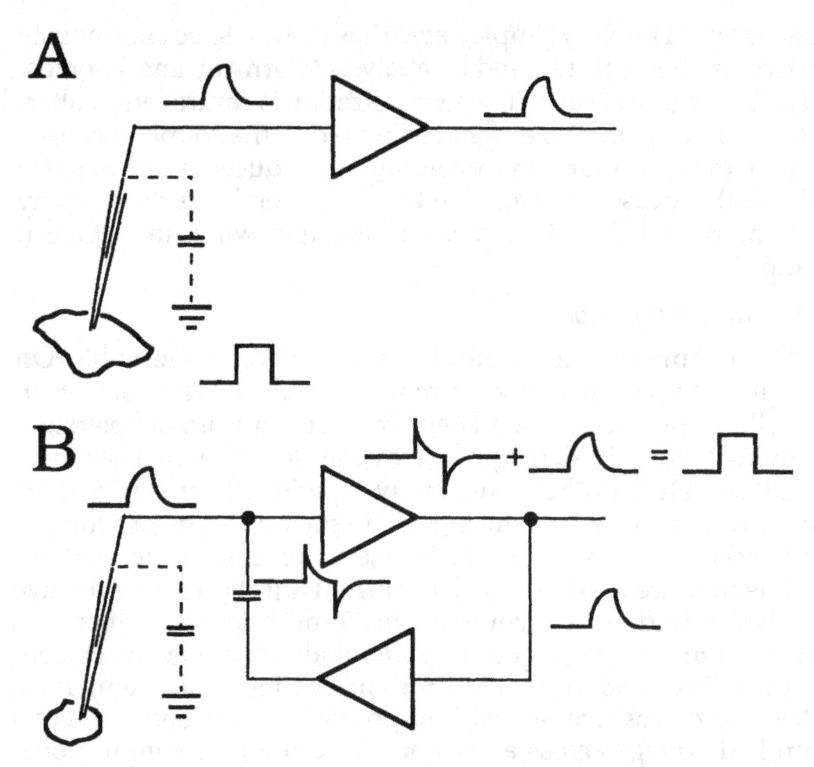

Fig. 3. (A) The input capacitance of the biological amplifier and the shunting capacitance of the recording micropipet combine with the electrode resistance to form a low-pass filter that markedly attenuates the high-frequency components of the measured signal. (B) A feedback amplifier whose output is coupled through a capacitor to the input node can restore the lost high-frequency components.

farads of input capacitance, still quite a low value, would cause 50% attenuation at 640 Hz, which is even less tolerable. Since the losses are because of the presence of what is essentially a capacitor between the amplifier input node and ground, it makes sense to try to restore them by injecting a remedying signal across a capacitor into the input node (Lettvin et al., 1958). This is the principle of capacitance neutralization (Fig. 3). The signal that is available is the intracellular one, since it appears at the output of the input amplifier. When this signal is amplified in voltage and coupled to the input node through a small capacitor, effective restoration of the signal occurs. This very simple device has proven to be enormously effective, both in practice and by analysis (Cornwall and Thomas, 1981). The degree of capacitive neutralization is controlled by altering the voltage gain of the signal being fed to the coupling capacitor. However, too large a voltage gain produces a net positive feedback that causes the amplifier to ring, generally destroying any cell being recorded from because of the large swings in voltage at the input.

2.1.3. Current Injection

Current injection in a controlled manner is often desirable. On one hand, hyperpolarizing current injected at the onset of an intracellular recording often keeps the neuron from depolarizing and dying before the damage done by the electrode tip has time to repair itself. On the other hand, many experimental manipulations of a neuron require current injection, such as to determine the input resistance of a neuron or to inject horseradish peroxidase.

The first circuits developed for current injection used a passive Wheatstone bridge. Consequently, the term "bridge amplifier" has come to signify any biological amplifier that is equipped for current injection. The modern method for current injection (Fein, 1966) makes use of positive feedback of precisely unity gain to place a controlled voltage across a resistor connected to the input node. With no signal at the current command input of the amplifier (Fig. 4 A), the unity gain geedback circuit maintains that voltage at precisely zero. Any voltage imposed on the current command input is duplicated across the series resistor, and a proportional current

$$I = (V_R - V_i)/R \qquad (2)$$

passes through the electrode and neuron (Fig. 4 B). These experimental currents produce a large voltage drop across the recording electrode. A tradition remains, stemming from the days of the Wheatstone bridge circuit, of using some means to remove that artifactual voltage. This is referred to as balancing the bridge. With the passive (i.e., Wheatstone) bridge circuit of the past, bridge balance was necessary just to obtain the most accurate current injection, which was still not very accurate. Today, with an active bridge amplifier, bridge balance is just cosmetic and in no way affects the magnitude of the current being injected. (*See* Park et al., 1983, for a review of bridge balance techniques.)

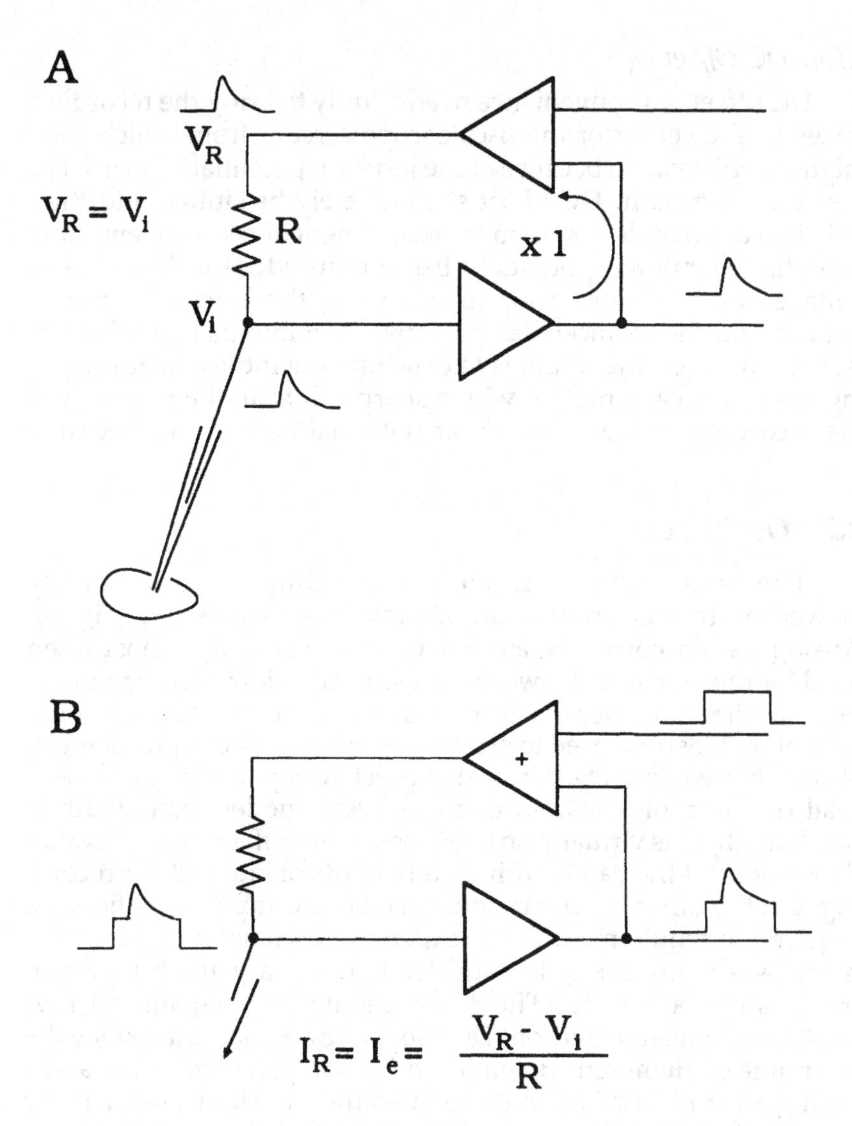

Fig. 4. Current injection through an active bridge. (A) When no current is being injected, a unity gain feedback loop (arrow) maintains the voltage at the upper end of the resister, R, at precisely the same value as is at its lower end, which is also the input node of the biological amplifier. (B) A command voltage is added to the signal contained in the feedback loop in order to create a voltage differential across R. The current injected is equal to that voltage drop divided by the value of R, which is typically 100 MΩ.

2.1.4. DC Offset

DC offset is a convenience needed only to bring the recording traces to the center of the oscilloscope screen, from which they might have deviated because of the junction potentials that exist in the recording path. DC offset should rarely be shifted and then only in ways that do not harm the assessment of resting membrane potential. Membrane potential is best measured as the difference in voltage recorded just prior to removing the electrode from a neuron and the extracellular potential seen immediately thereafter. The best routine, then, is to conclude the intracellular recording session of each neuron with a sharp exit from the neuron and the recording of the extracellular potential now at the electrode tip.

2.2. Oscilloscopes

The traces from intracellular recording are traditionally viewed on the screen of a cathode ray tube oscilloscope (Fig. 5). Analog oscilloscopes remain the type of instrument most often used for this purpose. However, those used in intracellular recording experiments differ in several respects from the more common type of oscilloscope used in electronics testing. Electrophysiological signals are relatively slow. The fastest sweep used might have a total duration of 2 ms, so that only very modest bandwidth is needed. There is virtually no oscilloscope manufactured today that does not meet the bandwidth requirements of intracellular recording. On the other hand, absolutely stable DC input amplifiers are required, a requirement that inexpensive oscilloscopes might not meet. Most physiologists consider it necessary to display two traces simultaneously. These are usually a low-gain (10 mV/oscilloscope screen division) DC coupled trace that can display the full range of membrane potential and action potential height and a high-gain (1 mV/division) AC coupled trace to show postsynaptic potentials in detail. Independent time bases for these two traces are an added feature that can permit the composition of more meaningful records during the recording experiment. Storage oscilloscopes preserve the image of a trace until purposefully erased and are a convenience. Recently, digital oscilloscopes (e.g., Nicolet 2090) have been improved to the point of being able to fully replace analog ones. In operation, they act like a conventional

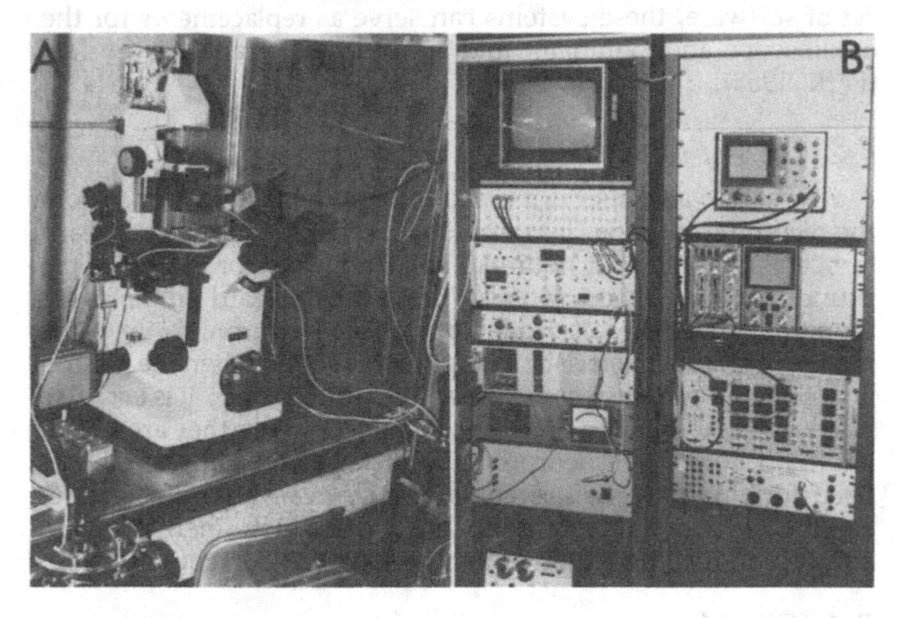

Fig. 5. The instruments used in intracellular recording. (A) In vitro chamber placed in an inverting microscope. This setup is suitable for intracellular recording and voltage clamp. (B) Photograph of the electronic instruments used in intracellular recording. The devices in the left rack are, from top to bottom: auxiliary monitor for the laboratory computer (which is not visible in this photograph), plugboard for interconnecting instruments, patch-clamp amplifier, biological amplifier, power supply, panel containing a calibrator and thermometer, power supply, and waveform generator. In the right rack are: a blank panel, analog oscilloscope for monitoring intracellular traces, digital oscilloscope, stimulus timer with pulse generator, and audio monitor.

storage oscilloscope, but have the added very important advantage of having made an internal representation of the trace in digital memory. From here it is possible to make an archival copy of the trace, such as to floppy disk, send the trace to a computer for further processing, or make a hard copy, such as to a plotter. The current forefront of technology involves the efforts of several firms to develop systems that convert a personal or laboratory computer into a digital data acquisition system. If well designed, made of quality components, and complete with a full and workable set of software, these systems can serve as replacements for the oscilloscope and stimulator and aid in the later analysis of data (Park, 1985).

2.3. Audio Amplifiers

An audio amplifier is used to convert the voltage signal from the recording amplifier into sounds that the investigator can learn to use as a monitor of neuronal spike activity. For example, the sound of distant action potential firing can be heard on the audio monitor as the microelectrode approaches a neuron, in the same manner as the characteristic change in this sound when the neuron is penetrated and the recording becomes intracellular. It is useful to familiarize oneself with these various sounds, since they are a great aid in approaching a neuron for recording. The audio amplifiers used for this purpose are commercially available from sources serving the scientific community (e.g., Grass Instrument Co.), but a simple hi-fi audio amplifier and speaker will serve just as well.

2.4. Stimulator

Physiological stimulators used for intracellular recording are usually employed to stimulate either peripheral nerves (directly or indirectly, as through the skin) or central nervous structures (e.g., a fiber bundle, a nucleus, and so on) in order to activate a neuron from which intracellular recordings are being made. Usually, a single stimulus pulse is delivered at intervals of 0.5–1 s as the experimenter searches for a neuron to record. The stimulation parameters are dependent on the structure being stimulated. The parameters used for deep central nervous system structures usually consist of a brief pulse (e.g., 0.05–0.1 ms) having a modest intensity (less than 1 mA).

There are many types of stimulators, but they all have the capability to control the intensity and duration of stimulus pulses and the frequency at which pulses are delivered. Some stimulators can deliver a paired pulse with varying interpulse intervals. Stimulators consist of a stimulus timer, usually a digital device, that is capable of generating the patterns of pulses needed to trigger nerve stimuli and the beginning of an oscilloscope sweep. The stimulus currents themselves are obtained from stimulus isolation units (SIUs) that provide the necessary control of current or voltage, and whose output is isolated from ground.

2.5. Micromanipulator

Micromanipulators are used to move the microelectrode through the neuronal tissue. They can be purely mechanical devices, usually based upon a micrometer advance mechanism, either directly coupled to a microelectrode holder (e.g., Narashige Canberra type) or coupled through a master–slave hydraulic system. Alternatively, they may be electromechanical (e.g., Inchworm). It is generally considered that, in order to penetrate neurons successfully, the microelectrode must advance in a stepwise fashion, rather than at a continuous velocity, through the brain. Continuous advancement causes the microelectrode tip to deform rather than sharply penetrate the membrane of a fortuitously placed neuron. For this reason, the mechanical and electromechanical classes of micromanipulators have proven to be the most suitable for intracellular recording. The Inchworm micromanipulator operates on a piezoelectric principle that causes an advance in small sharp steps. Other electromechanical manipulators achieve the same result by coupling a stepper motor to a mechanical screw advance. Controlled movements of as little as 1 μm are possible with both of these types. Purely mechanical manipulators require the skill of the experimenter to obtain the correct form of motion. If precisely made (e.g., Narashige), they are very effective and are preferred in our laboratories. Fluid coupled (hydraulic) manipulators generally have too much visco-elastic damping to achieve sharp movements of the microelectrode. We have found, too, that they are not stable positioners, but drift as much as 50 μm/h, probably because of thermal expansion or contraction of the hydraulic fluid in the long tube connecting the master and slave cylinders. These manipulators, however, do have the advantage of a small head-stage and are sometimes the only choice where space is limited, as in some in vitro preparations.

The large mass of manipulators, such as the Narashige or Inchworm, make them sensitive to vibrations. The way to overcome this is to choose an extremely rigid system of support for the micromanipulator that allows little or no movement relative to the preparation (Purves, 1981).

2.6. Mechanism for Storing Electrophysiological Traces

There are several ways of storing electrophysiological traces. Conventionally, they are photographed from the face of the oscilloscope screen using a camera (i.e., Grass Kimograph Camera). In addition, traces can be stored on FM tape as analog signals, or digitized and stored on a computer disk or tape. Resting membrane potential is often monitored during a recording experiment using either a digital voltmeter or a DC pen recorder.

3. The Preparation of Vertebrates

3.1. *In Vivo Preparation*

Animals are anesthetized according to accepted standards of animal anesthesia. The animal is placed in a stereotaxic frame that rests upon an air piston vibration isolation table. The bone and dura overlying the stimulation and recording sites are removed, and bipolar stimulating electrodes (e.g., epoxylite-insulated steel insect pins, bared 0.2–0.5 mm at the tips) are stereotaxically positioned in the structures to be stimulated (e.g., in the case of the neostriatum, the cortex, substantia nigra, and thalamus) and affixed in place with methyl metacrylic polymer (dental acrylic). A recording well is constructed of dental acrylic around the exposed recording site, and this recording well is sealed with low melting point (40°C) paraffin after each insertion of a microelectrode. The weight of this paraffin plug serves to reduce vascular pulsations in the brain. Prior to recording, the cisterna magna is pierced to relieve intracranial pressure, further reducing vascular pulsations. Vascular pulsations and respiratory movements can be reduced further by suspending the animal in a stereotaxic frame by means of a clamp to a lower vertebra. The temperature of the animal is maintained between 37–38°C by a feedback control device.

3.2. *In Vitro Slice Preparation*

For the experiments combining intracellular recording and labeling, slices are prepared from ether-anesthetized rats perfused through the heart with a small volume (20–50 ml) of cold (5–10°C) oxygenated Krebs Ringer. This perfusion serves to remove erythrocytes from the cerebral vasculature that would otherwise obscure the HRP-labeled neurons. The endogenous peroxidase activity of erythrocytes reacts the same way as HRP in producing labeling. Perfusion prior to decapitation does not seem to alter electrophysiological activities of the slice. Immediately after perfusion, the animal is decapitated and the brain gently removed from the skull. Slices are prepared using either a tissue chopper or a Vibratome for sectioning. With the tissue chopper method, the structure to be sectioned is initially placed on a small piece of filter paper. The tissue is sectioned with the tissue chopper in parasagittal planes at a thickness of about 350 μm. It is important to use a clean and sharp razor blade on the tissue chopper to ensure better survival of the slice. With the Vibratome sectioning method, the brain is placed immediately in a small Petri dish containing cold (ca. 5°C) oxygenated Krebs Ringer. With a razor blade, the brain is trimmed by hand into an appropriately oriented block. This block of tissue is then glued to the cutting stage of a Vibratome using a cyanoacrylic cement. The Vibratome bath is filled with cold Krebs Ringer solution and 350-μm sections are made. Four to six slices sectioned using a Vibratome or a tissue chopper are then stored in a beaker containing continuously oxygenated Krebs Ringer at room temperature, until each slice is ready to be recorded from. We have found it usually best to store the tissue in this manner at least 1–2 h prior to recording. For both tissue chopper and Vibratome sectioning methods, the entire operation from decapitation to placement of the slices in oxygenated Krebs Ringer should be carried out within 20 min.

4. The Microelectrode and Marker

The fabrication of microelectrodes is the same for in vivo and in vitro preparations. The characteristics of the microelectrode are determined by the need for acceptable intracellular recordings and only slightly influenced by the special requirements for injection of HRP. Glass microelectrodes with tip sizes of less than 1-μm tip diameter (probably in the range between 0.5–0.1 μm) are necessary to penetrate CNS neurons successfully (Fig. 6). The finest tips (less than 0.2 μm) cannot be filled with the slightly viscous HRP solution. We use micropipets that are pulled immediately before use from 10-cm long blanks of 2–3 mm Pyrex™ glass using a Narishige vertical puller or blanks of 1–2 mm glass using a horizontal puller (e.g., WPI C753). The modern style of horizontal microelectrode puller (Brown and Flaming, 1977) uses tungsten ribbon filaments

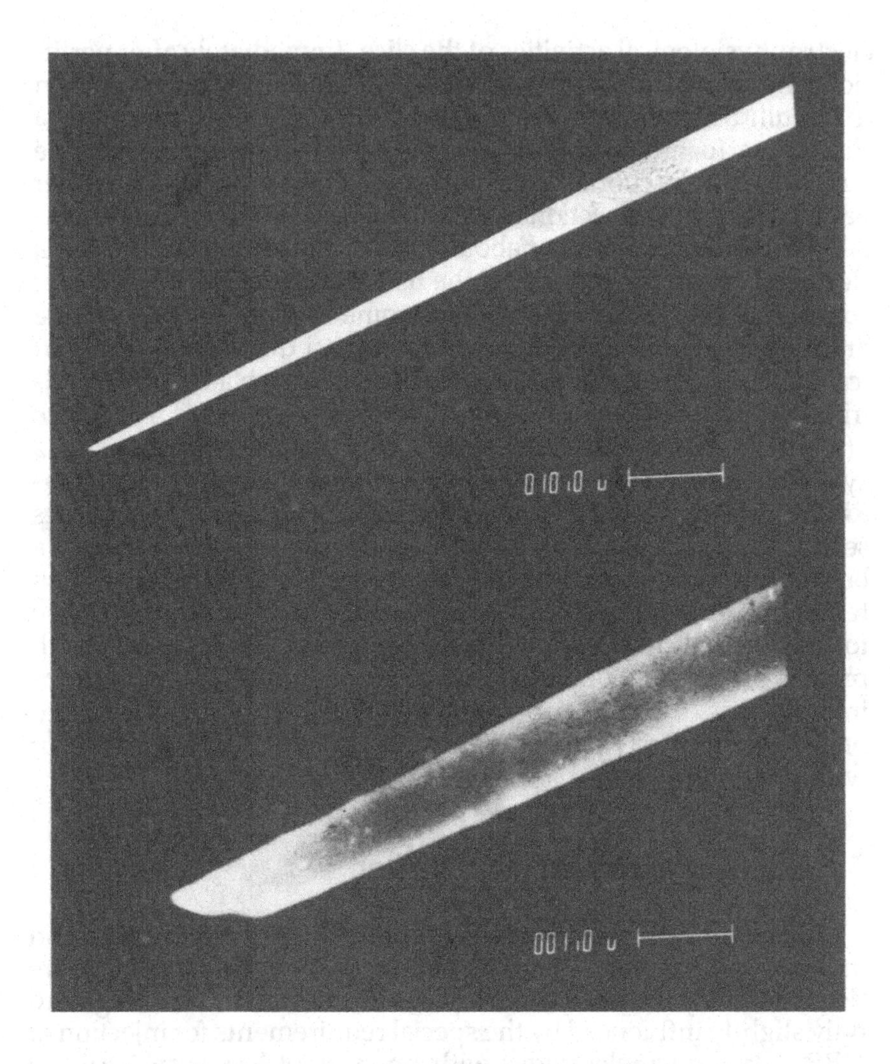

Fig. 6. Scanning electron micrograph of a glass micropipet that has been beveled by mechanical bumping. The scale is 1 μm.

for more controlled heating of the glass blank and a manifold for delivering a cooling puff of air during the pull, so that short shanked pipets that still have desirable tip characteristics can be produced.

The micropipets are then sharpened by one of two methods. The large pipets made from 3-mm capillary tube can be sharpened by the mechanical process of bumping the electrode tip against a roughened glass surface (Kitai et al., 1976; Park, 1987a). This is done under microscopic observation with the microscope stage being used to manipulate the microelectrode. Pipets fabricated from 1–2 mm glass must be beveled against a fine abrasive. Originally, microelectrodes were beveled by grinding with a disk that had been coated with a fine abrasive (Brown and Flaming, 1974; Baldwin, 1980). Several systems have been devised that use an abrasive slurry (Ogden et al., 1978; Corson et al., 1979). These have the advantage of being very inexpensive. Good results, however, can be obtained by combining the techniques by beveling the micropipets in a thin slurry that is made to rotate on the disk of a conventional beveling device.

The micropipets are then filled with 4% HRP in 0.5*M* potassium methylsulfate or 0.5*M* KCl and 0.05*M* Tris Buffer, pH 7.6, yielding electrodes with resistances in the range between 50– and 100 MΩ. The filling is done by placing several microliters of solution in the barrel of the microelectrode with a Hamilton syringe, and then using a fine glass filament to work the solution into the shank and tip of the electrode. Once the fluid meniscus has reached a sufficiently narrow portion of the pipet shank (ca. 30-μm diameter), capillary forces draw the electrolyte solution the rest of the way to the tip. The glass filament can then be manipulated to dislodge any remaining bubbles in the pipet shank and stem.

5. Recording and Injection Procedures

5.1. *In Vivo Procedure*

As the electrode is advanced through the tissue, a 0.5-nA, 10-ms hyperpolarizing or depolarizing square wave pulse is passed through the electrode once each sweep in order to monitor electrode impedance. An increase in electrode impedance, signaled by an increase in the amplitude of the voltage signal resulting from this constant current pulse, is often an indication that the electrode tip is in contact with the somatic membrane of a neuron. Although neurons can be successfully penetrated through just mechanical advancement of the microelectrode, a much higher degree of success results from use of a special circuit that places large but brief AC or DC signals through the electrode. For example, a 2–10-ms,

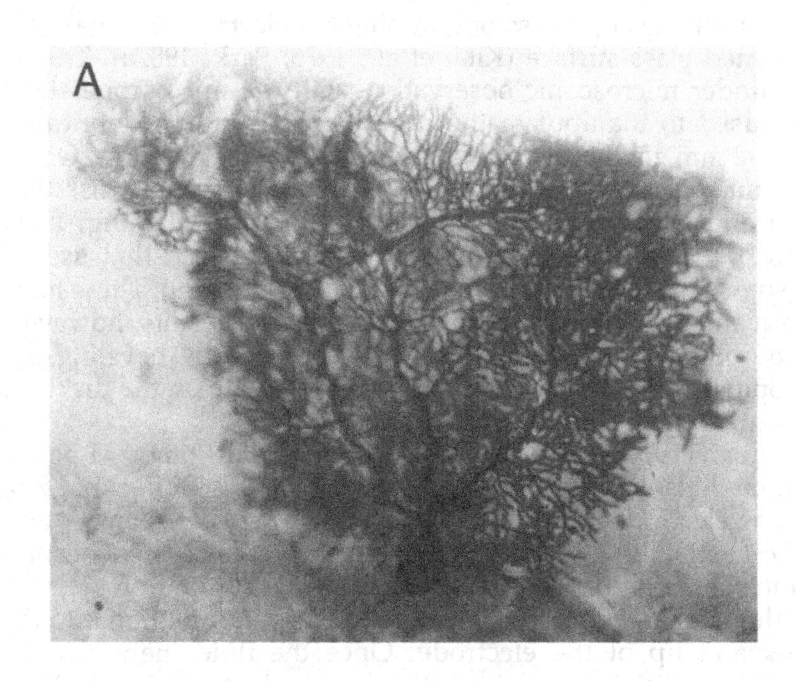
A

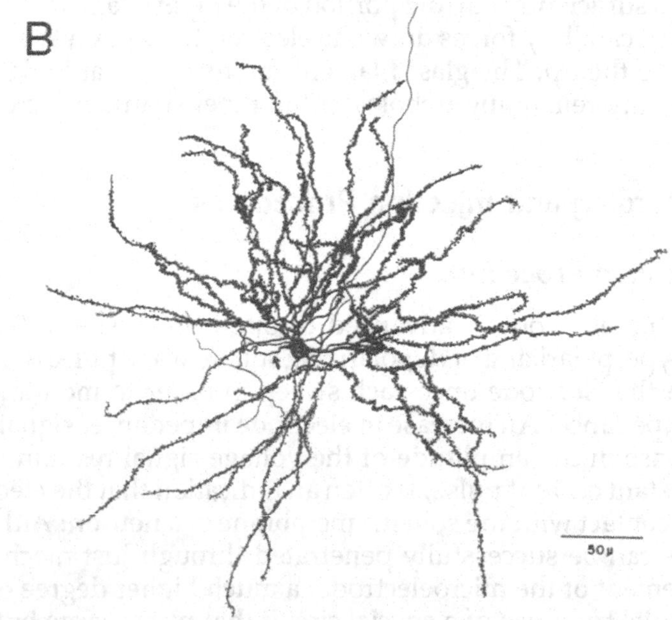
B
50µ

10–50-V DC pulse works quite well. Such circuits are now a part of many commercially available intracellular recording amplifiers. It is also possible to use a momentary excessive capacitance compensation ("ringing the electrode") to achieve penetration. Successful penetrations are signaled by a drop in recorded voltage to a stable resting potential that meets the criteria for the experimental design (generally above 40 mV), the presence of postsynaptic potentials in response to stimulation, and action potentials with overshoot. Recordings that fail to meet these standards are discarded and the probe continued. After completion of a protocol of neurophysiological analysis, the neuron receives an iontophoretic injection of HRP. In order to fill a rat striatal neuron, we use an injection current of 2–5 nA, given as square wave pulses of 100–150 ms in length and delivered at 5 Hz. A hyperpolarizing current of up to 1.0 nA may be maintained during the off period of the cycle. Typically, 2–5 min of injection is sufficient to recover a well-filled neuron (Fig. 7); 30 min facilitates the tracing of axons over longer distances, but injections as short as 1 min can yield lightly stained neurons. These latter, lightly stained neurons, are valuable for electron microscopic studies, as the cellular organelles are less likely to be obscured by a dense accumulation of osmophilic HRP reaction product.

For reliable recovery of an intracellularly labeled neuron, it is crucial that the penetration remain secure and stable during the iontophoresis of HRP. Consequently, the injection must be terminated upon the appearance of any signs of deterioration of the cell, such as a drop in resting membrane potential. The distribution of HRP through the dendrites and proximal axonal arborization occurs swiftly, and only a few minutes of survival are sufficient to recover a neuron that is well filled to this extent. However, in order to trace axons for more than a few millimeters, survival times of 12–24 h may be necessary. When neurons have been recovered

←

Fig. 7. Neurons labeled with intracellularly injected horseradish peroxidase. (A) A Purkinje cell has been filled with horseradish peroxidase and reacted with the diaminobenzidine method. The label is able to penetrate throughout the dendritic arbor. The axon can be seen emerging from the soma at 8 o'clock. (B) A reconstruction of an HRP-labeled striatal medium spiny neuron drawn from serial light microscopic sections. The arrow indicates the axon.

after such survival times, it has sometimes been possible to trace axons for long distances (Chang et al., 1981; Bishop et al., 1979; Donoghue and Kitai, 1981). A compromising factor, however, is that HRP-filled neurons often undergo soma-dendritic degeneration during prolonged survival experiments. Degeneration of the axon appears to proceed more slowly, and it can be possible to gather useful information about the axonal arborization of a neuron whose soma and dendrites are severely degenerated.

For those experiments that are designed to test the properties of evoked synaptic potentials, the questions that can nearly always be satisfactorily answered using intracellular recording techniques and the methods by which they are answered are outlined below. In all, five properties of postsynaptic potentials can be determined by manipulating two parameters: stimulus strength and intracellular current injection. The former is varied, usually by changing stimulus current, in what is referred to as the power test. Intracellular current may take the form of pulses or the injection of steady (DC) current. The five properties that can be discerned are:

1. Is the response graded?
2. What is the current-voltage (I/V) relationship of the response?
3. Can the response be reversed?
4. What is the reversal potential?
5. Is the response monosynaptic?

Items 1 and 5 depend upon the power test and items 2–4 upon current injection.

Item 1, testing whether the response is graded, is so basic that its description is often overlooked in the results sections of research reports. It is tested by power (changing stimulus strength). The amplitude of a postsynaptic potential must change with alterations of stimulus strength in a way that reflects spatial summation of afferent terminals (Fig. 8 A). All-or-none responses, some of which could conceivably by mistakenly interpreted as postsynaptic responses, are recognized as such at this stage of the analysis.

Items 2–4 are interrelated and are tested by current injection, most efficiently in the order in which they have been named. The current-voltage properties of a potential are estimated by injecting what is essentially DC current for the potential in question. This

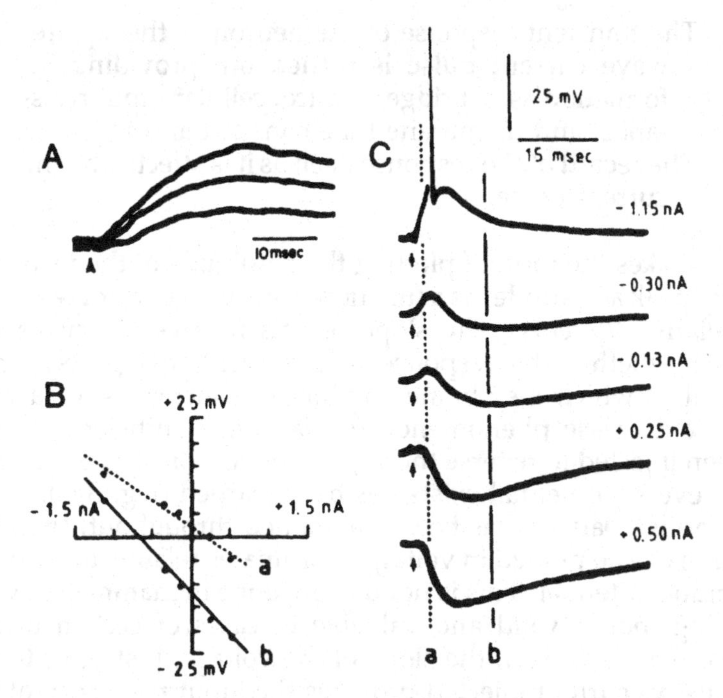

Fig. 8. Synaptic responses evoked following local stimulation of neostriatum in an in vitro slice preparation. (A) EPSPs induced by various stimulation intensities. Note the incremental amplitude of the EPSPs without a change in latency. (B) Relative current–voltage (I/V) curve based on the responses in C. The lines a and b correspond to measurements made along the line segments a and b shown in C. (C) EPSP-IPSP sequence and the effect of constant intracellular current injection on the response. The injection of hyperpolarizing current results in an increase in the amplitude of the EPSP and eventually elicits an action potential (–1.15 nA). Injection of depolarizing current decreases EPSP amplitude and increases IPSP amplitude. The arrows in A and C indicate the presentation of the stimulus.

can take the form of constant, truly DC, hyper- and depolarizing current or long current pulses that frame the stimulus and response. The latter is preferred as it provides, on the same trace

1. A record of quiescent membrane potential (where no current is injected) both before the response and after it

2. The transient response of the neuron as the square-wave current pulse is turned on, providing information as to bridge balance, cellular input resistance, and membrane time constant and
3. The record of the response itself as it is affected by the current pulse.

Analysis takes the form of plotting the amplitude of the response (usually peak amplitude) as a function of injected current resulting in a relative I/V curve.The slope of this relative I/V curve determines whether the response is a conventional postsynaptic potential in which membrane conductance increases, or a conductance-decrease phenomenon *(see below)*. If sufficient current has been injected to reverse the sign of the response, then the fact that a reversal potential exists has been learned (Fig. 8 C).

If bridge balance has been maintained throughout, then the I/V curve can be plotted in voltage coordinates relative to absolute membrane potential. This is not usually done in mammalian CNS recording, but is valid and valuable in cases of certain bridge balance. For such data, the slope of the voltage, just prior to the stimulus, vs current injected provides the input resistance of the cell. Similarly, input resistance during the response can be determined. If the response reverses, then the absolute membrane voltage at which that occurs can be known. This value can be compared with plausible estimates of the Na^+, Ca^{2+}, K^+, or Cl^- equilibrium potentials to provide a hint at the ionic species mediating the synaptic response.

Because absolute I/V plots are difficult to obtain, a different technique to measure conductance change during the response is often used. This consists of passing a train of brief square-wave current pulses through the electrode as the response is elicited. The aim is to measure a steady-state change in voltage (V) as a function of the step change in current (I) and from this determine input resistance (or its inverse, conductance) according to Ohm's law. Care must be taken that the current pulses are long enough to provide a sufficient approximation of the steady state. Capacitive current flow from the step change must have essentially ceased. Inspection of the voltage record, ensuring that the trace is flat or close to flat, usually suffices to guarantee this condition.

Responses that increase in amplitude as the membrane is brought away from its reversal potential have an I/V curve with a negative slope and are conductance-increase postsynaptic potentials (Fig. 8 B,C). Responses whose I/V curve has a positive slope are either conductance-decrease synaptic potentials or, more likely, network-related responses, that is, disinhibition if they are depolarizing or disfacilitation if they are hyperpolarizing. It is not possible to distinguish network-related responses from conductance-decrease postsynaptic potentials on the basis of current injection measurements alone.

The fifth point of interest is whether the synaptic response is monosynaptic or not. The data for this determination come from the power test. Monosynaptic inputs produce a potential whose latency is constant as stimulus strength is changed (Fig. 8 A). Latency is measured from the instant of stimulation to the point where the intracellular trace first deviates from a trace taken under identical stimulus conditions, but with the recording electrode in the extracellular space near the recorded neuron. These extracellular controls are generally taken at the end of an intracellular penetration. Observation of constant latency in these conditions is evidence in itself of a monosynaptic input. However, the observation of shifts in latency may not alone necessarily indicate a nonmonosynaptic (polysynaptic) connection (Park, 1987b). In evaluating records for shifts in latency (polysynaptic case), the amount of the shift is compared to any fluctuations in latency observed at constant stimulus strength. Cells having constant latency fluctuations larger than any shifts seen with changing stimulus strength are discarded.

There are sixth and seventh properties that are often asked of postsynaptic potentials: What ion or ions carry the synaptic current, and what is the neurotransmitter? Identification of the mediating ion is not usually an issue in tracing synaptic connectivity in vivo, since the experimental approaches for determining it are limited. The comparison of reversal potential with estimates of equilibrium potentials, mentioned above, is weak, since estimates of the latter are generally only approximate. The only direct experiment is if Cl^- is suspected. Intracellular (Cl^-) can be increased by passing negative (hyperpolarizing) current through a KCl recording microelectrode. This makes E_{Cl} less negative, so that Cl^- mediated IPSPs are reversed at resting membrane potential.

A pitfall in studying evoked responses is mistaking an action potential that is activated by orthodromic activation as being antidromic. In central nervous system studies, four tests for an antidromic action potential have evolved. First, the response must be an action potential, that is, all-or-none, stereotyped, and invarient between trials (Fig. 9 A,B). Secondly, the response must have constant latency for a range of suprathreshold stimulus strengths. Some variation in latency can be expected from certain well-characterized changes in axonal conduction velocity (Swadlow et al., 1978), but these can be eliminated if the stimulus is given at fixed intervals (i.e., 1 Hz). Thirdly, the response should follow high-frequency stimulation, or its equivalent, double shock (Fig. 9 G). Finally, the response should disappear when paired with an action potential triggered in the soma by depolarizing current injection for interstimulus intervals that are less than the sum of antidromic latency plus absolute refractory period (i.e., collision).

Collision of separately triggered action potentials is used in another experimental procedure that tests for branched axonal connections (Fig. 9 C-G). In this procedure, antidromic action potentials are triggered from two sites suspected of receiving an axonal branch from the neuron being recorded from. At certain interstimulus intervals, extinction of test evoked action potentials occurs in the segment of common axon proximal to the branching (Fig. 9 D,F). This is sufficient evidence to demonstrate the branching pattern and to prove that the two evoked action potentials are indeed antidromic.

Several procedures can be employed that bear on the problem of identifying neurotransmitters within a system. These include double-labeling procedures and the systemic application of specific antagonists, where available, to characterize the transmitter mediating a particular input. Double-labeling involves the intracellular injection of a marker (e.g., HRP) during recording and subsequent histochemical treatment (e.g., immunohistochemistry) for neurochemical identification (Kitai and Penny, 1987).

5.2. In Vitro Slice Procedure

In vitro recording requires a chamber in which the slice is maintained. Most brain slice chambers are constructed of acrylic plastic (Plexiglas)™ and comprise an inner chamber, containing the slice of brain tissue, and an outer chamber that forms a temperature-regulating waterbath (Fig. 10). The temperature of the waterbath is kept at a constant temperature (37°C for most

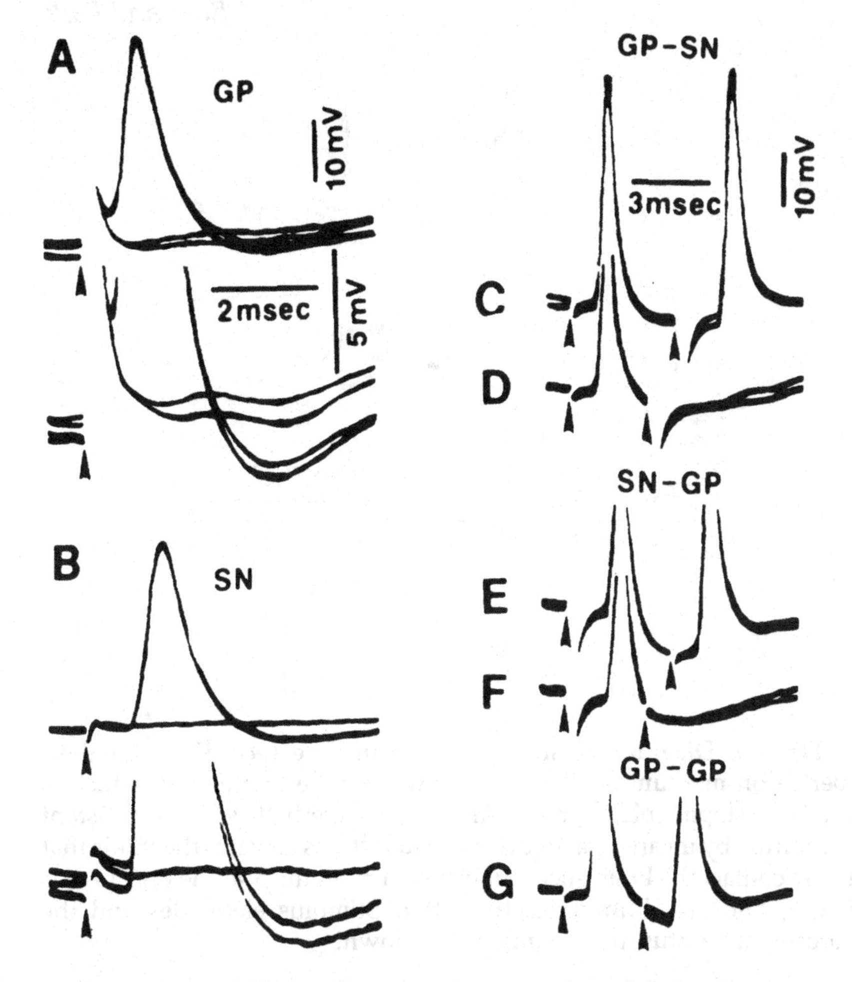

Fig. 9. (A and B) Antidromic responses of a subthalamic neuron to stimulation of globus pallidus (GP) and substantia nigra (SN). Top traces are DC recordings and the lower traces are high-gain AC recordings that demonstrate that there are no underlying synaptic potentials associated with the spikes. (C–F) Collision of antidromic spikes in response to paired stimulation of GP and SN. Interstimulus intervals of 4 ms for C and E and 3 ms for D and F. Note test SN and GP responses are blocked at 3-ms interstimulus intervals. (G) Paired stimulation of GP at 3-ms interstimulus interval to test the refractoriness of this neuron. Note that this neuron is capable of firing at the 3-ms interval, indicating that the failure of the test (second) response in C and F is the result of collision. The arrows indicate the presentation of stimuli.

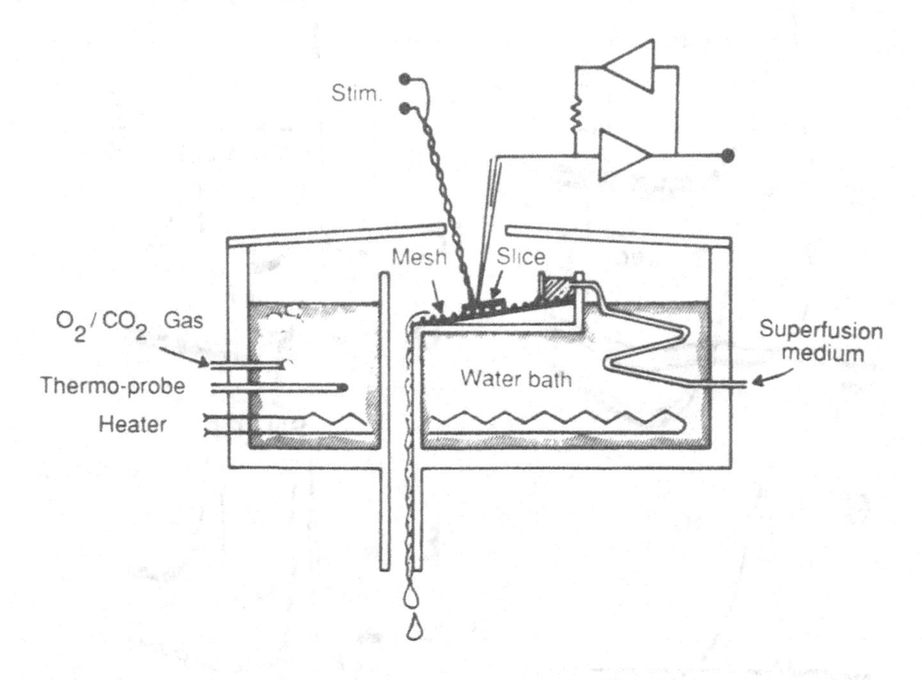

Fig. 10. Diagram of an in vitro brain slice recording chamber. Superfusion medium is allowed to flow over the tissue slice, which is placed on a sloping plastic mesh. An integral waterbath is held at constant temperature by means of a regulated heater. It acts to warm the fluids that directly contact the brain slice, superfusion medium, and the O_2/CO_2 gas mixture, and also humidifies the latter. Stimulus electrodes and the recording apparatus are schematically shown.

experimental designs) with a regulated heating element placed in the outer chamber. O_2/CO_2 gas is pumped continuously through the waterbath to create a warm, moist environment over the brain slice. The superfusing Krebs Ringer passes by gravity feed to the recording chamber through a heat-exchanging tube that traverses the waterbath. The recording chamber is tilted so that the superfusing of Krebs Ringer solution can run over the slice of brain tissue at a rate of approximately 1 mL/min. In this design, the upper surface of the slice is directly exposed to the warm and moist O_2/CO_2 gas mixture. A dissection microscope placed over the slice chamber allows visualization of stimulation and recording electrode placements. The microscope must have sufficient working distance to allow room for the recording and stimulating electrodes and the micromanipulator (Fig. 11).

The superfusing Krebs Ringer is composed as follows: NaCl 124.0 m*M*, KCl 5.1 m*M*, $MgSO_4$ 1.3 m*M*, CaCl 2.5 m*M*, KH_2PO_4 1.25 m*M*, $NaHCO_3$ 26.0 m*M*, and D-glucose 10.0 m*M*. The pH of the Krebs Ringer is maintained at 7.3–7.4 by bubbling a gas mixture of 95% O_2/5% CO_2 through the perfusate. The osmolarity of the Krebs Ringer is checked prior to each experiment with an osmometer and adjusted to 305 ± 5 mOsm. It is worthy of note to mention that the Ca^{2+} concentration of the Krebs Ringer heavily influences the patterns of synaptic activity recorded in the neostriatal slice preparation. Thus, it has been reported that no inhibition exists in rat neostriatal neurons in slice experiments that used 1.2 m*M* $CaCl_2$ (Misgeld and Bak, 1979). On the other hand, when the molarity of $CaCl_2$ was raised to 2.3–2.5 m*M*, inhibition was observed (Lighthall et al., 1981).

The following procedures are used for intracellular HRP labeling in slice. When a neuron is penetrated that has a stable resting potential of 40 mV or more, HRP is iontophoretically injected as in the procedure for in vivo experiments described above. After completion of the injection, the slice containing the injected neuron is carefully placed on a small piece of lens paper saturated with the Krebs Ringer solution. The paper and slice are then gently floated, slice up, on fixative solution containing 0.5–4.0% formaldehyde and 1.0–2.0% glutaraldehyde in standard isotonic buffer for 30 s to 1 min. The lens paper is then upturned, and the slice immersed in the fixative. This procedure of transporting the slice on the lens paper prevents it from becoming curled during the fixation. Fixation continues for 3–12 h. Following this, each slice is sectioned at 50 μm using either a sliding freezing microtome or a Vibratome. For Vibratome sectioning, the slice is embedded in warm agar (2%), and the agar is cut into a block prior to sectioning. The sections are processed according to the standard HRP procedures. For a detailed description of the histochemical procedures for visualizing intracellular horseradish peroxidase, alone or combined with immunohistochemical labeling readers are referred to Kitai and Wilson (1982) and Kitai and Penny (1987).

The in vitro slice preparation has become the preparation of choice for studying the properties of individual neurons. Without

Fig. 11. Photograph of an in vitro brain slice recording chamber. This is the same chamber that is diagrammed in Fig. 10. To the right of the stereo microscope is a Canberra-type mechanical micromanipulator. It holds a recording microelectrode. The head stage of the biological amplifier, and the coaxial lead connecting it and the microelectrode, can also be seen. The stimulating electrode is held in the manipulator that is to the right of the microscope.

the pulsations that arise from circulatory and respiratory movements in nearly all in vivo preparations, the intracellular penetrations that are obtained in the in vitro slice preparation are of extremely high quality and can be maintained for hours. The slice then becomes a perfect platform from which to study the membrane properties of neurons.

Acknowledgments

Supported by USPHS Grant NS20841 to M.R.P., USPHS Grants NS20702 and NS23886 to S.T.K., and the Neuroscience Center of Excellence, State of Tennessee. The authors thank Salmon Afshapour for his assistance in the photographic preparation of the figures.

References

Andersen P., Eccles J. C., and Sears T. A. (1964) The ventro-basal complex of the thalamus: Types of cells, their responses and their functional organization. *J. Physiol. Lond.* **174,** 370–399.

Baldwin D. J. (1980) Dry beveling of micropipette electrode. *J. Neurosci. Meth.* **2,** 153–161.

Bishop, G. A., McCrea R. A., Lighthall J. W., and Kitai S. T. (1979) An HRP and autoradiographic study of the projection from the cerebellar cortex to the nucleus interpositus anterior and nucleus interpositus posterior of the cat. *J. Comp. Neurol.* **185,** 735–756.

Bourque C. W. and Renaud L. P. (1983) A perfused *in vitro* preparation for hypothalamus for electrophysiological studies of neurosecretory neurons. *J. Neurosci. Meth.* **71,** 302–14.

Brown, K. T. and Flaming, D. G. (1974) Beveling of fine micropipette microelectrodes by a rapid precision method. *Science* **185,** 693–695.

Brown K. T. and Flaming D G. (1977) New microelectrode techniques for intracellular work in small cells. *Neuroscience* **2,** 813–827.

Chang H. T., Wilson C. J., and Kitai S. T. (1981) Single neostriatal efferent axons in globus pallidus: A light and electron microscopic study. *Science* **213,** 915–918.

Coombs J. S., Eccles J. C., and Fatt P. (1955) The specific ionic conductances and the ionic movement across the motoneuronal membrane that produce the inhibitory postsynaptic potential. *J. Physiol. Lond.* **130,** 326–373.

Cornwall M. C. and Thomas M. V. (1981) Glass microelectrode tip capacitance: its measurement and a method for its reduction. *J. Neurosci. Meth.* **3,** 225–232.

Corson D. W., Goodman S., and Fein A. (1979) An adaptation of the jet stream microelectrode beveler. *Science* **205,** 1302.

Cullheim S. and Kellerth J. O. (1976) Combined light and electron microscopic tracing of neurons including axons and synaptic terminals after the intracellular injection of horseradish peroxidase. *Neurosci. Lett.* **2,** 307–313.

Donoghue J. P. and Kitai S. T. (1981) A collateral pathway to the neostriatum from corticofugal neurons of the rat sensory-motor cortex: an intracellular HRP study. *J. Comp. Neurol.* **201,** 1–13.

Fein H. (1966) Passing current through recording glass micro-pipette electrodes, IEEE Trans. *Biomed. Eng.* **13,** 211–212.

Freygang W. H. (1958) An analysis of extracellular potentials from single neurons in the lateral geniculate nucleus of the cat. *J. Gen. Physiol.* **41,** 543–564.

Jankowska E., Rastad R., and Westman J. (1976) Intracellular application of horseradish peroxidase and its light and electron microscopical appearance in spino-cervical tract cells. *Brain Res.* **105,** 555–562.

Kita T., Chang H. T., and Kitai S. T. (1983) Pallidal inputs to subthalamus: Intracellular analysis. *Brain Res.* **264,** 255–265.

Kita T., Kita H., and Kitai S. T. (1984) Passive electrical membrane properties of rat neostriatal neurons in an in vitro slice preparation. *Brain Res.* **300,** 129–139.

Kitai S. T. and Bishop G. A. (1981) Intracellular staining of neurons, in *Neuroanatomical Tract-Tracing Methods* (Heimer L. and Robards M. J., eds.), Plenum, New York, pp. 263–277.

Kitai S. T. and Kita H. (1984) Electrophysiological study of the neostriatum in brain slice preparation, in *Brain Slices* (Dingledine R., ed.) Plenum, New York, pp. 285–296.

Kitai S. T., Penny R. G. and Chang H. T. (1989) Intracellular labeling and immunocytochemistry, in *Neuroanatomical Tract-Tracing Methods 2: Recent Progress* (Heimer L. and Zaborszky L. eds.), Plenum, New York, pp. 173–199.

Kitai S. T. and Wilson C. J. (1982) Intracellular labeling of neurons in mammalian brains, in: *Cytochemical Methods in Neuroanatomy* (Chan-Palay V. and Palay S., eds.), Alan R. Liss, New York, pp. 533–549.

Kitai S. T., Kocsis J. D., Preston R. J., and Sugimori M. (1976) Monosynaptic inputs to caudate neurons identified by intracellular injection of horseradish peroxidase. *Brain Res.* **109,** 601–606.

Kitai S. T., Tanaka T., Tsukunhara N., and Yu H. (1972) Facial nucleus of cat: Antidromic and synaptic activation and peripheral nerve representation. *Expt. Brain Res.* **16,** 161–183.

Lettvin J. Y., Howland B., and Gesteland R. C. (1958) Footnotes on a headstage, *IRE Trans. Med. Electronics* **ME-10,** 26–28.

Light A. R. and Durkovic R. G. (1976) Horseradish peroxidase: An improvement in intracellular straining of single electrophysiologically characterized neurons. *Exp. Neurol.* **53,** 847–853.

Lighthall J. W., Park M. R., and Kitai S. T. (1981) Inhibition in slices of rat neostriatum. *Brain Res.* **212,** 182–187.

Llinas R., Yarom Y., and Sugimori M. (1981) Isolated mammalian brain in vitro: new technique for analysis of electrical activity of neuronal circuit function. *Fed. Proc. Fed. Am. Soc. Exp. Biol.* **40,** 2240–2245.

Misgeld U. and Bak I. J. (1979) Intrinsic excitation in the rat neostriatum mediated by acetylcholine. *Neurosci. Lett.* **12,** 277–282.

Nastuk W. L. and Hodgkin A. L. (1950) The electrical activity of single muscle fibers. *J. Cell. Comp. Physiol.* **35,** 39–72.

Neale E. A., MacDonald R. L., and Nelson P. G. (1978) Intracellular horseradish peroxidase injection for correlation of light and electron microscopic anatomy with synaptic physiology of cultured mouse spinal cord neurons. *Brain Res.* **152,** 265–282.

Ogden T. E., Citron M. C. and Pierantoni R. (1978) The jet stream microbeveler: An inexpensive way to bevel glass micropipettes. *Science* **201,** 469–470.

Park M. R. (1985) A complete digital neurophysiological recording laboratory, in *The Microcomputer in Cell and Neurobiology Research* (Mize R. R., ed.) Elsevier, New York, pp. 411–434.

Park, M. R. (1987a) Intracellular horseradish peroxidase labeling of rapidly firing dorsal raphe projection neurons. *Brain Res.* **402,** 117–130.

Park M. R. (1987b) Monosynaptic inhibitory postsynaptic potentials from lateral habenula recorded in dorsal raphe neurons. *Brain Res. Bull.* **19,** 581–586.

Park M. R., Kita H., Klee M. R., and Oomura Y. (1983) Bridge balance in intracellular recording; introduction of the phase-sensitive method. *J. Neurosci. Meth.* **8,** 105–125.

Purves R. D. (1981) *Microelectrode Methods for Intracellular Recording and Iontophoresis* (Academic, New York).

Richerson G. B. and Getting P. A. (1987) Maintenance of complex neuronal function during perfusion of the mammalian brain. *Brain Res.* **409,** 128–132.

Snow P. J., Rose P. K., and Brown A. G. (1976) Tracing axons and axon collaterals of spinal neurons using intracellular injection of horseradish peroxidase. *Science* **191**, 310–311.

Stewart W. W. (1978) Functional connections between cells as revealed by dye-coupling with a highly fluorescent naphthalimide tracer. *Cell* **14**, 741–759.

Swadlow H. A., Waxman S. G., and Rosene D. L. (1978) Latency variability and the identification of antidromically activated neurons in mammalian brain. *Exp. Brain Res.* **32**, 439–443.

Wilson C. J. and Groves P. M. (1979) A simple and rapid section embedding technique for sequential light and electron microscopic examination of individually stained central neurons. *J. Neurosci. Meth.* **1**, 383–391.

From: ***Neuromethods, Vol. 14: Neurophysiological Techniques: Basic Methods and Concepts*** **Edited by: A. A. Boulton, G. B. Baker, and C. H. Vanderwolf**

Recording and Analysis of Currents from Single Ion Channels

William F. Wonderlin, Robert J. French, and Nelson J. Arispe

1. Introduction

In this chapter, we will give an overview of the methods for making electrical recordings from single ion channels in cell membranes and in planar lipid bilayers, endeavoring to point out the advantages and limitations of each approach. Since extensive treatments of various aspects of the techniques are currently available, including complete books devoted to patch clamping (Sakmann and Neher, 1983a) and to ion channel reconstitution (Miller, 1986), we attempt to present an integrated view, with references to other discussions of details of the theory and practice that provide additional information that might be necessary to establish these techniques in the laboratory. Since the techniques of single-channel recording are being applied to an expanding range of biological problems, numerous companies are beginning to address the needs of researchers in these areas. Consequently, we will list some sources of useful equipment, software, and supplies. Finally, we will attempt to highlight current developments in instrumentation and techniques, which have appeared since other reviews were written. Of particular interest are recent approaches to presentation, analysis, and modeling of single-channel data that have just appeared or are in press at the time of writing of this article.

1.1. Single-Channel Recording

There are two requirements that must be met in order to resolve current fluctuations that result from transitions between conducting (open) and nonconducting (closed) conformational states of an individual ion channel. First, background noise must be reduced to very low levels by minimizing current through nonspecific leakage pathways in parallel with the channel, and second, the recording must be focused so as to collect current only from one or a few channels. Two general experimental approaches have been used to resolve single-channel currents. In the first, one or a very few channel-forming molecules were introduced into a planar lipid bilayer and the current across the bilayer monitored (Bean et al., 1969; Ehrenstein et al., 1970; Hladky and Haydon, 1970). Lipid bilayers with resistances of >100 gigohms (1 gigohm = 1 GΩ = 10^9Ω) can easily be formed, even when the area of the bilayer is >10^4 μm^2. In the second method, a very small area of a cell or vesicle membrane is electrically isolated by forming a high-resistance seal (nowadays, generally >10 GΩ) between a glass pipet and the membrane. The current flowing into the pipet across the membrane patch is then monitored using a high-gain current-to-voltage converter (Neher and Sakmann, 1976; Sakmann and Neher, 1983a). In either approach, the end goal is to isolate one channel, or a very few, in the region of membrane from which current is monitored and to minimize the leakage conductance in parallel with the channel.

1.1.1. History of Single-Channel Recording

The first recordings of unitary conductance steps attributed to single ion channels were reported by Bean et al. (1969). EIM (excitability-inducing material) is a bacterial protein that confers a voltage-dependent conductance on lipid bilayers (Mueller et al., 1962). Bean et al. (1969) observed "incorporation jumps" as the conductance of a bilayer progressively increased in the presence of EIM. They also saw the conductance decrease in steps of a similar size during the degradation of EIM by proteolytic enzymes. Shortly thereafter, Ehrenstein et al. (1970) reported steady-state conductance fluctuations having the same amplitude. They were able to account for the negative conductance of bilayers "doped" with EIM on the basis of the voltage-dependent probabilities of occurrence of the open and closed states of the channels. A subsequent paper showed that the kinetics of approach toward a steady-state current level, following a voltage step, were determined by the rates of opening and closing of the individual EIM channels (Ehrenstein et al., 1974). Other ground-breaking studies were performed by Hladky and Haydon (1970, 1972), who showed that gramicidin, a polypeptide antibiotic, induced unitary fluctuations in the conductance of a lipid bilayer. Their experiments on ion conduction through the channel paved the way for future studies by suggesting (1) that occupancy of the channel was limited, perhaps to one ion at a time, and (2) that under different ionic conditions single-channel conductance might be limited either by the rates of ion entry into the channel from the aqueous solution, exit from the channel, or by translocation within the channel. These analyses of gating and ion conduction of channels formed by bacterial proteins laid the groundwork for more recent functional studies of ion channels that play central roles in a variety of living cells.

Use of the planar lipid bilayer facilitated these initial measurements of single-channel currents by providing a preparation in which the leakage conductance was extremely low. Bilayer conductances of less than 10 pS (i.e., resistances of greater than 100 GΩ) are easily attained. The first recordings of single-channel currents from a living cell were made by Neher and Sakmann (1976), who observed currents through extrajunctional cholinergic channels in denervated frog skeletal muscle. These records were achieved by gently pressing a polished glass recording pipet against the enzyme-treated cell membrane to obtain an electrical seal of a few 10s of megohms. The resolution of the single-channel recordings was dramatically improved when Sigworth and Neher noticed that much higher resistance seals ("gigaseals" > 1 GΩ and routinely in the range 10–100 GΩ in favorable preparations) were often spontaneously formed when gentle suction was applied to the pipet. The resultant decrease in noise and increase in useable bandwidth enabled them to make the first recordings from single sodium channels (Sigworth and Neher, 1980). Gigohm seal patch recording has enabled single-channel analysis in a variety of systems at a level of resolution that could not have been anticipated. "Gigaseals" also provided such a mechanically stable bond between membrane and pipet that they opened the possibility of three new recording configurations (e.g., Hamill et al., 1981)—from membrane patches ripped off from the cell having either the extracellular or the cytoplasmic surface facing the bath solution, and from the whole cell after disrupting the patch in the pipet tip while preserving the glass-to-membrane seal (*see* chapter by S. Jones, in this volume).

1.1.2. Choosing the Experimental Approach to Fit the Question—from Molecular Mechanism to Cellular Response

The variety of methods and recording configurations now available for recording single ion channel currents enables the approach to be tailored specifically to a particular problem. If one is interested in the activity of an intact cell, records from on-cell patches may be taken. Beating heart cells have been studied in this manner (Levi and DeFelice, 1986). The action of a suspected intracellular modulator may be tested by applying it to an inside-out patch. Properties of a purified channel protein may be examined in the simplest of physicochemical systems after reconstitution into planar lipid bilayers. Channels from intracellular membrane systems, or from cell surfaces made inaccessible *in situ* by a glial sheath, may be studied after membrane fractionation and insertion into bilayers. One of the variety of possible preparations can be adopted based on the experimental conditions and manipulations needed to answer the question that the investigator has in mind.

2. Methodology

2.1. Introduction

Our presentation of single-channel methods is organized along the three basic steps involved in the investigation of the activity of single ion channels, including:

1. The recording of single-channel currents
2. The detection and measurement of single-channel events and
3. The analysis and interpretation of the amplitudes and durations of these events.

The detailed discussion of these topics will be preceded by a brief overview of background material.

2.1.1. Glossary

The following notation will be used in the text.

Preparation/Headstage amplifier:

g_{chan} —conductance of an open ion channel
i_{open} —open channel current level

i_{closed} —closed channel (baseline) current level
i_f —feedback current maintained by headstage amplifier
i_{input} —sum of current flowing through R_{input} and C_{input}
i_{gate} —gate leakage current of headstage amplifier FET
C_m —membrane capacitance
C_f —stray capacitance across headstage feedback resistor
C_{stray} —stray capacitance at input of headstage amplifier
C_{input} —total input capacitance
R_{chan} —resistance of an open ion channel
R_f —feedback resistance of headstage amplifier
R_m —resistance across patch or bilayer membrane
R_{shunt} —shunt resistance between membrane and pipet or partition
R_{series} —access resistance in series with membrane
R_{input} —lumped input resistance of headstage
V_m —transmembrane potential
V_{com} —command potential for voltage-clamp
V_{out} —voltage output of headstage amplifier
V_{eq} —equilibrium potential of permeant ion(s)
$V_{in(-)}$ —signal voltage at inverting input of headstage
$V_{in(+)}$ —signal voltage at noninverting input of headstage
$V_{er(-)}$ —error voltage at inverting input
$V_{er(+)}$ —error voltage at noninverting input

Filter/Sampling:

f_{cut} —filter cut-off frequency (–3 dB)
t_{rise} —filter rise time (10–90%)
f_{samp} —sampling frequency, $1/t_{samp}$
t_{samp} —sampling interval, $1/f_{samp}$
s_i —i^{th} sample
BW —bandwidth (Hz)
$S^2_{(Hz)}$ —noise variance density, amp^2/Hz
$rms_{(\sqrt{Hz})}$ —root mean square noise density, $amp/\sqrt{Hz}$
$S^2_{(BW)}$ —noise variance in bandwidth BW (amp^2)
$rms_{(BW)}$ —rms noise in bandwidth BW (amp)
SNR —signal-to-noise ratio

Event detection:

i_{thresh} —current-level threshold for detection of a state transition
t_{st} —time at which a state transition occurs
FTC —false threshold-crossing rate
dur_{max} —duration at which FTC is 100-fold less than the transition rate
t_{dwell} —a continuous variable representing the durations of events
t_{all} —minimum true duration for which all events are detected
t_{dead} —event detection dead time
t_{min} —minimum duration of detected events
t_{max} —maximum duration of detected events
t_{flat} —duration of flat segment between opening/closing steps

Data analysis:

N	—number of observations in a data set
Θ	—set of parameters in a kinetic scheme
δt	—bin width of dwell-time histogram on linearly scaled abscissa
δx	—bin width of dwell-time histogram on logarithmically scaled abscissa
t_{low}	—lower limit of histogram
t_{high}	—upper limit of histogram
f_{bin}	—frequency of observations in a bin of a histogram
$f_{bin}\Theta$	—predicted frequency of observations in a bin of a histogram based on parameter set Θ
bin_{mid}	—midpoint of bin (abscissa)
$P(x)$	—cumulative probability (probability distribution function) of x
$p(x)$	—probability density (pdf) of x, 1st derivative of P(x)
τ_j	—j^{th} time constant of exponential function, $1/k_j$
k_j	—j^{th} rate constant of exponential function, $1/\tau_j$
a_j	—area under j_{th} component of exponential function
$L(\Theta)$	—maximum likelihood estimate for unbinned data with parameter set Θ
$L_{bin}(\Theta)$	—maximum likelihood estimate for binned data with parameter set Θ

2.2. Overview of Single-Channel Recording

2.2.1. Passive Electrical Properties of Membranes and Ion Channels

Pure lipid bilayer membranes are impermeable to ions because of the prohibitively large energy required for solubilization of charged molecules in the hydrophobic lipid phase of the membrane (Parsegian, 1969). Thus, with regard to electrical properties, bilayer membranes exhibit a very high resistance (R_m) to the passage of electrical current, and because they are also exceptionally thin, they exhibit a large capacitance (C_m). Evolution has produced a variety of membrane proteins that alter these electrical properties of bilayer membranes. One class of these proteins includes the ion channels, transmembrane proteins that possess a hydrophilic pore through which ions can cross the membrane. The energy required for movement of ions across the membrane is greatly reduced by these channels, because the environment of the pore is polar and entry of ions into the lipid phase is not required. The pore of an ion channel is usually selective in allowing certain species of either cations or anions to pass, and this selective permeation results in the generation of an electric current. Since the transfer rate, or turnover number, through an ion channel is very high ($>10^6$ ions/s), the ionic currents generated by the opening of a single ion channel can be measured. The ability of the pore of a channel to transfer ions is characterized by its conductance ($g_{chan} = 1/R_{chan}$), which is given here by the chord conductance:

$$g_{chan} = i_{open}/(V_m - V_{eq}) \quad (1)$$

where g_{chan} is the conductance of the channel (units of picoSiemens [pS]; a 1 pS conductance is equivalent to a $10^{12}\Omega$ resistor), i_{open} is the current amplitude (units of picoamps—pA—or 10^{-12} amps), V_m is the transmembrane potential (volts), and V_{eq} is the equilibrium potential (volts) of the permeant ion(s). Single-channel conductances vary greatly (≈0.5–>200 pS) among different types of ion channels, although for a specific class of channel, it is usually quite consistent and can be used to aid in the identification or classification of channels. The single-channel conductance may also be dependent on the species and concentrations of permeant ions that are present and the range of membrane potentials that is examined.

2.2.2. Gating of Ion Channels

For most ion channels, the conductance fluctuates with time among discrete levels. In the simplest example, the conductance alternates between two conductance levels, the fully open (conducting) and closed (nonconducting) levels. Each conductance level is termed a "conductance state" of the channel. Thus, for the example above, the channel exhibits two conductance states, closed and open. The open state generates a current, i_{open}, with a magnitude proportional to the "driving force," $V_m - V_{eq}$ (cf Eq. 1). The transitions among conductance states are termed the "gating" of the channel. The transient increases and decreases in ionic permeability resulting from gating are essential to the coordinated electrical signalling produced by populations of ion channels. The records shown in Fig. 1 show examples of channel gating. Gating probably results from changes in the conformation or structure of the channel protein. Thermodynamically, there is an energy barrier to these changes in conformation. The relationship between the magnitude of this energy barrier and the energy contained in the random molecular vibration of the protein determines the probability that the conformational change will occur. The probability of conformational change can also be influenced by other factors, such as the transmembrane voltage (Fig. 1a) and/or the binding of chemical ligands to the channel protein. Either of these factors might, for example, decrease the energy barrier, thereby increasing the probability of the conformational change.

Often, the analysis of transitions among conductance states is

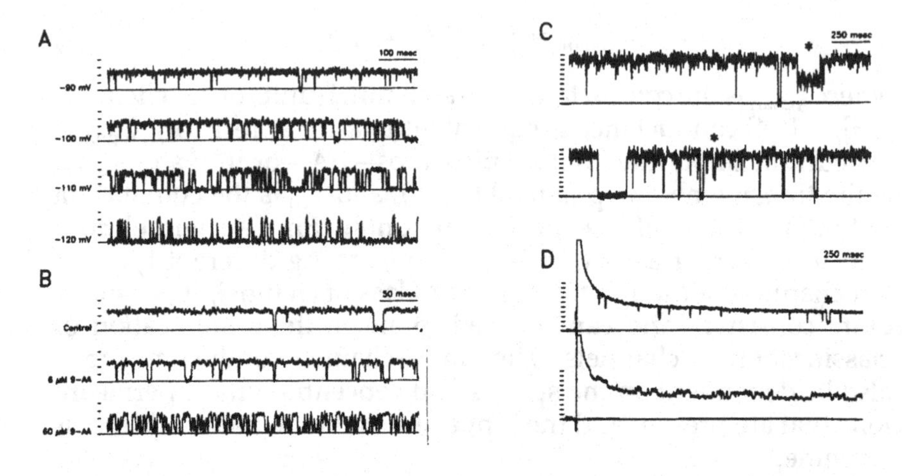

Fig. 1. Channel gating in isolated rat brain ion channels incorporated into lipid bilayers. In A–D, the orientation of channel opening is upward, the horizontal line represents the closed channel current level, and the horizontal tic marks at the left of the record are stacked in 1-pA increments. Vesicles were prepared from a rat forebrain homogenate and fused with lipid bilayers formed from PE : PS (50 : 50) in decane over a hole in a Teflon® partition (*see* Worley et al., 1986, for methods). (A) Voltage-dependent gating of a batrachotoxin-modified rat brain sodium channel. At –90 mV, the channel is almost always open. With increasing hyperpolization between –90 and –120 mV, the ratio of open and closed times is nearly reversed. The records in A and B were sampled at 10 kHz and digitally filtered at 1 kHz. The bilayer was formed over an ≈ 70 µm hole in A and B with a $C_{input} < \approx 50$ pF. The bilayer was bathed in symmetric 200 m*M* Na acetate. (B) Block of batrachotoxin-modified sodium channels by 9-aminoacridine (9-AA) at a holding potential of +100 mV. 9-AA produces brief closed events, some of which are too brief to reach full amplitude at a bandwidth of 1 kHz. Consistent with the C-O-B model, the average open dwell time, but not the closed dwell time, decreases with increasing 9-AA concentration (C) Conductance substates (marked by asterisks) and increased noise during the open state are present in a record from a large, Ca^{2+}-activated K^{+} channel. Note that both the fully open and the substate levels are noisier than the closed level, and that the noise excursions from the fully open level are asymmetric, tending to go towards the closed level. The extra noise presumably results from rapid, poorly resolved transitions among different conductance states of the channel. A rapid blocking reaction could also produce similar open-state noise. The bilayer was formed over a

complicated by the presence of conductance "substates" and multiple "kinetic states" for a single conductance level. Substates are states with conductance levels intermediate between the fully open and fully closed conductance levels (Fig. 1c). We refer to states that are distinguishable kinetically, but not on the basis of conductance, as "kinetic states." Within a set of kinetic states, each state probably corresponds to a different conformation of the channel protein that produces the same conductance level. For example, there may be several nonconducting states (e.g., normal closed state, inactivated state, and "blocked" state in the presence of a drug). Although these nonconducting states cannot be identified on the basis of conductance level, they can be identified if they exhibit markedly different rates of transition to and from other conductance states.

Some of the molecular vibrations of a channel protein are very rapid, occurring on the time scale of picoseconds. Because these rapid vibrations lead to random crossing of the energy barrier for conformational change, the changes in conformation and conductance are essentially instantaneous. The rapid "snapping" of the conformational change results in current fluctuations that are

---

175-μm hole, and the membrane was bathed in symmetric 200-m*M* KCl and 100-μ*M* free Ca^{2+}. The voltage was +40 mV, and the bandwidth was 300 Hz. (D) Pharmacological effects of a quaternary ammonium ion (tetrapentylammonium, TPeA) on a batrachotoxin-activated Na^+ channel in a lipid bilayer. The upper trace is a control; the lower was obtained with 1-m*M* TPeA bathing the inner end of the channel. The TPeA appears to affect the Na^+ channel in two ways: (1) by reducing the amplitude of i_{open}, and (2) by inducing long-lasting (tens to hundreds of ms) closed or blocked periods. A relatively long closure in the control record is marked by an asterisk for comparison with the amplitude of i_{open} in the presence of TPeA. The bilayer was formed on a 100-μm hole with symmetric 200 m*M* NaCl. The voltage was stepped from 0–+100 mV and the record digitally filtered at ≈ 175 Hz. Note the long, slow capacitive transient following the voltage step. This is an unavoidable result of the redistribution of solvent between the annulus and the bilayer when the voltage is changed and is an inherent problem of using voltage step protocols in this type of preparation. The duration of the transient can be reduced by reducing the size of the bilayer and/or the amount of material in the annulus.

essentially rectangular in shape with extremely small transition times between states. We discuss the transitions among states in terms of "events." Each event is a sequence of two state transitions. For example, an open event represents transitions from a closed to open to closed state, whereas a closed event represents transitions from an open to closed to open state. The duration of an event is the period of time between consecutive state transitions and is termed the "dwell time" of the event. Currently, detectable dwell times vary over a time scale that can cover several orders of magnitude (from microseconds to seconds) (e.g., Blatz and Magleby, 1986a). This is true for both the normal gating of a channel and the block of channels by various drugs.

An inescapable limitation in the study of channel gating is the poor resolution of very brief events, even with the best currently attainable recording bandwidths (≤30 kHz). Single-channel current records must be filtered to a level where the channel gating can be distinguished from random fluctuations in background noise. The effect of filtering on the resolution of brief events can be succinctly stated in terms of the rise time of the filtered record. If a rectangular test pulse is applied to the recording system, the rise time can be defined as the time required for the filtered output of the recording system to make the transition from 10–90% of the full amplitude of the step. Events of duration much greater than the rise time easily reach full amplitude and are clearly resolved. Events slightly shorter than the rise time do not reach full amplitude and are distorted, but may still be detected above background noise. Events shorter than ≈½ the rise time are not detected. Although these events are undetected, they can leave their mark on the current record, because a series of consecutive undetected events will be "time averaged." The current level during a time-averaged sequence will appear intermediate between the fully open and closed levels at a level proportional to the relative amount of time the channel spends in the conducting state. Time averaging is an important consideration in interpretation of channel conduction and gating. For example, in the case of channel-blocking drugs, an apparent concentration-dependent decrease in conductance might result from time-averaged, brief blocking events with the fractional open time decreasing with increasing blocker concentration (Coronado and Miller, 1979, 1980). In the case of normal channels, it can be tempting to segregate conduction and gating properties as if the conduction properties reflected only ionic permeation through the pore of a static protein molecule. In fact, very rapid conformational changes (i.e., gating) beyond the resolution of the recording system might be responsible for some apparent conduction properties (e.g., rectification) of channels (Läuger, 1985).

2.2.3. Inferring Kinetic Schemes from the Channel Gating

An important goal in single-channel analysis is the development of a kinetic scheme that can account for the conductance and gating behavior of each species of ion channel. The variables in such a model include the number of states, the pathways among states, and the rates for transitions among states. Prior to the advent of single-channel recording techniques, these variables could only be studied indirectly by examining the average simultaneous behavior of large populations of channels either at equilibrium or during the relaxation following a perturbation, such as a change in voltage. In contrast, single-channel recording techniques now enable us to examine more directly the number of states and the transitions among states in a single channel molecule. However, by limiting ourselves to one channel, we must now study the "average" behavior as a stochastic process continuing over a relatively long period of time.

By using established methods for event detection and measurement, we can analyze a single-channel current record and produce, for each conductance state, a distribution of dwell times in that state. Each distribution is summarized in histogram form, plotting on the ordinate the frequency with which the dwell times fall into time intervals or bins arranged along the abscissa. The expected shapes of these observed distributions can be predicted directly from kinetic schemes that include variables defining the conductance and kinetic states of the channel and the transition rates among different states. By comparing the relative goodness-of-fit of distributions predicted by several alternative kinetic schemes to our observed distribution, we can select the kinetic scheme that best predicts the behavior of the channel. The prediction of dwell-time distributions, even for relatively complicated kinetic schemes, is straightforward and exact. A far more substantial source of difficulty is the identification of biases or artifacts in our observed dwell-time distributions that arise from the inadequacies of the event detection and measurement methods (e.g., limited time resolution, digitization error). If these sources of errors are not controlled, the fitting of predicted distributions to our observed distributions may lead to erroneous conclusions.

Methods for the prediction of dwell time distributions from kinetic schemes are well-established (Colquhoun and Hawkes, 1981, 1982, 1983, 1987). The current recorded from a single ion channel can be treated as a series of successive transitions among two or more conductance or kinetic states. This series of transitions is usually modeled as a Markov chain process (*see* Horn [1984] for applications of Markov processes to single-channel analysis). There are two important assumptions of the Markov model. First, it is assumed that the channel gating results from transitions among a small number of discrete states. Second, the transition rates among states are assumed to be time homogeneous, i.e., they are independent of the length of time the channel has occupied a state. In essence, the channel has no memory, and the probability per unit time of leaving a state is independent of its history, including the previous occupancy of other states or the length of time spent in the current state. If these assumptions are valid, and if transitions among states are reversible, the distribution of dwell times in each state will be described by a positively skewed, monotonic exponential decay, with a decrease in the number of observations as the dwell time is increased. Irreversible transitions in the gating scheme may lead to nonmonotonic distributions (*see* Colquhoun and Hawkes [1983] for explicit derivations).

2.2.3.1. Two-States: Closed–Open Model. The fundamental building block of state transition kinetic schemes involves two states. For example, a simple, two-state channel with closed and open states (termed the C-O model) can be represented diagramatically as:

$$\text{closed} \underset{k_{-1}}{\overset{k_1}{\rightleftharpoons}} \text{open} \tag{2}$$

where k_1 and k_{-1} are rate constants (s^{-1}) for the opening and closing transitions, respectively. These rate constants are proportional to the probability that the transition will occur per unit time. For example, the rate constant k_{-1} is proportional to the probability per unit time that the channel, after entering the open state, will make the transition to the closed state. Therefore, it would be expected that, as k_{-1} is increased, i.e., the probability of the channel closing transition is increased, the average open dwell time will be decreased. The distribution of open dwell times is predicted by an exponential probability density function (pdf). The pdf is a continuous function, from which the probability density of a specific open dwell time, t_{dwell}, is given by:

$$p(t_{dwell}, \text{open}) = k_{-1} \exp(-k_{-1}t_{dwell}) \quad (3)$$

where t_{dwell} is a value of the open state dwell time, and $p(t_{dwell}$, open) is the probability density (s^{-1}) of t_{dwell} obtained from the pdf. The average dwell time in the open state is equal to the time constant (τ_{open}) of the exponential function, which is the reciprocal of the rate constant, k_{-1}. Thus, it may be useful to rewrite p(t_{dwell}, open) as:

$$p(t_{dwell}, \text{open}) = \tau_{open}^{-1} \exp(-t_{dwell}/\tau_{open}) \quad (4)$$

In a similar manner, the distribution of closed dwell times is given by:

$$p(t_{dwell}, \text{closed}) = k_1 \exp(-k_1 t_{dwell}) \quad (5)$$

with an average closed time equal to $1/k_1$ or τ_{closed}. It is important to note that, in the special case of a single exponential dwell-time distribution, the arithmetic mean of the dwell times is equal to τ under the restriction that there are no missed events. In practice, however, there is always a minimum detectable event duration, t_{min}, because of the limited bandwidth of the recording system. Taking into consideration t_{min}, the relationship between τ and the arithmetic mean is simply:

$$\tau = (\text{mean observed dwell time}) - t_{min} \quad (6)$$

2.2.3.2. Three States: Block of an Open Channel. Channel gating is usually more complicated than the simple C-O kinetic scheme diagrammed above. For example, a kinetic scheme that can account for the common observation that ion channels can be blocked by drugs that enter and block the channel during its open state is the C-O-B model. This blocking mechanism can be diagrammed as:

$$\text{closed} \underset{k_{-1}}{\overset{k_1}{\rightleftharpoons}} \text{open} \underset{k_{-2}}{\overset{k_2[B]}{\rightleftharpoons}} \text{blocked} \quad (7)$$

where [B] is the blocking drug concentration (molar), k_2 is the bimolecular rate constant for binding ($mole^{-1}s^{-1}$), k_2[B] is thus the transition rate (s^{-1}) for the blocking transition, and k_{-2} is the unblocking rate constant. As before, the average closed duration is $1/k_1$. The average blocked duration is $1/k_{-2}$. Since both the closed and blocked states are nonconducting states, they are kinetic states and are indistinguishable on the basis of conductance. They are combined in a single dwell time distribution for nonconducting states. This distribution can be described by the sum of two exponentials (Colquhoun and Hawkes, 1983):

$$p(t_{\text{dwell}}, \text{nonconducting}) = \left(\frac{k_{-1}}{k_{-1} + k_2[\text{B}]}\right) k_1 \exp(-k_1 t_{\text{dwell}}) + \left(\frac{k_2[\text{B}]}{k_{-1} + k_2[\text{B}]}\right) k_{-2} \exp(-k_{-2} t_{\text{dwell}}) \quad (8)$$

Estimation of the values of k_1 and k_{-2} is dependent on our being able to resolve the two components in the distribution of nonconducting dwell times. Identification of each component can be difficult, especially under the conditions where the rate constants are similar in value and/or the number of events is small. As a general rule, if there are m kinetic states at a specific conductance level, there should be m exponential components in the distribution of dwell times for that conductance level. The area under each component is proportional to the number of events in the distribution that belong to that kinetic state. The open state of the channel can be terminated by either normal closing, O-C, or by block, O-B. Therefore, the open dwell times in the C-O-B model are not only dependent on the closing rate constant, k_{-1}, but also on the transition rate for blocking, $k_2[\text{B}]$. The resulting distribution of open dwell times is described by a single exponential with an average open dwell time of $1/(k_{-1} + k_2[\text{B}])$. In general, the average dwell time in a given state is equal to the reciprocal of the sum of rate constants for transitions leading away from that state. Note also that in the C-O-B model, average open dwell time, but not the average nonconducting dwell time, is dependent on blocker concentration.

The number of identified states for channels has grown as the methods of recording and analysis have improved. Some kinetic schemes are considerably more complex than those shown above, with multiple open and closed kinetic states. Also, the pathways among the states can be linear, branched, cyclic, or a combination of all three patterns (e.g., Horn and Vandenberg, 1984; Colquhoun and Hawkes, 1987).

2.2.3.3. BURSTING MODEL. Often, the transitions between open and closed states appear in bursts separated by relatively long gaps during which the channel is silent. This is common in agonist-activated channels. A simple model that can account for bursting behavior in agonist-activated channels is:

$$R \underset{k_{-1}}{\overset{k_1[A]}{\rightleftharpoons}} AR \underset{k_{-2}}{\overset{k_2}{\rightleftharpoons}} AR^* \tag{9}$$

where R is the closed state, AR is the closed state with agonist bound, and AR* is the open state with agonist bound. If $k_1[A]$ and k_{-1} are slow relative to k_2 and k_{-2}, the channel activity will be divided into two epochs: (1) a quiet (nonconducting) period while the channel is in state R, and (2) a period of rapid opening and closing as the agonist-bound channel oscillates between the AR and AR* states before the agonist dissociates and the channel returns to the R state.

2.2.3.4. STATIONARY VS NONSTATIONARY ANALYSIS. Data analysis is easiest with stationary current records, i.e., records where the probability of occupancy of each state is independent of time. This type of record is common when long records are made at a constant potential or when a channel is allowed to reach a steady-state level of gating in the presence of an agonist. Alternatively, it is sometimes necessary to analyze nonstationary records. This is the case for channels that inactivate rapidly following a voltage step (e.g., Na^+ channels, Aldrich et al., 1983) or channels that desensitize in the presence of agonist (Brett et al., 1986). Methods for nonstationary analysis are discussed in Aldrich and Yellen (1983). It is important to emphasize that nonstationary probabilities of state occupancy observed, for example, after a voltage step do not contradict the Markovian assumption of independent state transitions and time homogeneous rate constants. We usually assume that voltage-dependent rate constants change instantly following a change in voltage. However, although the rate constants change instantaneously and are time homogeneous after the step, the probability of occupancy of each state will be nonstationary during a period of relaxation to a new steady-state. This concludes the overview of single-channel methods. In the next section (2.3), we will discuss in greater detail the actual recording methods, including the selection of a biological preparation, and the mechanical and electrical equipment with which single-channel recordings can be made.

2.3. Recording Single-Channel Currents

2.3.1. Recording Configurations

2.3.1.1. CHOOSING THE CONFIGURATION—REQUIREMENTS FOR RESOLUTION, MEMBRANE SIZE, AND ACCESSIBILITY. In a geometric sense, there are two solutions to the problem of isolating and recording currents from single channels. In the planar bilayer, regardless of the area of the membrane, the number of channels entering the membrane is limited to one by decreasing the concentration of vesicles in the suspension from which channels are incorporated. For patch clamping, the recording is focused, by the use of the pipet, on an area small enough to include only a single active channel. There are some practical limitations to these ideal situations. For example, when channels are incorporated into a bilayer from a membrane vesicle preparation, fusion of a single vesicle *(see below)* may insert multiple channels into the bilayer. Alternatively, with the high channel density in some native membranes (e.g., at a neuromuscular endplate), it may not be possible to make a pipet with a small enough tip for the patch to include only a single channel. Various tactics may be used in an attempt to circumvent these difficulties. Purely from the standpoint of minimizing noise, the smaller the membrane from which one records, the better, since high-frequency noise current increases directly with an increase in capacitance and, hence, with area. However, a number of other factors bear on the choice of the most appropriate method of single-channel recording to study a particular question.

As a path to an understanding of a channel's biological role and an analysis of the details of its function, the spectrum of available techniques includes recording configurations from on-cell patches, to excised patches and reconstituted systems that are highly complementary. The on-cell patch offers a way to observe function with minimal disturbance of the cellular system. Because a gigaseal provides an effective barrier between the solution outside the pipet and the patch, application of hormones to the bath can provide evidence for involvement of a diffusible intracellular messenger in hormonal effects on channel activity in the patch.

Excised patches and reconstituted systems offer ways to answer the question of what components are sufficient for such control systems to operate and, thus, to raise the possibility that these components may be functional in the intact cell. Exchange of solutions on each side of the membrane is possible in both patch recording and bilayer chamber methods, so the need to do this might not influence one's choice of method. Very fast solution changes *(see below)* are easiest in the small volumes bathing a pipet tip. However, open bilayer chambers offer an advantage for labile reagents, which are most conveniently added in solid form and immediately stirred into solution against the bilayer. These chambers also make possible drug or reagent application to either side of the membrane by simple addition of an aliquot of concentrated stock solution. Bilayer methods offer by far the most convenient and effective way of studying the influence of membrane lipid composition. Also, in some cases, membranes that are not normally accessible to a pipet tip may be concentrated by fractionation techniques and incorporated into planar bilayers or large vesicles for recording.

2.3.1.2. Patch Pipet Methods. We will provide only a brief overview of patch pipet making and recording since recent, detailed descriptions of the techniques are available in chapters of Sakmann and Neher (1983a), and in articles by Rae and Levis (1984) and Auerbach and Sachs (1985). Within this broad-ranging chapter, we cannot hope to do justice to all the foibles and intricacies of the art.

2.3.1.2.1. Pipet Fabrication. The relatively large-tipped (diameter about 1 μm) pipets used for patch-clamp recording can be easily pulled on a variety of micropipet pullers, though some added convenience may be provided by one of the recently available, microprocessor-controlled models, which can be programmed for multi-step pulling. Pipets are usually pulled in two or more steps, the first pull(s) to thin the glass and the last to separate the capillary into two pipets with a tip diameter of about 1 μm. A mechanical stop is easily made to terminate the first elongating pull, and on vertical pullers, the weight of the clamp assembly provides sufficient force for the pull. After pulling, the tip of the pipet is coated with Sylgard #184 to make the surface of the pipet hydrophobic and to increase wall thickness near the tip. This increases the wall resistance and decreases capacitative coupling to the bath both by decreasing wall capacitance and by decreasing the tendency of the aqueous solution to creep up the outside of the pipet. Sylgarding thus substantially decreases noise in the recording. One should apply Sylgard by hand while observing through a dissecting microscope. Rae and Levis (1984) suggest mounting the pipet with the tip up to decrease the tendency of a film of Sylgard to creep to the tip and thus prevent seal formation. A small chuck, allowing the pipet to be freely rotated by hand around its axis during application of the Sylgard, makes a convenient holder. After the application, the pipet is drawn back through a heated coil to cure the Sylgard. The tip of the pipet is then polished to its final diameter and configuration (see Sakmann and Neher, 1983b) while observing with a compound microscope at about 200–400×, using a heated platinum wire that has been coated with glass from a capillary similar to the pipet being made. A quick, convenient index of tip size can be obtained by attaching a syringe to the pipet with plastic tubing and noting the pressure required to bubble air through the pipet tip into methanol. Immediately before use, the tip of the pipet is filled by suction, after which the shank is backfilled, with a syringe, through a fine needle or a piece of pulled-out polyethylene tubing (*see* Corey and Stevens, 1983, Rae and Levis, 1984, for details and precautions). Any bubbles remaining in the taper are removed by tapping.

2.3.1.2.2. Seals and Recording Configurations. Regardless of the intended recording configuration, the first step is to establish a gigaseal on the cell or vesicle. Observation of the pipet and cell is easier using interference contrast (Nomarski) or modulation contrast (Hoffman) optics. For proper function, Nomarski optics require that recording chambers be constructed with only glass in the optical path. The Hoffman system, on the other hand, allows direct observation of cells in plastic culture dishes. Gentle positive pressure is applied to the pipet by mouth, syringe, manometer, or constant pressure pump during the approach to the cell in order to avoid clogging. The chance of clogging is also reduced by passing all solutions through a 0.2-μm filter before use. A gigohm seal, or gigaseal, has, by definition, a resistance >1 GΩ. For recording purposes, the higher the resistance, the better, and the resistance may reach well over 100 GΩ. Occasionally, a gigaseal forms spontaneously on contact with the cell. Usually, gentle suction must be applied to induce seal formation, which may occur in a few seconds or, occasionally a few minutes. A much greater membrane area is involved in making the seal than might, at first, be thought. The applied suction can draw the membrane into the pipet as much as 10–20 μm, and much of that membrane area is closely apposed to the glass and thus involved in making the seal. Once a gigaseal is formed, one has a choice of recording configurations from a patch *in situ* "on cell," or excised patches that are "inside-out" or "outside-out." Briefly, an inside-out patch is formed by withdrawing the pipet after seal formation until the membrane separates from the cell. Often, this results in a vesicle in the tip of the pipet. The outer layer of membrane is broken by briefly drawing the pipet out through the air-solution interface and then returning it to the solution, leaving a single layer of membrane adhering to the pipet tip. The cytoplasmic membrane surface faces the bath, and the extracellular surface faces the pipet contents. Outside-out patches are produced by breaking the patch in the pipet tip, while still attached to the cell, and then slowly withdrawing the pipet, allowing the membrane to break from the cell and reseal with the extracellular surface facing the bath. Immediately before separating the outside-out patch from the cell, one is in the "whole-cell" recording mode with the interior of the pipet in direct contact with the cytoplasm (*see* chapter by S. Jones, in this vol.). A clear description of the different patch configurations and their formation can be found in the article by Hamill et al, (1981), which is reprinted as an appendix in Sakmann and Neher (1983a). Seal formation is also possible using liposomes made from purified lipid, and the ability to record from isolated, excised patches allows patch recording from purified channels, reconstituted into freeze-thaw enlarged liposomes (Tank and Miller, 1983; Rosenberg et al., 1984).

2.3.1.2.3. Changing Solutions. The tiny dimensions of the patch-clamp preparations, particularly using an excised patch, offer another experimental advantage—the possibility of making quite rapid solution changes, at least at one membrane surface. The use of several parallel pieces of fine plastic tubing providing parallel laminar flow of a number of different solutions allowed Yellen (1982) to complete changes in the solution at the surface of a patch, in about 1 s, simply by moving the pipet bearing the excised patch from one solution stream to the next. A faster solution change (5–15 ms), easily controlled remotely by computer, has been achieved using a solenoid-driven pinch valve by Brett et al. (1986). An alternative method, which allows use of very small volumes of each solution (down to about 100 μL), is described by Kakei and Ashcroft (1987). This opens the possibility of studying the responses to step changes in agonist and other drug concentrations in the range of 10–100 ms. Although such rapid changes of the solution inside the pipet are not possible, one can change this solution during a recording, and this has been done in a few studies (e.g., Cull-Candy and Parker, 1983; Soejima and Noma, 1984; Kameyama et al., 1985; Fischmeister and Hartzell, 1987). Perfusion of the inside of the pipet allowed the latter workers to change solutions at the membrane surface within about 10 s.

2.3.1.2.4. How Inert Are Pipet Glasses? It is just becoming clear that patch recording pipets cannot be considered as an inert recording probe to convey a signal from cell to amplifier for analysis. Cota and Armstrong (1988) noted an inactivation-like block of potassium channels in whole-cell currents recorded from cultured pituitary cells using soft, lead-containing, or soda glasses. This block was prevented by adding 10–20 m*M* EGTA to the pipet solution. A study of the current–voltage relations of the cyclic GMP-activated conductance of vertebrate rods has shown that glass-dependent differences can largely be eliminated by inclusion of similar, high concentrations of EGTA in the pipet. Only for one glass, Corning ™0010, which contains minimal amounts of divalent and trivalent cations other than lead, was the form of the current–voltage relation independent of chelator concentration (Furman and Tanaka, 1988). Glass properties are generally controlled, in manufacture, by addition of substantial amounts of di- and trivalent metal oxides (Corey and Stevens, 1983). It is well to be aware that cations may be released from the tip of the pipet and accumulate to substantial concentrations. One can then carry out appropriate controls with different glasses, with high concentrations of chelating agents, or perhaps using bilayers formed on plastic partitions.

2.3.1.3. Bilayer Chamber Methods—Painted and Folded Membranes. Two general methods are commonly used to form planar lipid bilayers. Detailed accounts of each approach, plus many additional references may be found in various chapters of the book edited by Miller (1986). Both types of bilayer membranes have been used for studies on channels from biological membrane (*see* Section 2.3.1.5., *below*). In each method, formation of stable bilayers is facilitated, or enabled, by pretreating the hole in the plastic partition on which the membrane is to be formed with either the bilayer-forming solution, a dilute solution of petroleum jelly in pentane or hexane, or with hexadecane or squalene.

2.3.1.3.1. "Painted" Bilayers. A "painted" bilayer (Mueller et al., 1962; Miller, 1986) can be formed by spreading a dispersion of lipid in an organic solvent, usually decane, across a hole in a plastic partition between two aqueous solutions. Under the right conditions (White, 1986), the blob of lipid and solvent spontaneously thins to form a bilayer. If relatively low molecular weight solvents, such as decane, are used, significant amounts of solvent remain in bilayers formed by this method, resulting in membranes that are thicker than either native membranes or bilayers formed by apposition of two lipid monolayers. This is reflected by the lower specific capacitance (0.3–0.5 $\mu F/cm^2$) than for folded bilayers or native cell membranes. Solvent-free painted bilayers can be formed, however, by using a higher molecular weight solvent, such as squalene, which does not partition into the bilayer (White, 1986).

2.3.1.3.2. "Folded" Bilayers. "Folded" bilayers which are almost solvent-free, can be formed by apposition of two monolayers left by evaporation of a volatile solvent (pentane or hexane) from solutions of pure lipid spread on the surfaces of aqueous solutions in a two-compartment chamber. This was the method used by Montal and Mueller (1972). The solutions are raised, one at a time, past a hole in the partition between two compartments to bring the monolayers into contact. It should be noted that such membranes are not completely solvent-free, since an annulus of hydrophobic material is essential for formation of a stable bilayer (White, 1986), and this annulus is in a dynamic equilibrium with the bilayer. A recent paper notes that even "folded" bilayers are formed by a thinning process, in many respects similar to that seen in "painted" membrane formation (Niles et al., 1988).

2.3.1.3.3. Chambers. Two common types of bilayer chambers are used. In one type, a machined cup (usually 1–3 mL) of plastic (polystyrene, polycarbonate—Lexan™—or chloro-fluoro-polyethylene—Kel-F™) forms one compartment, and the hole (100–500 μm), on which the bilayer is formed, is drilled in a thinly machined area on one side of the cup. The cup slips snugly into one-half of a figure-eight-shaped hole in a plastic block. The second half of the hole forms the other compartment of the chamber and generally provides a window for convenient viewing of the bilayer during formation. It is useful to view the membrane through a dissecting microscope as it thins to form a bilayer, but it is important to note that membrane capacitance and the ability to incorporate functional channels provide the ultimate criteria that a bilayer has indeed been formed. The advantages of the cup-type chambers are that they are robust and are not prone to developing leaks, providing that no cracks develop during the machining and drilling of the hole. Disadvantages are that manufacture requires fairly delicate machining, and that it is difficult to mill very thin walls and drill small holes (<100 μm). A variant on this type of chamber uses, in place of the machined cup, a disposable plastic syringe or test-tube cap with a hole formed in a heat-thinned area of the wall, and there are many other possible improvisations on this theme. It is important in the making of small diameter holes that the thickness at the margin of the hole be minimized. We have found that small bilayers (<100 μm in diameter) are easier to form and more stable when the thickness of the margin is <≈25 μm (1 mil). This can be difficult to achieve if milling and drilling are required. We have experienced difficulty in attempting to incorporate channels into membranes formed on 100-μm holes drilled in cups. We suspect that this results, in part, from slowed access of vesicles to the bilayer through an unstirred tunnel formed by the walls of the hole (*see* section 2.3.1.5.1). A small diameter hole can be easily made in cups made of plastics that have a low melting temperature, such as polystyrene or polyethylene. A sharpened needle (tip diameter < 10 μm, tip angle ≈ 70–80°) is warmed and pressed part of the way through the wall of the cup from the inner surface. After cooling, the needle is removed and the plastic shaved away from the external surface with a disposable microtome blade. When the blade cut intersects the depression formed by the needle, a hole is formed. This method can be used to produce small diameter holes (≈30–50 μm diameter) that possess several ideal properties. First, the plastic at the margin of the hole is very thin, resulting in a reduced volume of solvent in the anulus and a decreased thickness of the unstirred layer, at least at the outer surface of the bilayer. Also, the thickness of the plastic increases rapidly beyond the margin of the hole, thereby providing mechanical strength and reducing stray capacitance across the partition.

An alternative type of chamber employs a thin Teflon™ partition clamped between two plastic blocks in which the chamber compartments have been milled. Teflon™ film is available in thicknesses from 0.0005-in, and holes of various sizes, down to about 30–50 μm, may be melted in the film using a fine, electrically heated wire or by dielectric breakdown using a spark from a high-voltage ignition coil (*see* Hartshorne et al., 1986). We use an 0.001-in platinum wire, pinched into a tight loop with watchmaker's forceps. For holes of 100–150 μm, we usually use 0.001-in film, whereas for smaller holes, the 0.0005-in thickness is used. To minimize the capacitance of the partition itself, the thin film is heat-bonded to a thicker piece (0.004 or 0.005-in) of Teflon™, with a prepunched hole of about 1 mm to expose a small area of the thinner film in which the small hole for bilayer formation is to be melted. Using these precautions, the total capacitance (including bilayer and partition) between the chambers may be reduced to 25–50 pF, or with the simpler construction of a single piece of Teflon™ and a hole of about 100 μm, total capacitance of ≈100 pF can be obtained. Instead of using Teflon™ film, it is also possible to cut film of the desired thickness from a block of high density polyethylene using an ordinary microtome.

2.3.1.4. Bilayers on Pipets

2.3.1.4.1. "Tip-Dip" Techniques. Very small folded bilayers may be formed on the tip of patch clamp pipets by passing the tip of a clean glass pipet, usually not fire-polished (but *see* Coronado, 1985), in, out, and in through a lipid monolayer spread on the surface of a saline solution (Coronado and Latorre, 1983; Hanke et al., 1983; Schuerholz and Schindler, 1983; Suarez-Isla et al., 1983). Under ideal conditions, this has the advantage of providing very low noise recordings, and for some preparations, the reduction in probability of channel incorporation because of the small membrane area is an advantage (e.g., Levitan, 1986). Also, very small volumes can be used, enabling reconstitution of channels from single, identified neurons (Levitan, 1986). However, even when a seal of 1–10 GΩ has been formed, this method seems prone to artifactual, intermittent, channel-like leakage, which can provide considerable annoyance and confusion. Stability and resistance of the seals between negatively charged lipids and glass are enhanced by divalent ions (Coronado, 1985), but the nature of the seal for all lipids is not completely understood. The method seems most generally useful for experiments in which substantial concentrations (0.01–0.1*M*) of divalent ions can be included in the solutions. If solutions must be free of divalents, phosphatidylethanolamines at neutral (or acid) pH are recommended. Many, though not all, phosphatidylcholines appear to seal poorly. Additional studies using different glasses and ionic conditions may well prove fruitful.

2.3.1.4.2. Solvent-Containing Membranes on Pipets. Decane-containing membranes can be transferred to the tip of a glass pipet (diameter about 10–30 μm), which has been made hydrophobic by silanization, by passing the pipet through a preformed painted membrane. This technique was mentioned by Mueller (1975) and has been described in detail by Andersen (1983). Membranes formed in this way can provide extremely high seal resistances and be stable enough to withstand very high voltages (Andersen, 1983). The method provides a way of picking one or a few channels out of a membrane into which many have been incorporated (*see* Green et al., 1987), as well as allowing low noise recordings. The total input capacitances achieved with this method differ only marginally from those attainable with the smallest bilayers formed on thin Teflon™ partitions, and a choice of the recording configuration may depend on considerations other than capacitance and, hence, noise levels.

2.3.1.5. Incorporation of Channels

2.3.1.5.1. Vesicle Fusion. The most adaptable method of incorporation of channels into a bilayer for electrical recording is fusion of vesicles containing functional channels (Miller, 1978; Cohen, 1986). Channel function may be verified by flux studies using the vesicles in suspension (e.g., Krueger and Blaustein, 1980). Clear evidence of fusion as the mechanism of incorporation of proteins into a bilayer has been obtained by Cohen et al. (1980) and Zimmerberg et al. (1980a). A variety of factors, including the presence of free Ca^{2+}, an osmotic gradient oriented to drive water into the vesicles, and the presence of lipids with negative charges appear to enhance the probability of incorporation in different preparations (e.g., Cohen et al., 1982; Zimmerberg et al., 1980b). There are, on the other hand, preparations that seem to break all of these rules. A convenient and generally successful method of establishing an osmotic gradient is to load vesicles with a solution isotonic or slightly (about 10%) hypertonic to the experimental solution, which is to be placed into the "cis" chamber (e.g., 100–200 m*M* salt) at the beginning of the experiment. The bilayer is formed with an hypotonic solution in the opposite ("trans") chamber (e.g., 10–20 m*M* salt), and the vesicles are added to the cis chamber. After incorporation, an aliquot of concentrated saline is added to the trans chamber to remove the gradient. This reduces the probability, but does not eliminate the possibility, of further incorporation. It simplifies the experimental protocol by not requiring perfusion of the cis chamber, which increases the likelihood of breaking the bilayer. If perfusion is required, it can be simply and effectively achieved by using two syringes in a push–pull arrangement, so that one removes solution from the chamber as the other introduces an equal volume, thus maintaining a constant level in the chamber. The chance of electrical transients breaking the membrane during perfusion can be reduced by shorting the cis and trans chambers together during perfusion (e.g., using a salt bridge).

Two processes contribute to the rate at which vesicles reach a bilayer after addition to the bath: (1) convective movement resulting from stirring, and (2) diffusion through the unstirred layer adjacent to the bilayer. Because convective movement is relatively rapid, diffusion through the unstirred layer is likely to be the rate-limiting process. The time for diffusion across an unstirred layer is proportional to the square of the thickness of the layer and to the radius of the vesicles. For example, diffusion of 0.1-μm diameter vesicles through a 100-μm unstirred layer would require ≈19 min, whereas diffusion of the same vesicles across a 20-μm layer would require only ≈0.8 min. The effect of the unstirred layer can be reduced by pressure application of vesicles from a pipet placed close to the bilayer (Niles and Cohen, 1987).

If incorporation from a given native membrane preparation is achieved without breaking and reforming the bilayer, channels usually incorporate with uniform orientation (an exception is noted by Latorre, 1986). Orientation of channels in, and incorporated from, vesicles from a purified, reconstituted preparation is usually random. Sometimes, however, it is worth deliberately breaking and reforming the membrane, since channels often appear in the bilayer after repainting. If it is difficult to achieve incorporation, the risk of nonuniform orientation may be worth taking.

Ease of incorporation of channels by vesicle fusion is preparation-dependent. For example, channels seem to incorporate more readily from rat sarcolemmal vesicles than from rat brain vesicles. Also, incorporation seems to occur more readily into painted decane-containing membranes than into folded membranes (*see* Cohen, 1986).

2.3.1.5.2. Incorporation of Channel Proteins from Monolayers. Channel proteins may also be incorporated into folded bilayers formed either on pipet tips or Teflon™ partitions by including the proteins in the monolayers from which the bilayers are formed. The monolayers are formed by equilibration with liposomes containing the purified channel protein (e.g., Nelson et al., 1980), or formed from a mixture of native membrane vesicles and exogenous lipid (e.g., Schindler and Quast, 1980; Levitan, 1986).

2.3.2. Current Amplifiers

The major challenge in the design of a current amplifier is to simultaneously maximize the signal-to-noise (SNR) ratio to permit resolution of small ionic currents (0.2–10 pA) that flow during the conducting state of an ion channel and to enhance the frequency response, or bandwidth (BW), of the amplifier such that rapid transitions between conductance states can be accurately identified. Unfortunately, in practice there is a considerable trade-off between SNR and bandwidth. The following discussion will focus on various factors that influence SNR and bandwidth, with special emphasis on factors under the control of the experimenter. For a more detailed description of the theoretical basis for the design of current amplifiers, the reader can consult several excellent reviews, including Hammill et al. (1981), Sigworth (1983a), and Rae and Levis (1984).

2.3.2.1. GENERAL DESIGN. A schematic of a generalized recording system consisting of the sequential steps of current-to-voltage conversion, high-frequency boost, and analog filtering is shown as a block diagram in Fig. 2. The principal component is the headstage, which includes the virtual ground circuit that produces the current-to-voltage conversion. This circuit is shown in greater detail on Fig. 3a. The input to the current-to-voltage converter is modeled as a parallel equivalent resistance, R_{input}, and capacitance, C_{input}. R_{input} consists of three parallel resistors:

1. R_m, the membrane resistance
2. R_{shunt}, the seal or shunt resistance and
3. $R_{chan}(=1/g_{chan})$, representing the resistance of the pore of the channel.

R_{chan} is in series with a voltage source, V_{eq}, the equilibrium potential of the permeant ions and a switch, sw_1, representing the gate of the channel. R_{input} is in series with R_{series}, representing the series resistance through which i_{input} must flow, including the resistance of the bulk and pipet solutions. C_{input} consists of two parallel capacitors, C_m, the membrane capacitance, and C_{stray}, the stray capacitance arising from various sources (e.g., amplifier input, bath chamber). Application of a command voltage, V_{com}, across R_{input} and C_{input} produces a current, i_{input}. When the channel is in the nonconducting state, the current level following charging of C_{input} is the baseline current, i_{closed}, equal to $V_{com}/(R_m||R_{shunt} + R_{series})$. A subsequent transition of the channel to the conducting state (closing of switch sw_1) then increases i_{input} to a new level, which is equal to the sum of i_{closed} and the additional current that flows through the open channel, i_{open}, or:

$$i_{input} = V_{com}/(R_m||R_{shunt} + R_{series}) + (V_{com} - V_{eq})/R_{chan} \quad (10)$$

The headstage amplifier, A_1, provides current feedback, i_f, across the feedback resistor, R_f, such that the potential at the inverting input, $V_{in(-)}$, is actively maintained at the same potential as the potential at the noninverting input, $V_{in(+)}$, in this case ground potential. The input impedance of A_1 is very high, and any current flowing across the membrane to the summing node at the (–) input must also flow across R_f. Under this condition, i_f is equal to i_{input}. Since $V_{in(-)}$ is held at ground potential, the voltage drop across R_f is equal to $i_f R_f$, thus providing a voltage output, V_{out}, proportional to i_{input}. The orientation of V_{com} and ground potential can be reversed, such that V_{com} is applied to the (+) input, and $V_{in(-)}$ is then actively maintained at V_{com} with the other side of the membrane grounded (Fig. 3B). In this case, V_{out} is the sum of V_{com} and $i_f R_f$. This circuit requires a differential amplifier (not shown) to subtract V_{com} from V_{out}. Because of the small size of the single-channel current (and, hence, i_f), the value of R_f must be quite large (1–100 GΩ) to produce a resolvable V_{out}. As a consequence of the large value of R_f, any capacitance, C_f, in parallel with R_f produces a low-pass filtering of the V_{out} response to a square current-pulse input, usually limiting the frequency response of V_{out} to a few hundred Hz. Under the conditions that the gain-bandwidth product of A_1 is large and that R_f and C_f act as a simple first-order, low-pass filter, the frequency response of the headstage can be restored by a frequency "boost" circuit (Fig. 2D). This circuit adds to V_{out} a scaled derivative of V_{out}, restoring the original square wave. Practical "boosted" bandwidths range from 1–30 kHz. The boost circuit is a critical determinant of the effective bandwidth of the amplifier and is discussed in detail in Sigworth (1983a) and Rae and Levis (1984).

A sequence of single-channel events can be idealized as a

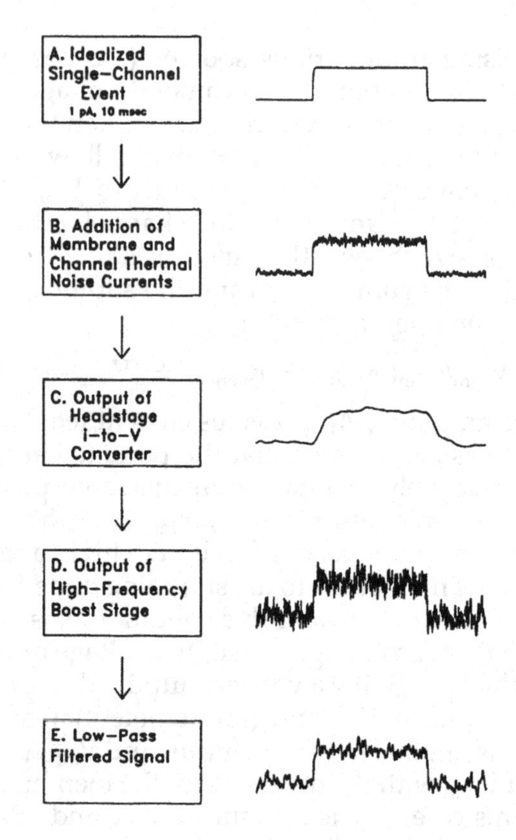

Fig. 2. Simulation of the processing by the recording system of an open event generated by a single ion channel isolated in a membrane. (A) The idealized event is a rectangular step with an amplitude of 1 pA produced by a 10 ms opening of a 25 pS channel held at 40 mV. The event is shown digitally filtered (a Gaussian digital filter is used in all records) with a bandwidth of 10 kHz. (B) The thermal noise currents generated by the channel resistance (R_{chan}, 40 GΩ) and the membrane resistance (R_m, 100 GΩ) have been added to the idealized event. The Gaussian noise currents added in B and C were generated by the whitenoise command of the DAOS (Ver. 7.0) software. The whitenoise array was digitally filtered at 10 kHz, normalized, and then scaled to an amplitude 6 $rms_{(BW)}$, where $rms_{(BW)}$ was calculated as the square-root of the sum of the variances of the independent noise sources. (C) The current pulse has been converted to a voltage (50 mV/pA) by the headstage current-to-voltage converter. The current record now also includes background noise generated by the thermal noise in R_f (50 GΩ) and R_{shunt} (100 GΩ), by shot noise resulting from i_{gate} (1 pA) and by interaction of the headstage input voltage noise (2

series of rectangular current pulses resulting from rapid transitions between conductance states. Ideally, kinetic analysis requires that we measure the duration and amplitude of events without distortion by the recording system. However, the limited bandwidth of current amplifiers distorts each current step, reducing the apparent rate of transition between states, and biasing the amplitude and duration of brief events. The distortion of the current steps is dependent on the rise-time (t_{rise}) of the amplifier, which is defined here for a squarewave input as the time required for the output to rise from 10–90% of its peak level. It is essential to maximize the bandwidth to minimize t_{rise}. Although the frequency responses of amplifiers are usually described in terms of bandwidth based on amplification of sinusoidal signals, t_{rise} is a particularly valuable characterization of a single-channel current amplifier because of the square-wave nature of single-channel current fluctuations. The t_{rise} of a headstage amplifier should be measured by applying a square current pulse to the amplifier input. This is easily done by capacitatively coupling a triangle wave to the amplifier input (e.g., through a small fixed capacitor or a wire placed adjacent to the input), because the coupling injects the derivative of the triangle wave, which is a square wave. The t_{rise} of a current amplifier (and the recording system) should always be measured, because it limits the ability of the recording system to record brief events.

←

nV/√Hz) with C_{input} (15 pF) (these noise sources are shown in greater detail in Fig. 4A). The current record has been digitally filtered at ≈ 150 Hz to simulate the low-pass filtering produced by the parallel feedback resistance and feedback capacitance (C_f, 0.02 pF). (D) The frequency response of the current record is "boosted" by adding to the headstage output a scaled derivative of the output. This restores the frequency response of the signal concomitant with increasing the high-frequency noise. To simplify simulation, we show the record that was constructed from signal and noise components before filtering as shown in C. For the sake of simulation, we assume that the boosting is 100% effective and that the effective boosted bandwidth is 10 kHz. Note that some peaks are large enough to cross the 50% threshold between the closed and open levels, which would be falsely interpreted as brief events. (E) The current record has been low-pass filtered to 5 kHz to reduce the background noise peaks to levels where the 50% threshold is not randomly crossed.

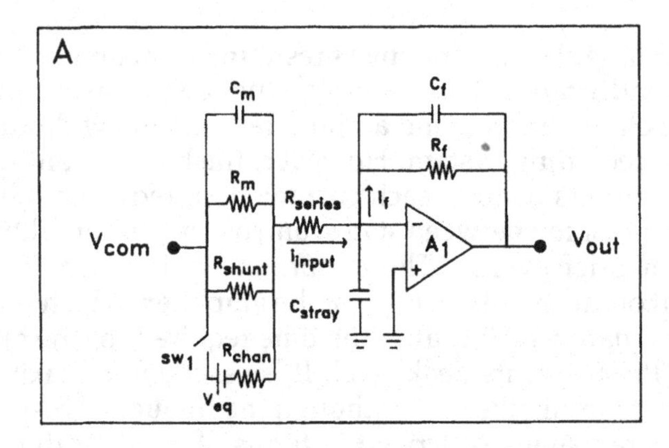

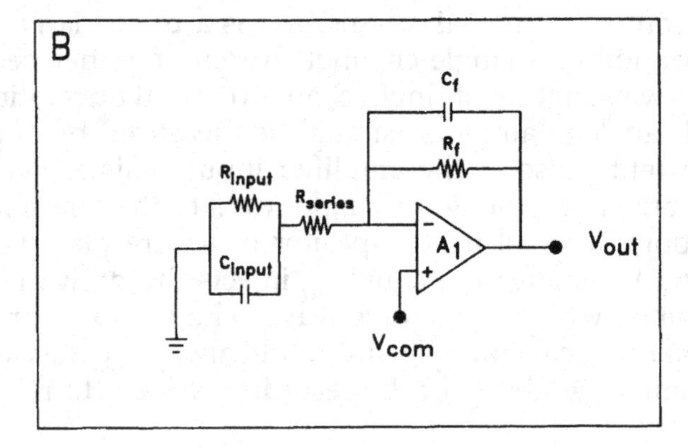

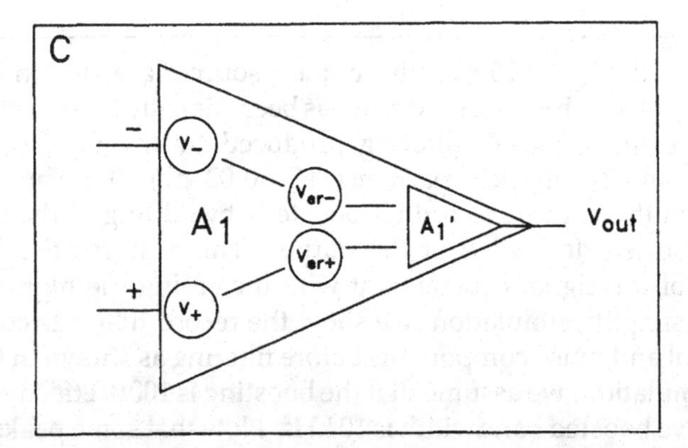

Fig. 3. (A) Schematic of headstage amplifier. The notation is defined in the text and glossary. This stage of the recording system provides the current-to-voltage conversion. The total input current, i_{input}, is equal to the amplifier feedback current, i_f. Because the summing node of

Two additional headstage circuit components that are not shown in Fig. 3 may be important under some conditions. The first is a transient cancellation circuit, which provides control of charging of C_{input} during changes in V_{com}. Step changes in V_{com} produce large transient currents during the charging of C_{input} that can both interfere with baseline identification and saturate the output of amplifier A_1 producing temporary loss of control of membrane potential. This is especially critical for voltage-clamp experiments with lipid bilayers where C_m can be quite large (>100 pF) and the charging current can reach several nA in amplitude. Transient cancellation circuits are discussed in detail in Sigworth (1983a). Briefly, the transient cancellation circuit uses an additional amplifier to provide charging current proportional to a scaled derivative of V_{com} through a capacitor coupled to the (–) input of A_1. Because the charging current is not supplied by A_1, this arrangement prevents saturation of the output of A_1 by a large current being passed across R_f. Transient cancellation is complicated by the fact that the charging of C_{input} must often be treated as if there were several RC circuits with different time constants. Thus, complete transient compensation requires injection of the sum of several derivatives of V_{com}, each with a different time constant. In practice, it may be difficult to completely cancel the capacitative current, especially if it is large and/or nonideal, such as with slow, electrostrictive changes in the capacitance of the annulus of a bilayer. Digital subtraction during computerized data analysis may prove useful under these conditions.

A second circuit is the series resistance compensation circuit. This circuit is critical where large currents might produce a significant voltage-drop across R_{series}, a resistance (e.g., bath solution, pipet tip) in series with R_{input}. This is especially important for the whole-cell voltage-clamp where, following formation of a high-resistance seal between pipet and cell surface, the patch of membrane is ruptured and the voltage of the entire cell membrane is

←

the amplifier is held at ground potential, the output voltage is proportional to i_{input} with a ratio of $R_f 10^{-12}$ V/pA. (B) Alternative arrangement obtained by reversing the ground reference and command voltage. In this case, V_{out} is the sum of $R_f i_f$ and V_{com}. (C) Schematic representation of A_1 showing addition of noise voltages ($V_{er(-)}$ and $V_{er(+)}$) to the noiseless signal voltages ($V_{in(-)}$ and $V_{in(+)}$) at input of ideal amplifier A_1'.

controlled. Compared to patch-clamp, R_m is much smaller and C_m is much larger because of the greater area of voltage-clamped membrane. With the larger currents required under these conditions, there is a voltage-clamp error because a larger portion of V_{com} is applied across R_{series}. This source of error can be reduced by using the output of the current monitor to scale V_{com} to compensate for R_{series}. These circuits are described in detail in Sigworth (1983a) (*see also* the chapter in this vol. by S. Jones).

2.3.2.2. NOISE. The full "boosted" bandwidth of the headstage amplifier is rarely usable, because an acceptable SNR usually requires filtering to remove noise such that legitimate transitions between conductance states can be resolved (Fig. 2E). Thus, the minimization of noise is an important step towards increasing the usable bandwidth. Noise can be represented quantitatively by several measures. Ultimately, the investigator is most interested in the peak-to-peak noise level, since this can be directly related to the single-channel current amplitude. However, from the viewpoint of circuit analysis, noise is typically expressed as a noise density in a 1-Hz bandwidth, either as the variance ($S^2_{(Hz)}$, volts²/Hz or amps²/Hz) of the noise density or its square root, the rms value ($rms_{(\sqrt{Hz})}$, volt/$\sqrt{}$Hz or amp/$\sqrt{}$Hz), which is simply the standard deviation of the noise density. Care should be exercised in measuring rms noise, because "approximate" rms meters produce an output assuming that the noise source is sinusoidal, whereas the output of a "true" rms meter is independent of the nature of the noise source. If the noise is frequency-independent, the variance of the noise in a larger bandwidth (BW) can be calculated as the product of the variance of the noise density and BW ($S^2_{(BW)}$, volts² or amps²). The rms noise in BW ($rms_{(BW)}$, volts or amps) can then be calculated as the square root of S^2_{BW}. If the noise density is frequency-dependent, calculation of the noise in BW requires integration of $S^2_{(Hz)}$ within BW. The total noise from several uncorrelated noise sources can be calculated directly as the sum of the noise variances, but not as the sum of the rms or peak-to-peak noise values. For normally distributed noise (Gaussian), the peak-to-peak noise in a specified bandwidth is ≈5–6 × the rms noise.

2.3.2.2.1. Thermal Noise. All conductors exhibit an inherent noise that results from the thermal agitation of the charged particles (electrons or ions) in the conductive medium. The noise occurs in the absence of applied current, and gives rise to a noise voltage across the conductor or a noise current through the conductor. This noise is frequency-independent and Gaussian, and is commonly known as thermal or Johnson noise. The physical basis of thermal noise is discussed in detail in DeFelice (1981). The thermal noise current density in a resistor is given by the equation:

$$S^2_{(Hz)} = (4kT)/R \tag{11}$$

where k is Boltzmann's constant (1.381×10^{-23}), T is absolute temperature, and R is the resistance. Thermal noise is generated in R_m, R_{shunt}, R_{chan}, and R_f, and places a lower limit on the SNR of the current amplifier. The values of $rms_{(BW)}$ for 1 GΩ, 10 GΩ, and 100 GΩ resistors with a bandwidth of 10 kHz are 0.4, 0.13, and 0.04 pA, respectively. One of the critical advances of patch recording technique has been the increase in the values of R_m and R_{shunt}, resulting from both a reduction in the area of membrane patch and an improved pipet-membrane seal. One advantage of the bilayer recording system is that very high values of R_m and R_{shunt} can be easily attained, greatly reducing these sources of thermal noise. Thermal noise currents are also generated in the feedback resistor (R_f). Thus, the SNR is increased by using high values of R_f (>10 GΩ), because the thermal noise currents are reduced at the same time that the scaling of the current-to-voltage conversion (mV/pA) is increased. Ideally, both R_{input} and R_f should have high values, because the thermal noise component is proportional to the equivalent parallel resistance ($R_{input} \| R_f$) and, therefore, dominated by the lower of the two resistance values.

2.3.2.2.2. Shot Noise. The conduction of current involves the movement of discrete charge carriers in a conductive medium. The discontinuous nature of charge movement results in the generation of a frequency-independent, Gaussian noise current termed shot noise. The shot noise current density is given by the equation:

$$S^2_{(Hz)} = 2Iq \tag{12}$$

where I is the current (amps) and q is the elementary charge (1.602×10^{-19}C). The source of shot noise in the headstage amplifier is the gate leakage current (i_{gate}) of the input FET transistors of amplifier A_1. The magnitude of i_{gate} typically ranges from ≈.5–10 pA. The minimization of i_{gate} and shot noise is an important consideration, because the shot noise generated by a 5-pA i_{gate} is equivalent to the thermal noise of a 10 GΩ resistor. The manufacturer's specification of i_{gate} for a given FET must be used carefully because (1) there is considerable variability among individual FETs, (2) i_{gate} is dependent on the biasing of the FET (i.e., the drain-source voltage), and (3) i_{gate} is temperature-sensitive with a doubling of i_{gate} for each ≈10°C increase.

2.3.2.2.3. Amplifier Noise Voltage. As discussed above, amplifier A_1 produces an output voltage (V_{out}) and feedback current (i_f) such that the (–) and (+) inputs are held at the same potential. For a noiseless amplifier, this process is described by:

$$V_{out} = G[V_{in(+)} - V_{in(-)}] \quad (13)$$

where G is the gain of the amplifier, and $V_{in(-)}$ and $V_{in(+)}$ are the potentials at the (–) and (+) inputs of A_1, respectively. A more realistic representation of A_1 is shown in Fig. 3C. $V_{in(-)}$ and $V_{in(+)}$ are now represented as being in series with noise voltages ($V_{er(-)}$ and $V_{er(+)}$) internal to A_1, followed by a noiseless amplifier (A_1'). With the addition of these noise voltages, Eq. (13) can be written as:

$$V_{out} = G\{[V_{in(+)} + V_{er(+)}] - [V_{in(-)} + V_{er(-)}]\} \quad (14)$$

The significance of the addition of the noise voltages is that V_{out} will now vary according to the magnitude of $V_{er(-)}$ and $V_{er(+)}$, even when $V_{in(-)}$ or $V_{in(+)}$ do not vary. Thus, in our current amplifier, even if i_{input} is constant, i_f will exhibit random fluctuations proportional to the error in $V_{er(-)}$ and $V_{er(+)}$. If C_{input} and C_f are neglected and only the equivalent resistance of R_f in series with R_{input} is considered in the feedback loop, the current noise produced by the noise voltage is small because the voltage is applied across a large equivalent resistance. However, with the addition of C_{input} to the circuit, the impedance of $R_{input}||C_{input}$ is reduced at high frequencies. Under this condition, the high-frequency components of the noise voltage in V_{out} must force larger currents to flow across R_f and R_{input} to maintain $V_{in(-)}$ at ground potential. The amplifier noise voltages $V_{er(+)}$ and $V_{er(-)}$ can be lumped as a single noise voltage, V_{er}. The current noise in bandwidth BW that arises from the interaction of V_{er} with C_{input} is equal to:

$$S^2_{(BW)} = (2\pi C_{input} V_{er} BW)^2 \quad (15)$$

where V_{er} is the rms noise voltage density ($rms_{(\sqrt{Hz})}$) of the amplifier. The noise voltage of a FET is approximately frequency-independent above ≈ 10 Hz, although this varies among different FET devices. As described by Eq. (15), this source of noise current is especially severe at high frequencies and with large C_{input}. Therefore, minimization of V_{er} is a major consideration in the selection of headstage amplifiers to be used in a system with a large C_{input} (e.g., bilayer).

2.3.2.2.4. Summary of Noise Sources. The noise current sources described above place a *lower* limit on the noise in actual recordings of single-channel currents. The relative contributions of these noise sources to total noise is shown in Fig. 4 for typical patch and bilayer recording conditions. We have not discussed the 1/f noise associated with many FET devices, because its characteristics vary greatly among different FETs and, although 1/f noise is a substantial source of noise at low frequencies, high-frequency noise dominates single-channel current records. In practice, the lowest overall noise level is usually greater than 2 × predicted level (Rae and Levis, 1984). This extra noise arises from a variety of sources, often resulting from the nonideal properties of circuit components. For example, the stray capacitance (C_f) across large values of R_f cannot usually be modelled by a simple $R_f||C_f$ circuit; rather, there is a distributed capacitance along the resistor that exhibits a decreasing impedance at higher frequencies, resulting in excessive high-frequency noise. Also, where patch pipets are used, the properties of thinly drawn glass can produce excessive noise currents (Rae and Levis, 1984). The sources of noise described above are particularly relevant to comparison of patch pipet and bilayer techniques in regard to primary sources of noise and concomitant bandwidth limitations (Fig. 4). For patch-clamping with pipets, the dominant noise current source is the seal resistance, a source of thermal noise currents. This noise current source cannot be compensated or corrected by any electronic circuitry (other than filtering), so it is imperative to establish good technique to maximize the pipet-membrane seal resistance. It should be possible with good patch-pipet technique to reduce C_{input} to a value (e.g., $<$15 pF) where amplifier noise voltage does not produce a substantial source of noise current. The investigator using the bilayer method is faced with a different balance of noise current sources. With the bilayer, $R_m||R_{shunt}$ is very large and an insignificant source of noise currents. Unfortunately, C_{input} is also quite large, and its interaction with amplifier noise voltage is the greatest source of noise currents, especially at high frequencies. This noise current source limits the bandwidth of bilayer recordings to ≈ 1 kHz. Therefore, every effort must be made to minimize C_m. We have successfully recorded channels with total bilayer/

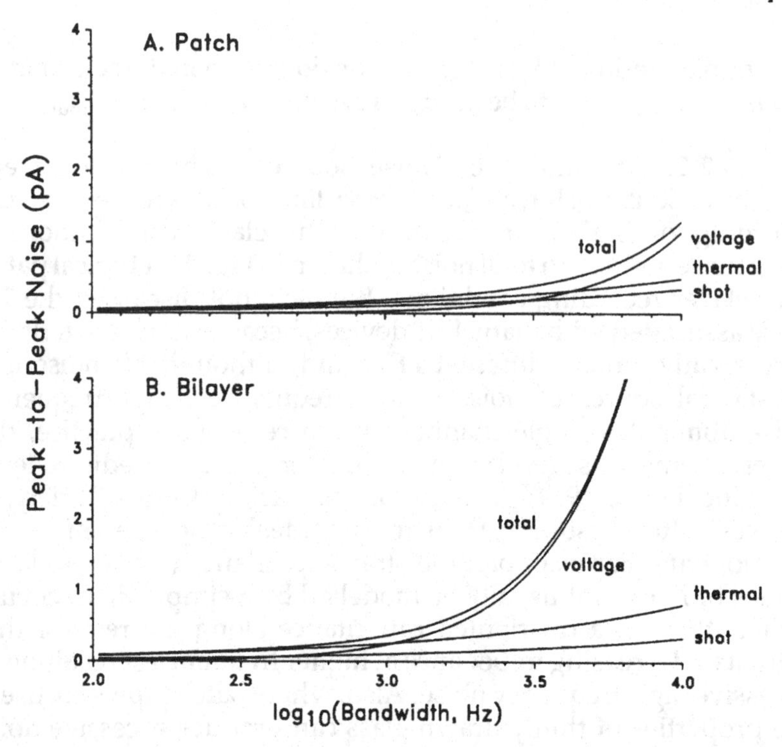

Fig. 4. Sources of noise currents in a headstage amplifer and preparation. The variance ($S^2_{(BW)}$) of the noise currents from each source (*see* equations in text) was calculated at 100 bandwidths equally spaced on a logarithmic scale between 100 and 10,000 Hz. The noise for each source is plotted as an approximate peak-to-peak value, $6\text{rms}_{(BW)}$. The total noise was calculated by summing the variances and then converting to a peak-to-peak value. (A) Noise currents in a nearly optimal patch-clamp arrangement. The values used in the calculations were $R_f = 50\ \text{G}\Omega$, $R_{input} = 50\ \text{G}\Omega$, $V_{er} = 2\ \text{nV}/\sqrt{\text{Hz}}$, $i_{gate} = 1$ pA, and $C_{input} = 15$ pF. The dominant noise source below ≈ 4.5 kHz is the thermal noise in the seal resistance and R_f. Above 4.5 kHz, voltage noise is the largest component. (B) Noise currents in a typical bilayer recording arrangement. The values used in the calculations are $R_f = 10\ \text{G}\Omega$, $R_{input} = 100\ \text{G}\Omega$, $V_{er} = 2\ \text{nV}/\sqrt{\text{Hz}}$, $i_{gate} = 1$ pA, and $C_{input} = 100$ pF. Above 1 kHz, there is a rapid rise in voltage noise because of the higher C_{input} in the bilayer system. The thermal noise is also higher than in the patch system, because R_f was made smaller to allow more rapid charging of the bilayer membrane. This would not be necessary if a switchable feedback resistor configuration were used, as described in section 2.3.2.4.

partition capacitances ranging from 25–50 pF (Fig. 1a, 1b). It is also possible that new developments in FET amplifier technology will reduce the noise voltage, such that its interaction with C_{input} will be reduced.

2.3.2.3. AMPLIFIER CONSTRUCTION. At the time of this publication, a number of high-quality patch-clamp amplifiers are commercially available. These amplifiers vary in regard to SNR and bandwidth levels and the availability of user convenience features, such as internal voltage command generators, rms meters, and transient cancellation and series resistance compensation. For many applications, very high-quality amplifiers can be "home-built." These amplifiers typically provide the current-to-voltage conversion and boost (Fig. 2), and varying degrees of transient cancellation and series resistance compensation. A simple head-stage amplifier can be constructed from a single operational amplifier that exhibits both a low bias current (<10 pA) and a high gain-bandwidth product(>10 MHz). The Burr-Brown OPA-102 op amp fulfills these requirements. Simple circuits using a single op-amp for the headstage are described in Hamill et al. (1981) and Alvarez (1986). In general, the highest quality headstage amplifiers are built as composite amplifiers consisting of a dual FET amplifier for the input followed by a high gain-bandwidth product amplifier for feedback. Examples of composite circuits with improved SNR and bandwidth are provided in Hamill et al. (1981), Sigworth (1983a), and Rae and Levis (1984).

2.3.2.4. NEW DEVELOPMENTS IN AMPLIFIER DESIGN. A critical element in present amplifier design is R_f, the feedback or gain resistor. As described above, R_f can be a significant source of thermal noise currents, and its large value severely limits the dynamic properties of the amplifier. One approach to avoid these problems has been to employ a very small capacitor (<1 pF) to integrate the feedback current and thereby provide the current-to-voltage conversion (Offner and Clark, 1985; Prakash et al., 1987). This approach is promising. An ideal capacitor is a noiseless device, and its use as an integrator should extend the recording bandwidth without requiring additional boost circuitry. However, this is not a simple alternative, because the integrator configuration requires a reset circuit to avoid amplifier saturation, and incorporation of the reset circuit within a feedback loop sensitive to sub-pA currents is difficult.

Recently, an alternative approach has been developed to aid in the charging of large values of C_{input}, such as in the case of lipid bilayers. This technique uses a logic-controlled FET to switch between two values of R_f. For example, the Axopatch-II (Axon Instruments) can provide logic-controlled switching between a 50 MΩ and a 50 GΩ R_f, permitting rapid charging of C_{input} through the lower-valued R_f followed by high-resolution recording, using the higher-valued R_f, after the command step. This method is effective, although it cannot speed the charging of the bilayer capacitance if a slow component of the charging results from electrostrictive changes in bilayer area and capacitance. This is most severe in the case of solvent-containing painted bilayers, where the annulus is relatively large. We have attained a settling time <1 ms following a 100 mV step applied to a 40 pF bilayer, by combining switching to a lower value of R_f and digital subtraction of the residual capacity transient.

The rate of charging of C_{input} can also be increased without modification of the headstage electronic circuitry using a "supercharging" method (Armstrong and Chow, 1987). Supercharging involves adding a brief, rectangular spike to the leading edge of the voltage command step. This spike briefly increases the current flow through the access (R_{series}) resistance and increases the rate of charging of C_{input}. This method works well for whole-cell voltage clamping with a patch pipet and, with appropriate choices of R_f, could also be applied in patch and bilayer recordings.

2.3.3. Monitoring and Storing Single-Channel Current Signals

During the course of an experiment, the current amplifier output is monitored on an oscilloscope. Both analog and digital oscilloscopes are suitable for this purpose, although digital storage oscilloscopes are especially useful. The ability to store a reference trace is useful for detection of baseline shifts, conductance changes, or thinning of a bilayer. Digital oscilloscopes also provide accurate on-line measurements of event amplitude and duration and peak-to-peak noise levels. Signals can also be monitored (and stored) on a strip-chart recorder. Strip-chart recorders provide a useful low-bandwidth (<100 Hz) overview of a long, continuous record. For steady-state recordings, an FM tape recorder provides a good storage medium, because a single tape can hold several hours of recording. A better, large-capacity storage medium is the hybrid audio digitizer and VCR cassette recorder. This technology provides the combination of highest quality recording (>90 dB SNR) with much lower cost. An audio digitizer modified for DC recording typically samples the signal on each channel at a sampling rate of 44 kHz (low-pass filter preset at 22 kHz), and the binary-coded data are then stored on VCR tape. For playback, the original signal is recovered by sending the binary-coded data to a digital-to-analog converter or by directly reading the digitized signal from the tape with a computer. This hybrid recording system provides a much higher quality signal storage than traditional FM tape recorders. However, there are two features of the hybrid system that are less convenient than an FM recorder. First, it is not possible to monitor the recorded signal on-line (as with a three-head FM tape recorder). Fortunately, the SNR is sufficiently high that the signal can be recorded with a lower relative gain, and monitoring for noise or off-scale transients is less important. Second, it is not possible to slow the playback speed to permit monitoring on a strip-chart recorder with a smaller bandwidth. One other alternative is to use a computer as the primary storage medium. This is highly desirable for many nonstationary protocols, e.g., the current response to a voltage step. If high precision is desirable in linking the stimulus and response, simultaneous digital-to-analog (voltage command output) and analog-to-digital (signal sampling) conversion within random-access memory (RAM) is required. Of course, the length of the sample is limited by the size of the RAM. Finally, for longer records, rapid access hard (Winchester) disks are currently available, capable of storing up to ≈ 5–65 million samples.

2.3.4. Signal Conditioning

The output of the headstage amplifier is a voltage proportional to the sum of two sources of currents: (1) the current flowing through the open channel, i_{open}, and (2) the baseline current, i_{closed}, including the ohmic leak current and noise currents arising from the biological preparation and the current amplifier. These noise currents can be large enough to obscure the event record and preclude accurate kinetic analysis. Because of this noise, the full bandwidth of the current amplifier usually cannot be used, and it is necessary to low-pass filter the output of the current amplifier to increase the SNR to tolerable levels. For a low-pass filter, the "cutoff frequency" (f_{cut}) is defined as the frequency at which the power of the output is reduced by ½, or the signal is attenuated by –3 dB. Above f_{cut}, there is increasing attenuation with increasing frequency, with the slope of the attenuation proportional to the number of poles in the filter. Low-pass filtering can be accomplished either with conventional analog filters or with programmed digital filters. At least some analog filtering will usually be required prior to digitization *(see below)*. Regardless of analog or digital type, the characteristics of the filter can significantly affect the current signal. The sharp-cutoff filters, including the Butterworth and Tschebycheff filters, have the steepest attenuation slope and provide the best reduction in high-frequency noise, but also exhibit a relatively slow t_{rise} and distort a rectangular current pulse by producing an overshoot or ringing response to a step input. A better choice is a Bessel or Gaussian filter (the responses of these filters become very similar with increasing number of poles, Colquhoun and Sigworth, 1983). These filters distort the signal less because they have a shorter t_{rise} and do not overshoot, but this is accomplished at the expense of a more gradual attenuation of noise above f_{cut}. The 10–90% t_{rise} of a Gaussian filter is $\approx 0.339/f_{cut}$.

The strategy behind using a low-pass filter is to select a value of f_{cut} that provides an appropriate SNR, while maintaining a high enough f_{cut} (and therefore small enough t_{rise}) such that distortion of the current events is minimized. Implementation of this strategy is made more complicated by the varying nature of background noise, dependent on relative contributions from different noise current sources (Fig. 4). For example, the noise density of thermal noise currents, such as the noise arising in R_m or R_f, is flat, or frequency-independent and the peak-to-peak noise therefore increases linearly with increasing bandwidth. This is the situation expected for ideal patch-recording conditions where C_{input} can be minimized. At the other extreme, the density of the noise currents arising from the interaction of amplifier noise voltage and C_{input} increases much more steeply, proportional to the square of the bandwidth (f^2) above a corner frequency. This is the situation expected for most bilayer recordings where C_{input} can be quite large. Empirically, the relationship between noise density and bandwidth is sometimes intermediate between the flat and f^2 profiles, with an approximately linear dependence on bandwidth (f) above a corner frequency. The importance of the dependence of noise density on bandwidth is discussed in detail in Colquhoun and Sigworth (1983). Briefly, the flat relationship offers the greatest flexibility in selection of f_{cut}, whereas the steepness of the f^2 relationship forces f_{cut} to be set at relatively low frequencies. The values of f_{cut} at which the balance between BW and SNR is optimized decrease as the dependence of noise on frequency changes from flat to f to f^2.

2.3.5. Digitization of the Current Record

Single-channel current records can be analyzed using either analog or digital techniques. In theory, direct measurements from analog signals might be preferable, because they offer a higher resolution in measurement of duration and amplitude than that available with digitized signals. However, a major limitation of analog measurements is that they cannot be used in automated measurement procedures, and they are too labor-intensive to be practical for measurement of the large number of events ($>>1000$) in a single record that are required for reasonably precise data analysis. Therefore, unless otherwise stated, digital measurement will be assumed in subsequent discussions. Digitization of the current record has the drawback that the continuous variables of amplitude and duration are transformed into discontinuously distributed variables because of the discrete sampling process. For current amplitude, the interval of resolution (pA) is:

$$\text{Resolution} = \frac{\text{Input range}}{(2^n)(\text{gain})} \tag{16}$$

where input range is in units of mV, gain is the current-to-voltage gain in units of mV/pA, and n is the number of bits in the analog-to-digital converter. If, for example, a gain of 100 mV/pA is used with a $\pm$ 5000 mV input range and a 12-bit A/D converter, the resolution is 0.024 pA. For duration, the interval of resolution is the sampling interval (t_{samp}, s), which is the reciprocal of the sampling frequency (f_{samp}, Hz). The problems resulting from the discretization of the analog record can be alleviated by interpolating the record between sampled points where critical measurements are to be made (e.g., transitions between states). Limiting the interpolation of the record to the intervals near these transitions is an effective means of reconstructing a signal from a digitized record and saves computation time compared to interpolation of the entire record. Cubic spline interpolation is discussed in greater detail, and sample programs are presented in Colquhoun and Sigworth (1983).

In most cases, it is appropriate to first set f_{cut} at a value low enough to minimize the detection of false events *(see below)*, but high enough to minimize distortion of the rectangular current steps. This choice of f_{cut} sets the lower limit of the sampling rate (f_{samp}), because the minimum value of f_{samp} is the Nyquist frequency, 2 f_{cut}. At this f_{samp}, aliasing of noise in the record is avoided. However, the Nyquist frequency is inadequate for accurate reconstruction or visualization of the record. Higher f_{samp} values, up to $\approx 5 f_{cut}$, are desirable. According to Colquhoun and Sigworth (1983; *see* Fig. 11-1 in Colquhoun and Sigworth), if interpolation is not utilized, f_{samp} values of 10–20 f_{cut} might be necessary to accurately reconstruct the signal. Such sampling rates may be prohibitive if the space for storage of digitized data is limited.

2.4. Event Detection and Measurement

The prototypical event is a rectangular current pulse that can be characterized by measurement of its amplitude and duration.

2.4.1. Baseline Identification

The baseline current level (i_{closed}) is the current present during the closed state of the channel. This baseline current is the sum of an ohmic leak current and noise currents from various sources. Several factors make baseline identification difficult, including high-frequency noise, low-frequency drift, and sudden shifts in level. It is essential that the baseline be accurately identified throughout a record. After initial identification, the baseline can be "updated" periodically by dividing the record into sequential blocks or frames, with the accuracy of the baseline "fit" reexamined in each block. A variety of algorithms can be used to identify or check the baseline. Initially, the baseline can be identified by averaging the current level during a period when the channel is closed. Updating is especially fast when a windowed average (e.g., ± 3 SD of the previous baseline) is used for each frame. An alternative method is the "zero-crossings" method (Sachs et al., 1982; Sachs, 1983), in which the baseline is identified as the closed channel current level at which the number of crossings is maximized. Although this method is significantly slower than the averaging method, it is less sensitive to outlying points. A fast, but less reliable, method is to identify the peak that corresponds to the baseline in the smoothed, point-by-point current level histogram. All of these algorithms are suitable for automated analysis programs. However, implementation can be difficult where records are badly corrupted by a noisy or drifting baseline. It is always a good idea to periodically examine, by eye, the accuracy of baseline identification by automated programs. An alternative to all of the methods described above is simply to position a cursor, by eye, on the apparent baseline. Although this method precludes automation, it may in cases of very "active" baselines be the only suitable method.

Identification of the baseline can be exceptionally difficult if the channel tends to occupy the open state much of the time. For example, batrachotoxin-activated Na^+ channels can be open > 98% of the time (French et al., 1984). In this case, there are very few lengthy periods of channel closed time suitable for baseline identification. One solution is to induce long channel closures by using a channel blocker whose blocking kinetics are distinctly slower than the gating process(es) of interest. In the case of batrachotoxin-activated Na^+ channels, the guanidinium toxins saxitoxin and tetrodotoxin are quite useful for this purpose, because their high affinity for the Na^+ channel receptor and slow off-rates provide long, recognizable baseline periods.

2.4.2. Event Detection

Event detection is usually based on the threshold-crossing method, in which a threshold (i_{thresh}) is set at a current level that, when crossed by the current record, indicates that a state transition has occurred. In the following discussion, we will assume that the baseline has been subtracted so that i_{closed} is zero and i_{open} is equal to the single channel current amplitude. The most common approach is then to set i_{thresh} at 50% of i_{open}. Other values of i_{thresh} are discussed in Colquhoun and Sigworth (1983) and in Magleby and Pallotta (1983a). For the remainder of this discussion, a 50% threshold will be assumed. Consider the detection of an open event within a series of N samples of the current record, S_1, S_2, . . . S_i . . ., S_n, with sample interval t_{samp}. The channel is assumed to be initially in the closed state. The amplitudes of consecutive S_i are sequentially compared with i_{thresh}, and, if an S_i is greater than i_{thresh}, a transition to the open state is detected, marking the beginning of the open event. Testing is then continued until a value of S_i less than i_{thresh} is detected, marking the transition to the closed state and the end of the open event. An important feature of the 50% threshold technique is that it provides a symmetric threshold for measurement of the time of the state transition, t_{st}. Because

i_{thresh} is the same for both opening and closing transitions, the values of t_{st} for opening and closing transitions are not differentially biased by the event detection procedure.

Event detection can be automated, using the 50% threshold method as an algorithm. It is important to verify the accuracy of the detection routine by comparing the actual record with an idealized record constructed from the measured amplitudes and durations of detected events.

2.4.2.1. Missed Real Events. Events are detected when the current level crosses the threshold, i_{thresh}, which we have assumed to be 50% of the open channel level, i_{open}. The true duration of an event that produces a filtered duration at i_{thresh} equal to t_{samp} is termed t_{all} (Fig. 5D). All events with a true duration greater than or equal to t_{all} will be detected. As the true event duration is decreased below t_{all}, the amplitude of the filtered event also decreases until a true duration is reached where the amplitude just reaches i_{thresh}. This duration is termed the dead time, t_{dead}, and events shorter than t_{dead}, cannot be detected since i_{thresh} is not crossed. This is true whether event detection is attempted on analog or digitized records. The dead time, t_{dead}, results from the filtering of the current record and, for a Gaussian filter (Colquhoun and Sigworth, 1983), is given by:

$$t_{dead} = 0.538 t_{rise} = 0.179/f_{cut} \quad (17)$$

Events with true durations between t_{dead} and t_{all} can also be missed if the transition times t_{st1} and t_{st2} occur within a single sample interval, t_{samp}. In other words, the detection process only "sees" the record at discrete, sampled points and ignores any activity within a sampling interval. With t_{samp} equal to t_{dead} (i.e., f_{samp} is approximately $5.6 \times f_{cut}$), the ratio of t_{all} to t_{dead} is ≈ 1.235 (McManus et al., 1987). If the time constant of a single-exponential dwell-time distribution is also equal to t_{dead}, then only 0.67 of the intervals with true dwell times between t_{dead} and t_{all} will be detected (McManus et al., 1987). McManus et al. (1987) discuss the detection error produced by discrete sampling and provide a correction that estimates the fraction of true events that are actually detected. This error is small if the fastest time constant in the data is at least $10 \times t_{samp}$, because under this condition most true event durations are longer than t_{samp}. The error is also small if t_{samp} is less than 10–20% of t_{dead}.

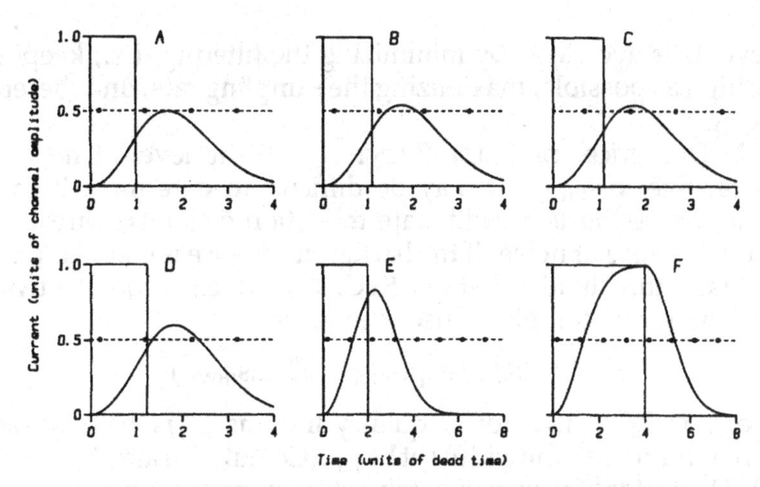

Fig. 5. Effect of filtering on amplitudes, durations, and detection of true intervals. Idealized (true) single-channel currents are plotted as rectangular responses starting at time 0, where time is expressed in units of t_{dead}. The responses observed with filtering are shown as the slowly rising and falling curves. The small filled circles on the dashed 50% threshold line indicate samples taken once every dead time. (A) An interval with a true duration equal to 1.0 t_{dead} just reaches the 50% threshold level. Intervals with true durations less than 1.0 t_{dead} are not detected. (B,C) Only a fraction of the intervals with true durations between 1.0 and 1.235 t_{dead} are detected. In B and C, an interval with a true duration of 1.1 t_{dead} and a filtered duration at the 50% level of 0.676 t_{dead} has a 67.6% chance of being detected when the sampling period is 1.0 t_{dead}. The interval is not detected in B and is detected in C. (D) With a sampling period of 1 t_{dead}, all intervals with true durations of 1.235 $t_{dead} = t_{all}$ or greater are detected, because the filtered duration at the 50% level is equal to or greater than the sampling period. (E,F) Intervals with true durations greater than 2 t_{dead} have observed durations at the 50% level equal to the true durations. Intervals with true durations from 1–2 t_{dead} have observed durations at the 50% level less than the true duration. (reproduced with permission from McManus et al., 1987).

Missing brief events can introduce several biases into the dwell-time distributions (*see* section 2.5.2.1.3.). Also, if several consecutive brief events are missed, there will be an apparent change in current level because of the averaging of the current during the sequence of missed events. The probability of missing real events is decreased by minimizing the filtering, i.e., keeping f_{cut} as high as possible, maximizing the sampling rate, and thereby reducing t_{samp}.

2.4.2.2. DETECTION OF FALSE EVENTS. If the level of noise is large relative to i_{thresh}, it may be difficult to determine if each threshold crossing is a valid state transition or a false threshold crossing because of noise. If the background noise is Gaussian, the rate of false threshold crossings (FTC, s^{-1}), which is equal to twice the frequency of complete false events, is:

$$FTC = 2kf_{cut} \exp(-i_{thresh}^2/2rms_{(BW)}^2) \tag{18}$$

where f_{cut} is the –3 dB cutoff frequency and $rms_{(BW)}$ is the rms noise current within the bandwidth set by f_{cut} (Colquhoun and Sigworth, 1983). The factor k is approximately unity, varying with the characteristics of the noise, with $k = 0.849$ when the noise spectrum is flat and $k = 1.25$ when the noise is proportional to f^2 (Colquhoun and Sigworth, 1983). FTC decreases with decreasing f_{cut}, in large part because of the consequent reduction in $rms_{(BW)}$. The ratio $i_{thresh}/rms_{(BW)}$, with i_{thresh} being fixed (e.g., ½ i_{open}) is a simple predictor of the severity of FTC and can be varied by adjusting f_{cut}. The maximum FTC should be one or two orders of magnitude smaller than the slowest transition rate of interest to minimize the error resulting from event durations being shortened by false threshold crossings. Normal values of $i_{thresh}/rms_{(BW)}$ are 3–5 for relatively active and inactive records, respectively. If the noise is distributed symmetrically above and below the baseline, an estimate of FTC can be made by setting a threshold with magnitude equal to i_{thresh}, but on the side of the baseline opposite from i_{open}. The rate at which this inverted i_{thresh} is crossed provides an estimate of FTC (McManus et al., 1987). One advantage of this method is that it does not require calculation of $rms_{(BW)}$.

The effect of detection of false threshold crossings is to improperly terminate real events, thereby shortening the dwell times of the real events. The mean duration of a given state of a channel is simply the reciprocal of the sum of transition rates leaving that state. Thus, if it is desired that this sum be 100-fold larger than FTC, the longest dwell time (dur_{max}) of a given state that can be accurately measured is given by:

$$dur_{max} = 100/FTC \tag{19}$$

Also, false threshold crossings add false brief events to the dwell time distributions. The FTC can be reduced by increasing the filtering (decreasing f_{cut}), but at the cost of increasing the likelihood of missing brief events, as described above.

2.4.2.3. EFFECT OF NOISE ON DETECTION OF TRUE EVENTS. For an ideal, noiseless current record, the random exponential distribution of true dwell times may result in many events with true dwell times equal to $\approx t_{dead}$ whose amplitudes will be near i_{thresh}. Addition of random noise to the record affects the detection of these events by pushing the amplitude of some subthreshold events over i_{thresh} and dropping the amplitude of some suprathreshold events below i_{thresh}. This contribution of noise to event detection error has been discussed by McManus et al. (1987). Because of the exponential distribution of dwell times, the addition of noise will produce a net increase in detected events, because there are more subthreshold events with dwell times slightly less than t_{dead} than suprathreshold events with dwell times slightly greater than t_{dead}. The effect of noise on event detection is negligible when the time constant of the distribution is $>10\ t_{dead}$, because a smaller fraction of events have true dwell times near t_{dead}. The effect of noise on event detection is also negligible if $rms_{(BW)}$ is less than 0.02 of i_{open} (McManus et al., 1987).

The effect of noise in increasing the detection of events and the effect of discrete sampling in decreasing the detection of events (section 2.4.2.1.) affect the detection of the same group of events, those with dwell times near t_{dead}. Because the errors in event detection introduced by the effects of noise and discrete sampling are opposite, they can counteract each other, resulting in no net error in event detection. A method for adjusting the sampling error to balance the noise detection error is provided in McManus et al. (1987).

2.4.3. Measurement of Event Amplitude

Let us assume that the baseline has been subtracted from our record and that the zero current level corresponds to the baseline. Measurement of event amplitude is then relatively straightforward. For events longer than $\approx 2.5\ t_{rise}$, the event amplitude (i_{open}) is measured by averaging the current level within a window beginning $\approx 1\ t_{rise}$ after the opening and ending $\approx 1\ t_{rise}$ before the end of the event. This window corresponds to the "flat" region of the event, and its duration is designated t_{flat}. Exclusion of the regions surrounding t_{flat} is necessary because of the distortion of the current step by the finite t_{rise} of the recording system.

Qualitatively, it is expected that the measurement of the amplitude will be most accurate when i_{open} exhibits a flat top and a generally rectangular shape. Quantitatively, the number of sampled points to be averaged, n, is equal to t_{flat} times f_{samp}, and the accuracy of the estimate of i_{open} increases according to $\sqrt{n}$. For events less than $\approx 2\ t_{rise}$ in duration, the current step is markedly distorted by the filtering and only the peak of the event can be measured, which underestimates i_{open}.

2.4.4. Measurement of Event Duration

The duration or dwell time of an event is the interval of time between the two sequential state transitions that bound the event, e.g., C-O-C. Each transition can be considered instantaneous and to occur at a specific time, t_{st}. Thus, for an open event, the duration is the interval between t_{st1}, the time of transition from the closed to the open state, and t_{st2}, the time of transition from the open to the closed state; i.e., the duration is equal to $t_{st2} - t_{st1}$. If measurements are made on an analog record or a digitized record that has been interpolated, the values of t_{st1} and t_{st2} are determined directly as the time at which i_{thresh} is crossed.

In the case of digitized records that are not interpolated, the duration of an open event is calculated as the product of the number (N) of s_i above i_{thresh} and the sampling interval, t_{samp}. The precision with which the duration of an event, t_{dwell}, is measured without interpolation is limited to $\pm\ t_{samp}$, since the value of t_{st}, a continuously distributed variable corresponding to the time at which each state transition occurs, is forced to take discrete values corresponding to the times at which the record is sampled, the s_i. Thus, we can only say that, where a threshold crossing is detected at s_i, the true t_{st} occurred between $s_i - t_{samp}$ and s_i.

Events with true durations between t_{dead} and $2\ t_{dead}$ will be measured with an apparent duration less than their true duration because of the limited rise time of the recording system. The true dwell time can be calculated from the observed dwell time (t_{dwell}) by:

$$\text{true dwell time} = t_{dwell} + a_1 \exp(-t_{dwell}/a_1 - a_2\, t_{dwell}^2 - a_3\, t_{dwell}^3),\ (t_{dwell} > 0) \quad (20)$$

with $a_1 = 0.5382\ t_{rise}$, $a_2 = 0.837\ t_{rise}^{-2}$, and $a_3 = 1.120\ t_{rise}^{-3}$ (Colquhoun and Sigworth, 1983).

2.4.5. Multiple Channels

Thus far we have limited our discussion to the detection and measurement of events in records where only a single channel is active. This is the ideal recording situation from the standpoint of data acquisition and analysis. However, in both patch (Patlak and Horn, 1982) and bilayer (Garber and Miller, 1987) recording configurations, there is often a problem in obtaining recordings from single channels. This can occur in patch recordings from native membranes where the density of channels is high. In the case of bilayers, multiple channel recordings result from single vesicles containing multiple channels that incorporate on fusion with the bilayer. The occurrence of multiple channels produces two problems. First, if a single channel of interest is present against the background activity of several extraneous channels, the activity of the channel of interest may be obscured by the other channels. Such activity can preclude the use of automated event detection and measurement routines. Background channel activity can be reduced by several approaches. First, the number of channels in the membrane might be reduced by decreasing the area of the patch or the size of the vesicles. Second, the activity of the extraneous channels might be selectively eliminated by blocking drugs. Third, if the ionic selectivity of the extraneous channels is different from that of the channel of interest, their activity might be eliminated by limiting the ions present to those that are permeant only in the channel of interest. A more serious problem is the occurrence of multiple channels of the species of interest. Again, this problem might be remedied by reducing the size of the patched membrane or the vesicle size. Also, the probability of multiple channels being simultaneously open can be gradually reduced by titration with a high affinity, selective blocker (e.g. TTX for Na^+ channels). As long as the kinetics of the block are distinctly different and independent of the process(es) under investigation, this approach should produce epochs with only one active channel. If multiple channels persist, information regarding the conductance of the channels can be obtained, but in general, it is too complicated to extract the kinetic parameters. This results from the fact that the successive state transitions cannot be unambiguously linked to a specific channel. For example, if several channels are simultaneously open and there is a single closing, the channel that has closed cannot be identified. Thus, the open and closed dwell times cannot be unambiguously measured. For the special case of a simple, two-state kinetic scheme and multiple channels that gate independently, it is possible to estimate the rate constants (Labarca et al., 1980; French et al., 1984).

An automated method for the detection of events in multichannel records has been reported by Vivaudou et al. (1986). Jackson (1985) has presented an interpretive framework with which some kinetic information can be extracted from multichannel recordings under the condition that the number of channels can be determined and the recordings are stationary. A maximum-likelihood approach generally applicable to multiple channels, but limited in practice to four-channel recordings because of the increasing complexity of the calculations, is presented in Vandenberg and Horn (1984). An example of an analysis of multi-channel recordings can be found in Patlak and Horn (1982).

2.5. Analysis of Single-Channel Events

At this point, we have identified the open and closed events in our current record and stored measurements of amplitude and dwell time in a sequential file from which an idealized version of our original record can be reconstructed as a series of rectangular pulses. Our next step is to examine the distribution of dwell times and amplitudes. Because the number of events is usually quite large (>1000), it is necessary to use histograms as a data reduction technique to help reveal the distribution of observed amplitudes and durations. In our data analysis, we are principally interested in dwell-time distributions because they usually provide the greatest insight into the structure of the kinetic schemes that can account for the channel gating. Therefore, in the following discussion of single-channel data analysis, we will focus on the analysis of dwell-time distributions. Similar principles will apply to the construction and fitting of amplitude histograms.

2.5.1. Constructing Histograms

A histogram is constructed by first dividing the range of values of a dependent variable into discrete intervals or "bins," which are plotted on the abscissa. We then count the number of values that fall within the limits of each bin and plot this number above the respective bin. The result is a frequency histogram plot where the amplitude of each bin is the number of observations whose values fall within that interval. Alternatively, we can make a histogram by counting the number of observations whose values are smaller than the upper limit of a bin, which is a cumulative frequency histogram. Suppose we make a histogram of open state dwell times (t_{dwell}) (e.g., Fig. 6 A). In the case of a linearly scaled abscissa and constant bin width, δt, the lower limit (t_i) of the ith bin in a series of m bins is:

$$t_i = t_{low} + i(\delta t) \ (i = 0 \text{ to } m - 1). \quad (21)$$

where t_{low} is the lower limit of the leftmost bin. For each ith bin, we plot the frequency of occurrence of dwell times such that:

$$t_i \leq t_{dwell} < t_i + \delta t. \quad (22)$$

If the range of values for t is 0–200 ms, we might set δt equal to 5 ms. Then the amplitude of the first bin is the number of values 0 ms $\leq t_{dwell} <$ 5 ms, the amplitude of the second bin is the number of values 5 ms $\leq t_{dwell} <$ 10 ms, and so on. When this binning process is complete, the relative magnitudes of the bins provide a graphical representation of the frequency distribution of open dwell times. We could also construct a cumulative frequency histogram by plotting for the ith bin the frequency of occurrence of dwell times such that:

$$t_{dwell} < t_i + \delta t \ (i = 0 \text{ to } m - 1). \quad (23)$$

Thus, the amplitude of the first bin is the number of values < 5 ms, the amplitude of the second bin is the number of values < 10 ms, and so on. Both of these histograms summarize the event record with regard to the frequency of occurrence of various open dwell times. However, neither histogram contains information regarding the sequence of events. Sequential information requires joint probability analysis (*see* section 2.5.5.).

Digitization of the current record transforms the continuous variables of amplitude and duration into discontinuous variables, with the set of possible values for these variables determined by the intervals of resolution of the analog-to-digital converter and the sampling interval, respectively (section 2.3.5.). Because of the discretization of these variables, the bin width should be an integral multiple of the interval of resolution to avoid artifactual cyclic distortions of the histogram (Starmer et al., 1986). This problem can also be avoided by interpolation of points within the sampling interval to restore the continuous nature of the variables.

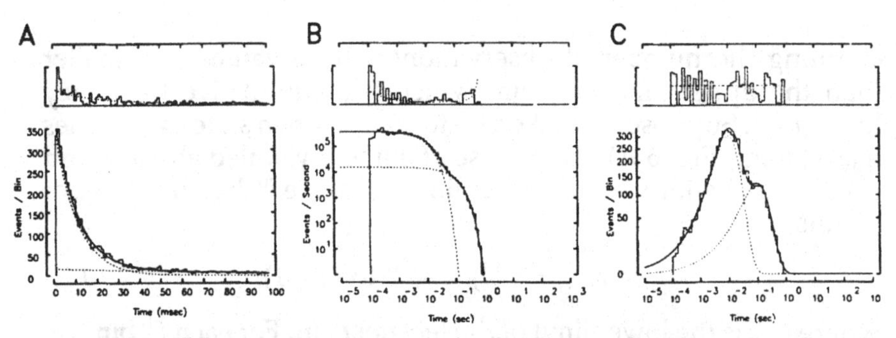

Fig. 6. Three representations of a dwell-time distribution with two exponential components. 5120 random numbers were generated according to a distribution with time constants of 10 ms (70% of the events) and 100 ms (30%), and binned for display as histograms in the lower panel of each part of the figure. Superimposed are the theoretical probability density functions for each component (dashed curves) and their sum (continuous curve). In each part of the figure, the upper panel plots the absolute value of the deviation of the height of each bin from the theoretical curve, with dashed curves showing the expected value of the standard deviation for each bin. The upper panels were plotted with vertical expansion factors of 2.1, 5.4, and 4.9, respectively. (A) Linear histogram. Events are collected into bins of 1-msec width and plotted on a linear scale. The 100-ms component has a very small amplitude in this plot. (B) Log-log display with variable-width (logarithmic) binning. The number of entries in each bin is divided by the bin width to obtain a probability density in dimensions of events/s on the ordinate. (C) Display with logarithmic time scale. Events are collected into bins of width $\delta x = 0.2$, and superimposed on the histogram is the sum of two functions. The ordinate is scaled with a square-root transformation with units of events/bin. Note that the scatter about the theoretical curve is constant throughout the display (reproduced with permission from Sigworth and Sine, 1987).

2.5.1.1. PROBABILITY–DENSITY FUNCTIONS AND PROBABILITY DISTRIBUTIONS. In a set of N dwell times, the expected number of observations of a specific dwell-time value ($t_{\text{dwell}i}$) is:

$$\text{frequency of } t_{\text{dwell}i} = \text{Probability } (t_{\text{dwell}i})\, N \qquad (24)$$

A continuous function that describes the relative probability of each value of t_{dwell} is the probability density function, or pdf. The pdf is used because, on a continuous time scale, the absolute probability of any single value of t_{dwell} is infinitesimal. The pdf has an area of unity and, for dwell times, dimensions of s^{-1}. For a dwell-time distribution, the pdf describes the limit of a probability per unit time that an observed dwell time falls in an interval (δt) centered on t_{dwell},

$$p(t_{dwell}) = \lim_{\delta t \to 0} \frac{\text{Probability}\ (t_{dwell} - \delta t/2 < t_{dwell} < t_{dwell} + \delta t/2)}{\delta t} \quad (25)$$

Under the restrictions that channel gating is a Markovian process and the state transition rates are time homogeneous, we can calculate the theoretical pdf for various kinetic schemes. For example, the pdf for the open dwell times of the C-O model is:

$$p(t_{dwell}) = k_{-1} \exp(-k_{-1} t_{dwell}) \quad (26)$$

where k_{-1} is the rate constant for channel closing. The probability that an observed dwell time t is less than or equal to some time t_{dwell} is given by the probability distribution function:

$$P(t_{dwell}) = \text{Probability}\ (t \leq t_{dwell}) \quad (27)$$

and is calculated by integrating the pdf from 0 to t_{dwell}. The values taken by the probability distribution are true probabilities and are dimensionless. (Note that the pdf is designated by a lowercase p and the probability distribution is designated by an uppercase P.) The probability distribution function for the open dwell times of the C-O model is:

$$P(t_{dwell}) = 1 - \exp(-k_{-1} t_{dwell}) \quad (28)$$

Because of the limited number of observations that we can make, we are usually interested in predicting the number of dwell times that fall within each bin of a histogram. The predicted frequency of observations in each bin (f_{bin}) can be calculated approximately as:

$$f_{bin} \approx N \delta t p(t_i + \delta t/2) \quad (29)$$

This method simply calculates, from the pdf, the probability density at the midpoint of the bin and is, therefore, subject to error if the pdf is not symmetric around the midpoint. An alternative method for calculating f_{bin} is

$$f_{bin} = N[P(t_i + \delta t) - P(t_i)], \quad (30)$$

where t_i is the lower limit of the ith bin. In this calculation, the probability is calculated as the difference between the values of the probability distribution at each edge of the bin. This is equivalent to integrating the pdf across the width of the bin and is, therefore, insensitive to the shape of the pdf as it crosses the bin (*see* Sigworth and Sine, 1987).

2.5.1.2. OPTIMIZING HISTOGRAMS FOR FITTING AND DISPLAY. Histograms have usually been plotted with a linearly scaled abscissa and ordinate. Recently, other axes scalings have been convincingly proposed (McManus et al., 1987; Sigworth and Sine, 1987). These variants have important implications for viewing and fitting histograms, and are discussed below.

2.5.1.2.1. Variable Bin Width and the Scaling of the Abscissa. It is important in the viewing and fitting of histograms that the bin width be set at a value that provides good resolution of detail in the distribution without also producing bins with very low frequencies. For a Gaussian-distributed variable, such as event amplitude, a linear scaling of bin width along the abscissa, as discussed above, usually meets these requirements. The exponentially distributed dwell times, however, are not well-portrayed with linear scaling, because in these highly skewed distributions, most of the expected dwell times occur at very brief durations, but a significant fraction of the dwell times extend over a much longer time scale. In constructing histograms of dwell-time distributions, we are therefore faced with the dilemma of using either: (1) a narrow bin width to provide resolution of brief events at the expense of empty bins at longer durations, or (2) a wider bin width to ensure adequate bin-filling at longer durations with a concomitant loss of resolution of brief events. One solution to this problem is to use a relatively narrow bin width and to combine bins where f_{bin} values are too low. The rebinning requires a rescaling of the amplitude of a bin, since the dimensions of δt^{-1} should be retained. The rescaling factor is $\delta t/\delta t'$, where δt is the original bin width and $\delta t'$ is the binwidth after rebinning.

A more systematic approach to varying bin width for dwell-time histograms is a logarithmic scaling of the abscissa (Fig. 6 B,C). A logarithmic time-axis provides the advantage that dwell times ranging over 5–6 orders of magnitude, e.g., 10^{-4}–10^{2} s, can be binned and conveniently displayed in a single histogram plot without loss of resolution or excessive zero or low-frequency bins. This is discussed in Blatz and Magleby (1986a,b), McManus et al. (1987), and Sigworth and Sine (1987). On a log scale, bin widths are scaled according to a dimensionless width (δx) rather than an absolute width (i.e., δt) as on a linear scale. For example, as presented by Sigworth and Sine (1987):

$$\delta x = 2.303/m \qquad (31)$$

where m is the number of bins per decade. Therefore, with log binning, the bin width is constant on a log scale, corresponding to a constant bin width ratio for successive bins on a linear scale. For m bins, the lower limit (x_i) of the ith bin on the log-scaled axis is:

$$x_i = \ln(t_i) = \ln(t_{low}) + i\delta x,\ (i = 0 \text{ to } m - 1) \qquad (32)$$

where t_{low} is the lower limit of the leftmost bin. Therefore, we log-bin according to:

$$\ln(t_i) \leq \ln(t_{dwell}) < \ln(t_i) + \delta x \qquad (33)$$

or, in terms of actual time:

$$t_i \leq t_{dwell} < t_i \exp(\delta x) \qquad (34)$$

Two approaches have been used to display a log-scaled histogram. Blatz and Magleby (1986a,b) and McManus et al. (1987) scale the amplitude of each bin according to the bin width, using a decoding file to calculate the bin width and bin midpoint (*see* McManus et al., 1987 for details). This results in dimensions on the ordinate of s^{-1}. Sigworth and Sine (1987) do not scale according to bin width; rather, they calculate the appropriate pdf by differentiating the log-transformed cumulative probability distribution. This results in dimensions on the ordinate of events/bin. The method described by Sigworth and Sine (1987) is most appropriate for continuously distributed dwell times, e.g., interpolated dwell times. If dwell times are not continuously distributed, there may be frequent empty bins at dwell times near the sample interval. The distribution can be "smoothed" using the approximate correction for sampling promotion error proposed by Korn and Horn (1988; *see* section 2.5.2.1.1.).

2.5.1.2.2. Ordinate Scaling. In determining the goodness-of-fit of a theoretical pdf to a histogram, it is desirable to weight most heavily the bins that provide the most reliable fit. The distribution of observations within each bin of a histogram is multinomial and should follow Poisson statistics. Poisson statistics predict that the variance of f_{bin} should be proportional to the magnitude of f_{bin}. With linear scaling this means, unfortunately, that the bins containing the most observations provide the least reliable fit (Fig. 6 A). A logarithmic ordinate has been used (McManus et al., 1987) to better resolve the detail in the dwell-time distribution, where f_{bin} is relatively small (Fig. 6 B). With a log-transformed ordinate, the variance again is not constant, but is greater for small f_{bin}. Sigworth and Sine (1987) have proposed a square-root transformation of the ordinate that produces a constant variance in each bin of a dwell-time histogram. This simplifies evaluation of the fit by ensuring that the variance of f_{bin} is independent of the magnitude of f_{bin} (Fig. 6 C).

The combination of logarithmic scaling of the time scale and a square-root transformation of the ordinate produces an histogram with properties that make it markedly preferable to a linear histogram, including:

1. Each exponential component of a distribution appears as a peak positioned at its time constant
2. The height of the peak reflects the number of events in that component and
3. the variability of f_{bin} is constant and independent of the magnitude of f_{bin} (Sigworth and Sine, 1987).

The peak(s) in the histogram result from the fact that the narrow bins at brief durations collect few events, whereas at longer durations the rate at which the expected number of events decreases is more rapid than the rate at which the bin width increases. This results in a maximum at the position of the time constant along the abscissa (Sigworth and Sine, 1987).

2.5.2. Dwell-Time Histograms

Dwell-time histograms can be constructed for each conductance state present in the single-channel current record. As described in the overview, the expected pdf for a single kinetic state is a single exponential. This can be extended to the general case of m kinetic states, where there should be m exponential components in the dwell-time distribution. The identification of the components in multi-component dwell time distributions can be very difficult and prone to error or ambiguity, depending on the degree of separation of the time constants and the number of events from which estimates are made. For example, consider the simple case of two components with equal areas and time constants separated by a factor of 10. To estimate these time constants with $<10\%$ error (using linear axes), more than 5,000 events must be collected (Auerbach and Sachs, 1985). The separation of individual components in multi-component distributions should be aided by the logarithmic scaling described in section 2.5.1.2.1. Several potential sources of error exist and are described below.

2.5.2.1. ERRORS IN DWELL-TIME DISTRIBUTIONS

2.5.2.1.1. Sampling Promotion Error. In the case of a channel opening, the dwell time of the opening measured from sampled data is equal to the product of the number of samples (N) above threshold and t_{samp}. Because the measurement error is $\pm t_{samp}$, the measured duration (Nt_{samp}) might be observed for events with true durations between $(N-1)t_{samp}$ and $(N+1)t_{samp}$. In effect, the durations are "binned" into intervals with midpoints (bin_{mid}) that are integral multiples of t_{samp}. The probability that the dwell time will be measured as N sample intervals is 1 when the true dwell time is equal to N sample intervals and the probability decreases linearly to zero as the true dwell time either decreases to $N-1$ sample intervals or increases to $N+1$ sample intervals (Sine and Steinbach, 1986b). The binning of dwell times can distort the dwell time distribution. McManus et al. (1987) used simulated records to show that, when t_{samp} is greater than $\approx 20\%$ of the time constant of the exponential distribution, the number of events (f_{bin}) will be greater than that predicted at bin_{mid} by the true pdf. Sine and Steinbach (1986b) termed this the sampling promotion error. For example, when t_{samp} is equal to the time constant of the distribution, the ratio of the observed f_{bin} to the true value at bin_{mid} is 1.086. Sampling promotion results from the exponential distribution of dwell times, which produces an asymmetric distribution of dwell times within $\pm 1\ t_{samp}$ on either side of bin_{mid}. Thus, for a bin centered at N sample intervals, more $N-1$ dwell times are added than $N+1$ dwell times are lost during the binning process. Sampling promotion results in an overestimate of the intercept of an exponential fit, but does not affect the time constant of the fit since the ratio is constant for all bins.

Sampling promotion can be prevented by maintaining t_{samp} less than 20% of the time constant of the fastest component of the distribution, because this reduces the asymmetry of the pdf around bin_{mid}. Sampling promotion can also be prevented by interpolation of the sample intervals containing the threshold crossings, which restores the continuity of the measurement. If these remedies are not possible, the error introduced by sampling promotion in the estimation of kinetic parameters using either least squares or maximum likelihood methods can be corrected as shown in McManus et al. (1987).

An approximate correction for sampling promotion error has been proposed by Korn and Horn (1988). As described above, the probability $p(N|t_{dwell})$ of an event covering N sample intervals, given a true duration t_{dwell}, is a triangular function between $N-1$ and $N+1$ sample intervals. If it is assumed that the true density of dwell times is constant over each bin in the histogram (i.e., the slope of the pdf is approximately flat across the bin), then the conditional probability density of t_{dwell}, given N, or $p(t_{dwell}|N)$, is also a triangular function given by:

$$p(t_{dwell}|N) = \begin{cases} [(t_{dwell}/t_{samp}) + 1 - N]/t_{samp} & (N-1)t_{samp} \le t_{dwell} < Nt_{samp} \\ [N + 1 - (t_{dwell}/t_{samp})]/t_{samp} & Nt_{samp} \le t_{dwell} < (N+1)t_{samp} \\ 0 & \text{elsewhere} \end{cases} \tag{35}$$

The error in the distribution of dwell times produced by sampling promotion can be approximately corrected by probabilistically redistributing the dwell times in each bin, using $p(t_{dwell}|N)$ to calculate the probability for events in the Nth bin to be located in any bin between $(N-1)t_{samp}$ and $(N+1)t_{samp}$. The probability of the true dwell time falling within the lower limit (t_1) and the upper limit (t_2) of each histogram bin is calculated as follows. For dwell times between $(N-1)t_{samp}$ and Nt_{samp}, the probability of a dwell time falling in the interval (t_1, t_2) is:

$$t_{samp}^{-1}[(t_2^2 - t_1^2)/2t_{samp} + (1-N)(t_2 - t_1)] \tag{36}$$

For dwell times between Nt_{samp} and $(N+1)t_{samp}$, the probability of an event falling in the interval (t_1, t_2) is:

$$t_{samp}^{-1}[(N+1)(t_2 - t_1) - (t_2^2 - t_1^2)/2t_{samp}] \tag{37}$$

For a linearly scaled histogram, this correction is simple. One-fourth of the dwell times is moved out of the Nth bin with one-eighth going to the $N-1$ bin and one-eighth going to the $N+1$ bin. Because of the varying ratio of the sample interval to the bin width, application of the correction to log-binned histograms requires that the limits t_1 and t_2 must be calculated for each bin and the dwell times redistributed among all log bins falling between $N-1$ and $N+1$ sample intervals. For the log bin containing the Nt_{samp}, the probability is calculated as the sum of the individual probabilities calculated from the areas under the ascending and descending limbs of the triangular conditional probability function.

2.5.2.1.2. Binning Promotion Error. When dwell-time histograms are constructed, dwell times with values between $bin_{mid} - \frac{1}{2}\delta t$ and $bin_{mid} + \frac{1}{2}\delta t$ are combined in each bin with width δt and midpoint bin_{mid}. McManus et al (1987) have shown that the binning process produces an overestimation of the magnitude at bin_{mid} relative to the magnitude predicted by the true pdf. This error is termed binning promotion error and, like sampling promotion error, results from the asymmetric distribution of the exponential pdf around bin_{mid}. Binning promotion error becomes severe when the bin width approaches the time constant of the distribution, because then the distribution of the pdf around bin_{mid} becomes markedly asymmetric because of its exponential decay. For example, when δt is equal to the time constant of the distribution, the ratio of the observed f_{bin} to the true value at bin_{mid} is 1.0422. Binning promotion error is negligible when the bin width, δt, is less than 20% of the fastest time constant of the distribution. Corrections for binning promotion error are provided in McManus et al (1987). The error introduced by binning promotion can be avoided in maximum likelihood calculations by using the area within the bin (calculated as the difference between the value of the probability distribution function, $P[t_{dwell}]$, at the upper and lower bin limits) rather than the value of the pdf, $p(t_{dwell})$, at bin_{mid} in the calculation of the expected bin frequency (Sigworth and Sine, 1987).

2.5.2.1.3. Missed Events. Events with a duration less than t_{dead} are not detected by the 50% threshold-crossing method. For each component in an exponential distribution, the fraction of true events that are undetected because they do not reach threshold (i.e., $t_{dwell} < t_{dead}$) is:

$$f_{miss} = 1 - \exp(-t_{dead}/\tau) \tag{38}$$

and the fraction of true events that crosses the threshold (i.e., $t_{dwell} \geq t_{dead}$) is:

$$f_{det} = 1 - f_{miss} = \exp(-t_{dead}/\tau) \tag{39}$$

where τ is the time constant of the component. If f_{miss} is large, the component may be eliminated from the multi-component distribution. In addition to the loss of this component, other significant distortions of the observed dwell-time distribution can result, including:

1. Addition of artifactual or "phantom" components to the distribution (Roux and Sauve, 1985)
2. Increases in the time constants of other components or
3. Either increases or decreases in the areas (i.e., number of events) under different components (Blatz and Magleby, 1986b).

Phantom components occur when more than 40–50% of the open and shut events are missed, and result from the formation of compound events consisting of one detected event combined with one or more contiguous missed events (Blatz and Magleby, 1986b). The presence of phantom components may lead to overestimation of the number of kinetic states in the distribution (Roux and Sauve, 1985). Since the effective duration of phantom components is less than 2 t_{dead}, phantom components can be avoided by only fitting intervals greater than 2 t_{dead} (Blatz and Magleby, 1986b).

The number of separate open and closed events actually observed is usually less than f_{det} as defined by equation 39, because, when a brief event is missed, the preceding event, the missed event, and the following event are detected as a single event with a duration equal to the sum of the three events. If this occurs often, the average dwell time and, therefore, the time constant(s) of the distribution will be lengthened and will require correction. As an example, let us again use the two-state, C-O kinetic scheme. We will assume an opening rate constant (k_1) of 1000 and a closing rate constant (k_{-1}) of 20. This should yield an average open-time (τ_{open}) of 50 ms and an average closed-time (τ_{closed}) of 1 ms. The overestimation of τ_{open} resulting from missed, brief closings can be estimated (Sachs et al., 1982; Blatz and Magleby, 1986b) under the restricted conditions of a two-state kinetic scheme (i.e., single-exponential distributions of open and closed t_{dwell}s) and that t_{dead} is much less than τ_{open} according to the equation:

$$\text{observed } \tau_{open} = (\text{true } \tau_{open}) \exp(t_{dead}/\tau_{closed}) \qquad (40)$$

With values of t_{dead} equal to 179 μs (BW = 1 kHz) and 17.9 μs (BW = 10 kHz), the observed τ_{open} would be 59.8 ms and 50.9 ms, respectively. Under the restrictions mentioned above, Eq (40) can be rearranged to calculate an useful approximation of the true τ_{open} from the observed τ_{open}. If the time constant is estimated as the arithmetic mean of the observed dwell times, Eq (6) should be used to correct the mean dwell time for the effect of t_{dead}. A more rigorous approach applicable to a two-state kinetic scheme that is not restricted to the situation when t_{dead} is much less than τ_{open} is discussed in Blatz and Magleby (1986b). With their approach, the "true" rate constants (k_1 and k_{-1}) are adjusted using an iterative procedure to obtain the best fit of predicted pdfs to the observed open and closed dwell-time distributions. The pdfs are calculated using "effective" rate constants k_{eff1} and k_{eff-1}, according to the equations:

$$p(t_{dwell}, \text{open}) = k_{eff-1} \exp(-t_{dwell}\, k_{eff-1}) \tag{41}$$

$$p(t_{dwell}, \text{closed}) = k_{eff1} \exp(-t_{dwell}\, k_{eff1}) \tag{42}$$

The effective rate constants represent the change in the true rate constant predicted to result from missed events. For example, the effective rate constant for channel closing is calculated from the true rate constant according to

$$k_{eff-1} = \frac{k_{-1}\, f_{det(C)}}{1 + f_{miss(C)}\, t_{miss(C)}/\tau_{open}} \tag{43}$$

where $t_{miss(C)}$ is the mean duration of all undetected true closings and is given by:

$$t_{miss(C)} = \frac{\tau_{closed} - (\tau_{closed} + t_{dead}) \exp(-t_{dead}/\tau_{closed})}{1 - \exp(-t_{dead}/\tau_{closed})} \tag{44}$$

With appropriate substitution, Eqs (43) and (44) can be rewritten to calculate the effective opening rate constant. This method forms the basis for corrections that can be applied to more complicated kinetic schemes with more than two states, but the details are beyond the scope of this discussion (*see* Blatz and Magleby, 1986b). An alternative method that provides an exact solution to the correction of the rate constants has been presented by Roux and Sauve (1985). However, because of the computational complexity of their exact solution, they have presented an approximate solution that should be adequate for most purposes.

2.5.3. Amplitude Histograms

2.5.3.1. Point-by-Point Current Level. A point-by-point current level histogram is constructed by binning all of the sampled points (current levels), regardless of the state of the channel. A point-by-point current amplitude histogram should show a peak corresponding to the baseline and an additional peak at the current level corresponding to each conductance level. For example, under the recording conditions of wide bandwidth and with a ratio of $i_{open}/rms_{(BW)} > \approx 6$, the current amplitude histogram of a channel that exhibits C-O kinetics with nearly equal opening and closing rate constants should be bimodal with clearly separated peaks centered at the baseline and i_{open}. The distribution of points around these peaks will appear approximately Gaussian, because the variability is dominated by the normally distributed background noise. A filling-in of the region between the peaks can occur under several conditions. First, as the ratio $i_{open}/rms_{(BW)}$ decreases, the overlap of the tails of the two normally distributed noise distributions increases. Second, with heavy filtering (increasing t_{rise}), the probability of sampling the signal at intermediate levels during a transition between states increases. Third, brief events with a duration $< t_{rise}$ will not reach the open current level and, if they occur frequently, may produce an additional "peak" between the peaks. Fourth, legitimate subconductance states may exist, resulting in additional small peaks between the fully open and closed levels. Finally, if not carefully controlled, baseline drift can smear the distribution by shifting the positions of the closed and open levels with time. A point-by-point current level histogram as described above is valuable, because it provides a rather complete description of the distribution of both noise and signal (including resolved and unresolved events). However, unless obtained under ideal recording conditions, it is not an accurate method for measuring the amplitude of different conductance states.

2.5.3.1.1. Extending the Bandwidth. Channel blocking drugs with relatively slow blocking and unblocking rates produce clearly resolved steps between the fully open and fully closed (blocked) conductance levels and do not affect the amplitude histogram. Block of batrachotoxin-activated Na^+ channels by tetrodotoxin is a good example. Other blockers, such as some inorganic and organic cations (e.g., Fig. 1 D), have such high rates of blocking and unblocking that the individual blocking events cannot be resolved and there is only the appearance of a decrease in conductance (Coronado and Miller, 1979, 1980; Yellen, 1984). These fast blockers distort the point-by-point amplitude histogram by moving the peaks closer together (or making a single, central sharp peak), but with only a moderate distortion of the shape of the peaks. For channel blockers with blocking and unblocking rates intermediate between these two extremes, the current record may appear to be dominated by a "flickery" noise that results from many poorly resolved blocking events. This flickering block is difficult to analyze using the traditional threshold-crossing method, because many blocking events do not cross the threshold and the durations of the events cannot be accurately measured anyway. The presence of flickering block does, however, produce a distortion of the amplitude histogram that can yield estimates of the rates of blocking and unblocking (Fig. 7). This method was developed by Yellen (1984) in his analysis of the flickering block of Ca^{2+}-activated K^{+} channels by Na^{+}. The distortion of the amplitude histogram by the poorly resolved events is predictable, consisting of a broadening of the peak corresponding to i_{open} and a shift of this peak towards the baseline level with increasing degree of block. A predicted point-by-point current level pdf for this distorted distribution can be calculated based on the effective time constant of the filter and the blocking and unblocking rates (*see* Yellen, 1984 for details). Comparison of this pdf with the observed distribution allows estimation of the blocking and unblocking rates. This method depends on two assumptions: (1) the flicker occurs between two states of known conductance, and (2) the blocking reaction is a two-state Poisson process (Yellen, 1984). This is a useful method for extending the effective bandwidth of our analysis by allowing us to extract kinetic information unattainable with traditional methods.

2.5.3.2. Event Amplitude. An alternative method for examining the distribution of event amplitudes is to construct a histogram of the amplitudes of events greater than $\approx 2.5\ t_{rise}$ duration. This event amplitude histogram will show a single peak centered at the average current level of the state. The pdf of the event amplitude distribution is approximately Gaussian, usually exhibiting a sharper peak and broader tails (Colquhoun and Sigworth, 1983). This deviation in the distribution results from the bias introduced by measurement of different t_{flat} durations. The longer events provide a more accurate measurement and contribute to the sharp peak,

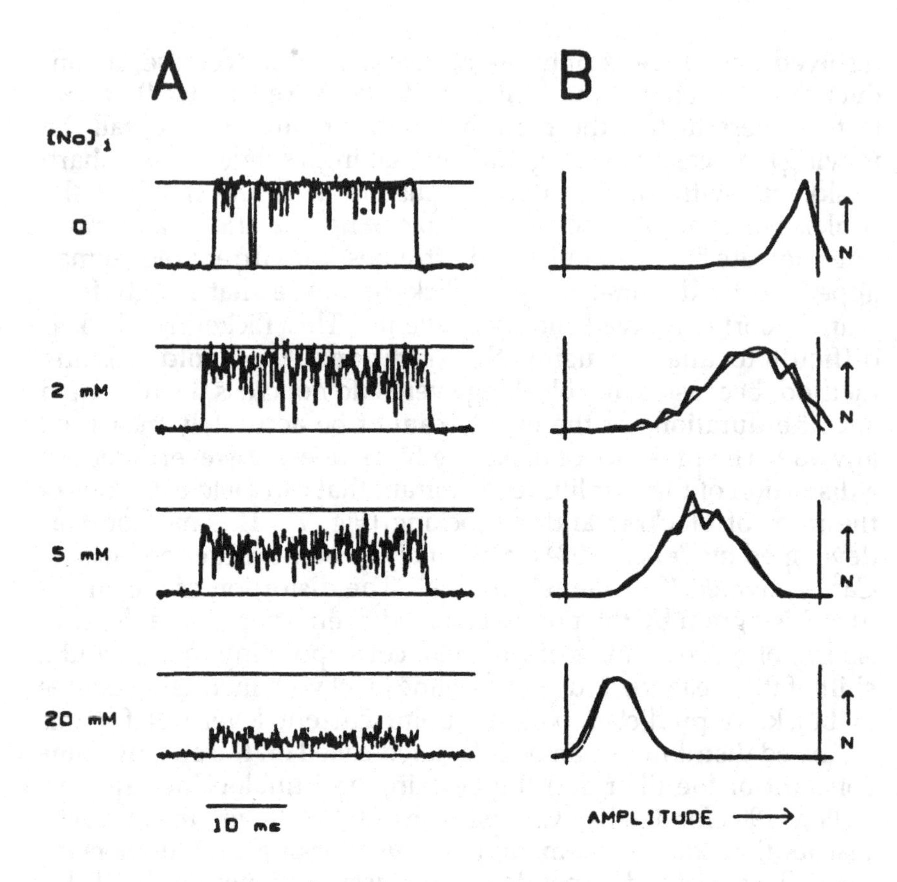

Fig. 7. Flickery block of a large, Ca^{2+}-activated K^+-channel by internal Na^+. (A) Single-channel records with the indicated concentration of Na^+ added to the internal solution containing 160 m*M* K^+ (the external solution contains 160 m*M* Na^+). The records are 40-ms long, and the lines indicate the baseline and open channel current (difference of 24.4 pA). The membrane voltage was +80 mV, and the data were filtered at 4 kHz. (B) Amplitude histograms for each of the individual channel openings at left, plotted as number of occurrences (N) on the ordinate vs amplitude on the abscissa. The left vertical line corresponds to the baseline amplitude and the right vertical line to the open channel amplitude (24.4 pA). Vertical scaling is arbitrary. The smooth curves superimposed on the lower three histograms are the theoretical distributions fitted empirically to the average histograms for many records (reproduced with permission from Yellen, 1984).

whereas the shorter duration events provide a less accurate measurement and contribute to the spread of the tails of the distribution. The distribution of event amplitudes is indistinguishable from a Gaussian distribution if only events with $t_{flat} \geq ½$ the time constant of the dwell-time distribution are included (Colquhoun and Sigworth, 1983).

2.5.4. Fitting Pdfs to Histograms

The following notation will be used in the subsequent discussion. A set of N events, each with dwell time t_{dwell}, will be considered. A histogram of m bins will be constructed from these data. The observed frequencies in these m bins are $f_{bin1}, f_{bin2}, \ldots, f_{bini}, \ldots, f_{binm}$. The set of parameters that describe the pdf (e.g., rate constant, area) will be represented by Θ. The expected frequencies in the m bins are calculated according to Eq (29) or (30), and are $f_{bin1}\Theta, f_{bin2}\Theta, f_{bini}\Theta, \ldots, f_{binm}\Theta$. An exponential pdf with k components will be fit with a set of parameters (Θ), including rate constants (k_j) and the area under each component (a_j):

$$p(t_{dwell} \mid \Theta) = \sum_{j=1}^{k} a_j k_j \exp(-k_j t_{dwell}) \tag{45}$$

It is preferable to fit the a_j rather than the initial value, because the a_j are less strongly correlated with the rate constants or time constants (Colquhoun and Sigworth, 1983).

The predicted pdf and the distribution of observations can be compared visually by overlaying the two with appropriate scaling on a histogram plot. The pdf is a smooth function with dimensions of s^{-1} and an area under the curve of unity. The observed frequency histogram, on the other hand, is a discontinuous function because of the finite bin width, δt. The dimensions of the frequency distribution are δt^{-1}, not s^{-1}, and the area is equal to the number of observations, not unity. Therefore, the pdf must be rescaled to be superimposed on the histogram. For a linear histogram, this rescaling involves multiplying the pdf by the total number of events (N) and then dividing the pdf by $1/\delta t$ (Colquhoun and Sigworth, 1983). The predicted (smooth) pdf should intersect the amplitude of each bin at its midpoint although, if δt is greater than $\approx$ 10–20% of the time constant of the distribution, there may be some discrepancy resulting from binning promotion error (*see* section 2.5.2.1.2.). The scaling of histograms with a logarithmic abscissa is discussed in section 2.5.1.2.1.

A qualitative evaluation of the goodness-of-fit of the predicted pdf to the distribution of observations can be performed visually. However, we are generally interested in using more automated and quantitative fitting methods. These quantitative methods provide a basis for iterative procedures that can be used to optimize our selection of the values of the parameters for a kinetic scheme. Discussion of these iterative procedures is beyond the scope of this chapter. References describing several iterative procedures can be found in Colquhoun and Sigworth (1983). In the next section, we will discuss several measures used in fitting the distributions, with the major focus on the maximum likelihood method.

2.5.4.1. MEASURES OF BADNESS-OF-FIT: χ^2 AND LEAST SQUARES. The quality of fit of the theoretical pdf to the histogram can be evaluated using methods that calculate a statistic proportional to the deviation of observed frequencies from expected frequencies. This includes the χ^2 and the least squares methods. For example, χ^2 is calculated as:

$$\chi^2 = \sum_{i=1}^{m} \frac{[f_{\mathrm{bin}i} - f_{\mathrm{bin}i}\Theta]^2}{f_{\mathrm{bin}i}\Theta} \tag{46}$$

Since the magnitude of χ^2 is proportional to the deviation of observed from expected frequencies, the parameter optimization methods attempt to minimize χ^2. Use of the χ^2 method requires that the data be binned in a histogram. This calculation of χ^2 is similar to a weighted least-squares method, because the magnitude of $f_{\mathrm{bin}i}\,\Theta$ in the denominator gives a greater weight to bins that are less variable in the calculation of the χ^2 statistic.

2.5.4.2. A MEASURE OF GOODNESS-OF-FIT: MAXIMUM LIKELIHOOD. The method of maximum likelihood treats the set of data as fixed and attempts to identify the parameter set, Θ, that has the greatest likelihood of predicting the data. For a set of N dwell times, we can calculate the probability density of each value of t_{dwell} based on a theoretical pdf with a specified set of parameters, Θ. From basic probability theory, we know that the probability of occurrence of a set of N independent observations is simply the product of the probability of each of the N observations. Likewise, if the dwell times are independent, then the probability or, more properly, the likelihood of observing all of the dwell times is simply the product of the individual probability densities:

$$\text{Likelihood (dwell times}\,|\,\Theta) = p(t_{\mathrm{dwell1}})\, p(t_{\mathrm{dwell2}}) \ldots p(t_{\mathrm{dwell}i}) \ldots p(t_{\mathrm{dwell}n}) \tag{47}$$

where each $p(t_{dwell})$ is calculated from the pdf using each observed value of t_{dwell}. This method of calculation is not practical, however, because it may involve calculating the product of thousands of extremely small probabilities. A more useful approach is to use the log-likelihood, $L(\Theta)$, which is the sum of the logarithms of the individual probability densities:

$$L(\Theta) = \sum_{i=1}^{N} \log p(t_{dwell\,i} \mid \Theta) \tag{48}$$

Because the set Θ that best predicts our distribution of dwell times is also the one that results in the largest value of $L(\Theta)$, we vary the values of Θ to maximize $L(\Theta)$. The set of Θ that maximizes $L(\Theta)$ will be considered our best estimate of Θ. Compared to χ^2 or least-squares, the maximum likelihood method is guaranteed to provide the least variable estimate of Θ (Horn and Lange, 1983).

For the general case of k exponential components, each $p(t_{dwell})$ can be calculated from the theoretical pdf as:

$$p(t_{dwell} \mid \Theta) = \sum_{j=1}^{k} a_j k_j \exp(-k_j t_{dwell}) \quad (0 < t_{dwell} < \infty) \tag{49}$$

Since the a_j must sum to unity, there are k–1 area parameters and k rate constants to be estimated. Dwell times cannot be measured over the entire range $0 < t < \infty$; rather, there is a minimum duration (t_{min}) and a maximum duration (t_{max}). This limitation requires that a conditional pdf be written that accounts for the restricted range of observable dwell times. This pdf is:

$$p(t_{dwell} \mid \Theta) = \frac{\sum_{j=1}^{k} a_j k_j \exp(-k_j t_{dwell})}{P(t_{min}, t_{max} \mid \Theta)} \quad (t_{min} < t_{dwell} < t_{max}) \tag{50}$$

where the denominator, the probability that each t_{dwell} lies between t_{min} and t_{max}, is defined as:

$$\begin{aligned} P(t_{min}, t_{max} \mid \Theta) &= \text{Probability } (t_{min} < t_{dwell} < t_{max}) \\ &= \sum_{j=1}^{k} a_j [\exp(-k_j t_{min}) - \exp(-k_j t_{max})] \end{aligned} \tag{51}$$

After Θ is estimated, the true number of events, including both observed and undetected events, can then be calculated according to

$$\text{Number of true events} = N/\text{Probability}\ (t_{min} < t_{dwell} < t_{max}) \tag{52}$$

where N is the number of observed events.

The calculation of $L(\Theta)$ as described above for unbinned data has an advantage over methods that use binned data, such as χ^2 and least squares, in that empty bins are not a problem and no information is lost or error introduced by the process of binning. However, if N is large, the procedure described above is computationally expensive. Sigworth and Sine (1987) and McManus et al. (1987) have demonstrated that the set of kinetic parameters, Θ, can also be estimated using a likelihood calculated from bin frequencies, rather than individual dwell times, with a considerable savings in computation time and no significant loss of accuracy. The method of the calculation of the log-likelihood for binned data, $L_{bin}(\Theta)$, described by Sigworth and Sine (1987) offers the advantage that it avoids binning promotion error in parameter estimation, whereas in the approach of McManus et al. (1987) a correction for the binning promotion error is provided. Sigworth and Sine (1987) use the probability distribution rather than the pdf to calculate $f_{bin}\Theta$. The probability of a particular value of f_{bin} is calculated as the difference between the probability distribution evaluated at the left (t_i) and the right (t_{i+1}) edges of the bin. The probability of t_i for k exponential components is:

$$P(t_i | \Theta) = 1 - \sum_{j=1}^{k} a_j \exp(-t_i\, k_j) \tag{53}$$

The log likelihood for m bins is then calculated as:

$$L_{bin}(\Theta) = \sum_{i=1}^{m} f_{bini} \log \frac{P(t_{i+1} | \Theta) - P(t_i | \Theta)}{P(t_{min}, t_{max} | \Theta)} \tag{54}$$

where the denominator, $P(t_{min}, t_{max} \mid \Theta)$, is the conditional probability that the dwell times fall within the range of the histogram with lower limit t_{min} and upper limit t_{max}, given the set of parameters, Θ. The numerator of this function is the difference between the probability distribution at the lower and upper limits of the bin, which is equivalent to integrating the pdf over the same interval. When bin width is small, the maximum likelihood estimates based on binned and unbinned data are very similar. With as few as 16 bins/decade on a log-transformed x-axis, there is no loss of accuracy compared with unbinned data, and acceptable behavior is observed with as few as 8 bins/decade (Sigworth and Sine, 1987).

Pairwise comparisons of alternative nested kinetic schemes can be evaluated using the likelihood ratio test (Rao, 1973; Horn and Lange, 1983). Comparisons of pairs are made using a ratio of the maximum likelihood estimates for each scheme. Twice the logarithm of the ratio of likelihoods is distributed as χ^2 with degrees of freedom equal to the difference in the number of independent parameters in the two schemes.

2.5.5. *Joint Probability Analysis*

The methods of analysis described above provide a wealth of information regarding putative kinetic schemes responsible for the conductance and gating properties of single ion channels. The variables in these kinetic schemes include the number of conductance states, the number of kinetic states within each conductance state, the pathways for transitions among states, and the transition rates along these pathways. Point-by-point amplitude histograms provide an estimate of the number of conductance states, whereas dwell-time distributions provide an estimate of the minimum number of kinetic states within each conductance state and estimates of some of the transition rates among states. However, these methods do not provide a means of testing alternative schemes that differ in the arrangement of pathways among the states. Recently, methods for joint probability analysis, including autocorrelation and autocovariance, have been proposed (Fredkin et al., 1985; Jackson et al., 1983; Labarca et al., 1985; Colquhoun and Hawkes, 1987; Bauer et al., 1987; Steinberg, 1987). Also, Horn and Vandenberg (1984) have applied the maximum likelihood method to a joint probability analysis of alternative kinetic schemes that vary in structure. These methods promise to add a new dimension to single-channel data analysis by providing a method of testing for the pathway structure within kinetic schemes. These methods differ from the traditional dwell-time analysis in that they use information regarding the sequence of events in the current record.

As in the dwell-time distribution analysis, joint probability analysis is based on the assumption that the transitions among states result from a Markovian process, i.e., transitions are independent and future transitions are not affected by past events. Let us assume that the analysis of a hypothetical channel record has led us to conclude that the kinetic scheme includes two conductance states (closed and open) and, within each conductance state, an aggregate of three kinetic states identified by the presence of three exponential components in both the closed and open dwell-time distributions. These states can be arranged in many different patterns by varying the arrangement of interconnecting pathways. A linear arrangement is a particularly simple scheme:

$$C_1 \rightleftharpoons C_2 \rightleftharpoons C_3 \rightleftharpoons O_3 \rightleftharpoons O_2 \rightleftharpoons O_1 \tag{55}$$

In this scheme, the aggregate of closed states (C_1, C_2, C_3) is linked to the aggregate of open states (O_3, O_2, O_1) by a single gateway, C_3–O_3. If we examine the sequence of either the closed or open dwell times, we expect to observe no correlation of sequential dwell times within either aggregate. This results from the presence of only one gateway between the aggregates. For example, in a sequence of channel opening, closing, and then reopening, the channel has no memory during its second opening of the open state in which it resided during its first opening. An absence of correlation is expected for any variant of this scheme as long as only one gateway between the two aggregates is allowed. An alternative scheme includes the same two aggregates of states, but includes two gateways (C_2–O_3, C_1–O_2):

$$\begin{array}{ccccc} C_3 & \rightleftharpoons & C_2 & \rightleftharpoons & C_1 \\ & & \downarrow\uparrow & & \downarrow\uparrow \\ & & O_3 & \rightleftharpoons & O_2 & \rightleftharpoons & O_1 \end{array} \tag{56}$$

Let us also suppose that the bidirectional transitions in the pathways C_2–C_1 and O_3–O_2 are slow relative to the bidirectional transitions in the gateways. Under these conditions, we expect to see correlations within the sequences of both open and closed dwell times. For example, if a channel in closed state C_2 opens to state O_3, and then closes again to C_2, it is more likely to reopen to O_3 than to make the transition to C_1 and then reopen to O_2. Thus, the successive open dwell times will not be independent and will tend to be grouped into "bursts" during which the closed-open-closed transitions are limited to one or the other gateway. The observation of correlations in sequential dwell times provides evidence that there are multiple gateways between the closed and open aggregates. Joint probability analysis should allow testing of alternative kinetic schemes that differ in the arrangement of pathways connecting aggregates of kinetic states.

2.5.6. Fractal Model

Our discussion of the kinetic analysis of single-channel gating has focused on the widely accepted paradigm of a Markov chain model, which is based on two fundamental assumptions. First, channel gating is assumed to involve transitions among a small number of discrete states. Second, the rates of transition among states are assumed to be time homogeneous, i.e., the probability of leaving a kinetic state is independent of the length of time the state has been occupied. The Markov model has been recently challenged as a paradigm for the interpretation of single-channel gating (Liebovitch and Sullivan, 1987; Liebovitch et al., 1987). These investigators have applied an alternative model, termed the "fractal model," which attempts to combine fractal mathematical theory with recent, more detailed studies of the dynamic conformational fluctuations of proteins. The fractal model can be clearly differentiated from the Markov model on the basis of the same two assumptions. First, unlike the Markov model, the channel is assumed to have a very large number of conformational states separated by relatively small energy barriers. This proposition is motivated by recent studies of the kinetics of conformational changes in proteins (e.g., *see* review by Frauenfelder et al., 1988), which have demonstrated that some proteins are very dynamic, rapidly fluctuating among many (perhaps thousands) conformational states. Thus, each of the conductance states is seen as a family of a very large number of kinetic states that can be approximated by a continuum. Second, the rate constants for transitions among states are time inhomogeneous, i.e., the rate constant for leaving a conductance state is inversely proportional to the length of time the state has been occupied. Therefore, unlike the Markov model, the channel exhibits memory. The physical interpretation of the time dependence of transition rates between open and closed states is that, after entering a new conductance state, the channel will continue to "wander" through many additional conformational states. The model implies that the longer the conductance state is occupied, the more the channel structure will have changed and the lower will be the probability (per unit time) of returning to the conformation from which the state can be exited. The quantitative modeling of this time dependence of transition rates has been based on the mathematical framework of fractal scaling. With this model, the effective rate constant (k) for leaving a conductance state is:

$$k = At^{1-D} \tag{57}$$

where A is the kinetic setpoint, t is the time the channel has resided in the current state, and D is the fractal dimension. The kinetic setpoint simply determines whether all of the transitions will be slow or fast. The fractal dimension, D, on the other hand, determines the dependence of k on t and is restricted to the range $1 \leq D < 2$. The inverse relationship between the transition rate and the time scale requires that the fractal dimension must be ≥ 1, and normalization of the probability density function requires that $D < 2$ (*see* Liebovitch and Sullivan [1987] for details of the pdf). When $D = 1$, the transiton rate is not dependent on t, and the dwell times are distributed as a single exponential, which is the same as predicted by the Markov model. As D increases from 1 to 2, the steepness of the inverse relationship between the time a channel remains in a conductance state and its probability of leaving the state increases.

Proponents of the fractal model have argued that a sensitive test of the validity of the fractal model for a given set of data is an examination of the relationship between the effective transition rate constant and the time scale of measurement. The details of this method of analysis are beyond the scope of this review, but, in brief, the fractal model predicts that the measured effective rate constant should increase as the minimum resolvable dwell time is decreased (e.g., the recording bandwidth is increased). When plotted with log–log scaling, this dependence appears as a linear increase in the effective rate constant with decreasing time scale (i.e., increasing resolution). However, as pointed out by Millhauser et al. (1988), this relationship leads to the improbable situation of infinitely large rate constants at zero time. Using this graphical analysis, Markov models are distinguished by the presence of plateaus in this relationship. The distributions of dwell times predicted by Markov and fractal models can also be clearly distinguished when plotted as described in section 2.5.1.2.2. (logarithmically transformed abscissa and square-root transformed ordinate). With this method, the distribution of dwell times predicted by the Markov model exhibits a peak corresponding to the time constant of each kinetic state (Korn and Horn, 1988), whereas the distribution predicted by the fractal model is always unimodal. The choice between the Markov and fractal models has stimulated much lively debate. At the present time, the quality of the fit of the two models to the gating of several species of ion channels has been compared using graphical and statistical methods of analysis (McManus et al., 1988; Korn and Horn, 1988). In most tests, the best Markov model for a given case has fit the distributions better than, or at least as well as, the fractal model. However, there has been a report suggesting that the best model might be some combination of the two models (French and Stockbridge, 1988). The challenge put forth by the fractal model has forced a more careful examination of the relationship between quantitative models of ion channel gating and the underlying molecular or physical processes (e.g., Lauger, 1988).

2.6. Summary of Recording, Event Detection, and Analysis

2.6.1. Useful Guidelines

The methods discussed above can be summarized in the form of some guidelines for the collection and analysis of single-channel current records.

2.6.1.1. CHOICE OF RECORDING SYSTEM. The bandwidth and noise levels of a patch-recording system are generally superior to a bilayer recording system. However, we wish to emphasize that the disparity between the two methods is almost completely dependent on the size of C_{input}. If careful attention is given to mimimizing C_{input} in the bilayer system, the bandwidth can be markedly improved and, as shown in Fig. 1, the rapid gating of relatively small conductance channels (≈ 25 pS) can be resolved at a bandwidth of 1 kHz. We believe that the choice between patch and bilayer methods should be principally based on the requirements of the experimental protocol, including factors such as accessibility of solutions and control of the composition of the membrane.

2.6.1.2. FILTERING AND SAMPLING. A value of f_{cut} should be selected that minimizes the false threshold-crossing rate while avoiding an excessive number of missed events. This value can be calculated based on the magnitudes of the $rms_{(BW)}$ noise, i_{open}, and the average event dwell time. Sampling promotion error is minimized if f_{cut} is chosen such that t_{dead} is less than 10–20% of the shortest time constant in the dwell-time distribution. After selection of f_{cut}, the sampling rate should be set $\geq \approx 5 \times f_{cut}$. Two-stage filtering is particularly useful. The first stage is the analog (Bessel) filtering of the record prior to digitization, with an f_{cut} set at 2–3 $\times$ the expected final f_{cut}. The second stage is the filtering of the digitized record using a programmed, Gaussian digital filter (*see* Colquhoun and Sigworth, 1983, for examples of programs). The effective f_{cut} of a two-stage filtering process (assuming high-order Bessel or Gaussian filters) is:

$$1/f_{cut}^2 = 1/f_1^2 + 1/f_2^2 \tag{58}$$

where f_1 and f_2 are the f_{cut} of the first and second stages (Colquhoun and Sigworth, 1983). Two-stage filtering is the most convenient method for adjusting f_{cut} to optimize the control of FTC and sample detection error. It is also useful to record a square current pulse test signal (applied to the headstage as described in section 2.3.2.1.), because the response to this pulse allows direct measurement of t_{rise}.

2.6.1.3. EVENT DETECTION AND MEASUREMENT. The 50% threshold-crossing method simplifies measurement of event duration by providing a symmetric method for calculation of durations in each conductance state and facilitates the use of some correction processes, such as the missed-event correction proposed by Blatz and Magleby (1986b). Cubic spline interpolation should be used in the measurement of event duration to avoid sampling promotion error.

2.6.1.4. HISTOGRAMS. The display and fitting of dwell-time histograms is facilitated by using a logarithmically transformed time axis and a square-root transformed ordinate. Amplitude histograms can be constructed using linear scaling of both the abscissa and ordinate because of the symmetric, Gaussian distribution of event amplitudes.

2.6.1.5. ESTIMATION OF PARAMETERS OF KINETIC SCHEMES. The maximum likelihood method is preferred for estimating kinetic parameters from observed dwell-time distributions. This method can be used with either unbinned or log-binned dwell-time data, with a large reduction in calculation time for log-binned data. The binning promotion error discussed by McManus et al. (1987) should be avoided by using the maximum likelihood calculation presented by Sigworth and Sine (1987) (Eq. 54). The relative goodness-of-fit of alternative kinetic schemes can be tested using the likelihood ratio test.

The missed events resulting from the limited bandwidth of the recording system can lead to several serious sources of error in the fitting of dwell-time histograms. Blatz and Magleby (1986b) have presented a method that corrects for the overestimation of time constants produced by missed events. Another source of error, the phantom states (Roux and Sauve, 1985) produced by missed events, can be avoided by not fitting dwell times $< 2t_{dead}$.

2.6.1.6. SIMULATION. Simulation of dwell-time distributions with known parameters provides a means to evaluate the accuracy of parameter estimates. For multi-component distributions, the sensitivity of parameter estimates to the degree of differences among rate constants, the relative areas under each component, missed events, or the total number of events can be determined. Good examples of the use of simulation can be found in Clay and DeFelice (1983), Blatz and Magleby (1986b), McManus et al. (1987), and Sigworth and Sine (1987).

2.7. Data Acquisition—Interfacing Hardware and Software

2.7.1. Studying Open-Channel Properties—Use of Voltage Ramps

Transitions between open and closed states can be easily identified not only at a constant voltage, but also when the voltage is steadily changing. Under a voltage ramp command, an opening appears as a step in the current level as well as an abrupt change in the slope of the record. Open channel current-voltage (i-E) relations can be efficiently collected by applying a voltage ramp over the desired range (e.g., Yellen, 1982, 1984; Eisenman et al., 1986). This approach offers a number of features that can be used to advantage. First, in principle, a complete i-E curve can be collected in a single sweep. Second, in cases where closing of a channel carrying inward current might be confused with opening of a channel carrying outward current at a constant voltage, channel openings are clearly identifiable as segments of increased slope in the ramp-evoked current. Third, for voltage-dependent channels, it may be possible to obtain i-E data over a wider range than in steady-state recordings, by starting the ramp at a voltage where there is a high probability of the channel being open. For success, this depends on being able to change the voltage fast enough that the nonstationary open probability remains reasonably high throughout the sweep.

In practice, multiple sweeps are generally used to increase the signal-to-noise ratio, and to enable removal of leakage and capacitative components to yield an uncontaminated single-channel i-E relation. Unambiguous open and closed periods are identified in each sweep, and these are averaged, piecewise, to give total current with the channel open or with the channel closed. The averaged "channel-closed" sweep represents only the leakage and capacitative current flowing during the voltage ramp. The "channel closed" sweep is subtracted from the "channel-open" sweep to give an uncontaminated single-channel i-E relation. Brief events, corresponding to openings too short for the full amplitude to be registered, are excluded from the averaged curves. A couple of tricks based on the properties of the channel under study can be useful in applying this technique. A reasonable proportion of time with the channel open can be assured by appropriate agonist concentrations for chemically activated channels, or by choice of the holding potential and the starting voltage and direction for the ramp for voltage-activated channels. Definition of the closed level may be aided by use of drugs that produce long duration blocked events, such as guanidinium toxins for sodium channels, or Ba^{2+} or charybdotoxin for certain Ca^{2+}-activated K^+ channels.

2.7.2. Data Acquisition Hardware and Software

2.7.2.1. OVERVIEW. The choice of data acquisition hardware and software invariably represents a compromise between requirements for absolute performance and flexibility, and the minimization of development time and cost. Recommendations about different options can be made with the certainty only that they will be out of date by the time of printing. That qualification notwithstanding, we proceed. Available software ranges from laboratory-oriented languages or development systems that facilitate writing of powerful applications programs to integrated hardware and software packages with ready-to-run experimental and analysis programs. We have not had firsthand experience with all of these packages, so we will attempt to provide only an overview of general features of some packages of which we are aware. Thus, we hope to convey an idea of the range of options available.

Some general design features are of importance if performance of the interfacing hardware is to be optimized. A high-quality, "deglitched" D/A converter is of course important for the production of a low-noise stimulus signal. In general, quieter operation will be possible if the analog circuitry is remotely boxed and powered separately from the computer using a high-quality linear power supply. The computer's logic circuitry can then be optically isolated from the analog components in the remote front-end box. Single-board systems, in which analog signals pass directly to and from on-board connections in the computer, will in general be significantly noisier. The increased performance of the distributed systems, of course, comes only at an increased cost. Speed of operation is increased by delegating certain operations to dedicated microprocessors and memory on the interfacing boards. For example, stimulus generation can be accomplished by downloading pulse parameters into the D/A board's memory, then commanding it to begin, and having all of the timing and pulse generation witout demanding CPU time from the controlling microcomputer. Mapping of buffer memory on D/A or A/D boards as part of the computer's random access memory (RAM) can also speed data capture or stimulus production. For many purposes, single board systems may be adequate and they are substantially cheaper. However, it is worth bearing these points in mind if experiments requiring high-speed data acquisition are anticipated.

With the release of the IBM PC and its many clones, computer-controlled data acquisition and analysis moved out of the province of a wealthy and specialized elite to become as routine a necessity as an oscilloscope in the electrophysiology laboratory. The financial accessibility of computing and interfacing hardware stimulated the development of a variety of software solutions to the problem of control of experiments and display and analysis of data, both in leading laboratories and in laboratory-oriented companies. It is reasonably certain that these systems, running under one or another version of the MS-DOS operating system, will be the mainstay of a majority of laboratories for several years to come. Even so, changes are coming rapidly, and interfacing hardware and software are now appearing for the new generation IBM PS/2 and Apple Macintosh machines, in parallel with continuing refinement of, and support for, the MS-DOS systems. All of the systems that we discuss offer 12-bit A/D conversion (–2048 to 2047 in bipolar mode); there is some variation in resolution of the D/A converters (we would not settle for less than 12-bit, particularly where divided pulse protocols are contemplated).

Two of the systems that we mention run on Digital Equipment Corporation (DEC) machines under the RT-11 operating system. Although more expensive than the PC-based systems, these systems offer some advantages, and retain a number of users and advocates. A well-integrated system running the laboratory-oriented BASIC-23 language is offered by Indec Systems. This offers faster sampling (125 kHz) than a number of the systems currently implemented for PC-compatibles, though this is not an inherent limitation, at least for the PC AT-compatibles. A further advantage of the BASIC-23 system is exceptionally convenient cursor-controlled data manipulation implemented via the superb HP-1345 display. This display offers 2048 × 2048 resolution and very high speed through its memory-mapped, DMA-linked operation, and has few, if any, peers in the microcomputer world for realtime display of data or composition of figures prior to output to a hardcopy device. By the standards of PC hardware, however, it is expensive, and this is probably the major reason that it has not been incorporated into more systems. Other than initial cost, the main disadvantage of RT-11-based systems is the paucity of cheap, general-purpose software, such as spreadsheet and database programs, which are available in profusion to run under MS-DOS and provide a valuable adjunct to specialized data acquisition and analysis software.

Although our focus is on single-channel data, a laboratory computer system will usually be required, at different times, to digitize and collect data using a variety of different protocols in the following modes:

1. Stimulus-driven or interval-driven acquisition, in which a trigger pulse and a command signal is provided by the computer and data collection proceeds for a preprogrammed time after the trigger
2. Event-driven acquisition, in which data are taken continuously into a circular buffer and saved to disk only when the signal crosses a specified threshold (*see* Sigworth, 1983b) and
3. Continuous digitization and acquisition for extended periods—the most satisfactory mode of collection for steady-state data and many exploratory experiments.

Long duration continuous acquisition requires either a rapid procedure for writing data to disk, or copious quantities of memory. DOS systems are often limited to 64 kbytes of absolutely continuous acquisition because of the partitioning of memory. However, this limitation can be circumvented by using dual-ported memory on the A/D board to provide double-buffering of the input, as implemented in Axolab. Continuous acquisition may be selected as a special case of event-driven acquisition by a suitable setting of threshold. Event detection may, in principle, be performed by a hardware device, reducing the demands on the computer, or by software examination of the data in the circular buffer. All of these modes of data acquisition can be useful, and one should be sure that a data acquisition system is capable of meeting all anticipated experimental demands. Maximum rates of acquisition in general depend on the computer, the software, the acquisition mode, and the interfacing boards. This may be as fast as 100 kHz during simultaneous stimulus output and data acquisition, though single board systems are generally slower. Again, specific requirements should be checked against systems specifications and, if possible, with an experienced user.

2.7.2.2. Some Available Hardware/Software Packages. Integrated "ready-to-run" systems, including both interfacing hardware and software for IBM PC AT or XT or compatibles running under MS-DOS, are available from the following companies: Axon Instruments—pCLAMP, Indec Systems—BASIC-FASTLAB and C-LAB, Intracel—SATORI and ACTIVE. All of these packages include some software for analysis of single-channel data. Experimental and analysis programs for BASIC-23 (Indec Systems, RT-11 operating system on DEC LSI-11/23 and LSI-11/73 computers) are also available. In some cases, source code for the applications programs is supplied; in others, it is not—an important consideration if the user anticipates the need to customize a program. A version of IPROC (Sachs et al., 1982; Sachs, 1983) is available for automated batch analysis of continuous single-channel records under DOS. Inexpensive (or free—*see* Robinson and Giles, 1986) programs are available for digitization of data using the Data Translation DT-2801 and DT-2801-A boards. A more comprehensive software package for controlling stimulus and data acquisition (supporting the DT-2801, DT-2801A or DT-2821 boards) and providing analysis of both voltage-clamp and single-channel data is available free of charge to academic and noncommercial users from Dr. J. Dempster, University of Strathclyde (Dempster, 1988).

Powerful and efficient tools for program development are offered by BASIC-FASTLAB, C-LAB, BASIC-23, and The Real Environment (TRE, Laboratory Software Associates). Each of these packages provides interpretive operation for immediate execution of individual commands or routines from the keyboard, and easy debugging of programs. Speed of execution, something distinctly lacking in more primitive interpretive systems, is increased by provision of many high-level single commands that enable operations on an entire array of data. This not only speeds execution, but also simplifies programming greatly, since most of the need to write loops for repetitive operations is eliminated. Straightforward commands control acquisition, manipulation, and display of data, allowing an investigator to relatively rapidly generate an experimental or analysis program tailored precisely to specific needs. However, the flexibility of one or the other of the available applications programs may make even this development time unnecessary for many purposes. Features of the software systems that we have introduced are summarized for ease of comparison in Table 1.

2.7.2.3. Compatibility Between Packages. One safe statement about software is that no single program will do everything that you want—not even the ones that you write yourself—since demands and tasks are forever changing. Consequently, it can be extremely useful to have convenient ways to link functions of various programs in sequence. For example, if your laboratory software does not offer database and spreadsheet functions, does it produce files that are compatible with a program that does? If analysis programs do not output publication-ready figures, do they enable data to be passed easily to a good graphics package? Companies are beginning to address this issue by generating output files either in a standard form or to be compatible with software offering complementary functions. For example, pClamp outputs data in ASCII files that can be processed by a variety of other programs, and AQ and ADCIN produce files that can be processed with IPROC. Since software is constantly under development, it is best to inquire carefully about compatibility and links between software before you make a purchase, preferably from someone who made the connection that you want to make.

3. Applications

3.1. Overview

The ability to resolve the currents through single ion channels has added a new dimension to our accessible view of the electrical responses of cells. Instead of identifying functional components of total transmembrane current based only on their amplitude and time course as a function of ionic conditions, voltage, and their dependence on pharmacological agents, we can view a single-channel record naturally chopped into discrete unitary events. Two general characteristics are immediately added to the "fingerprint" by which we can identify the current resulting from a particular ion channel—the amplitudes and the lifetimes of the single-channel openings and closings. It is no coincidence that the number of recognized distinct channel types has burgeoned since single-channel recording has been possible. The numbers of identifiable types of Na^+, K^+, and Ca^{2+} channels continues to grow steadily. Beyond the simple recognition of discrete dwell times in conducting and nonconducting states, it is often immediately obvious that openings appear in groups or "bursts" (e.g., Colquhoun and Hawkes, 1982; Magleby and Pallotta, 1983b). This kinetic superstructure imposed on the open and closed time distributions reflects particular subsets of the conformational states that the channel protein can assume. The functional fingerprints of particular channels are often so characteristic that merely examining a set of single-channel records will allow identification of several different molecules contributing to a cellular response.

However, single-channel recording offers more than an elaborate basis for a molecular taxonomy. In the sections below, we outline some studies illustrating some diverse problems for which single-channel recording has contributed uniquely to our understanding. These applications include:

1. Detailed analyses of the kinetic fingerprints mentioned in the preceding paragraph to give clues to the molecular transitions involved in channel gating (section 3.2.1.)
2. Kinetic analysis of pharmacological actions, where the object is to gain an understanding of the mechanism of action of a drug (section 3.2.2.), although the analytical approach follows the same principles as one directed at an understanding of intrinsic channel gating
3. An approach to a functional understanding of molecular structure of ion channels, including study of reconstituted, purified channels (section 3.2.3.), of effects of specific chemical modifications

Table 1
Software/Hardware Systems for Data Acquisition and Analysis

Package	Company	Source code available	Computer, operating system	Interface hardware	Notes
pClamp	Axon	Yes	IBM PC or PC/AT, MS-DOS	Labmaster	Modify or enhance using MS-QuickBASIC
Basic-Fastlab	Indec	For applications only	IBM PC or PC/AT, MS-DOS	Labmaster with Indec opto-isolation	Development system with applications
C-Lab	"	"	IBM PC or PC/AT, MS-DOS	"	"
Basic-23	"	"	RT-11	Cheshire (Indec)	"
Satori, Active	Intracel	No	IBM PC or PC/AT, MS-DOS	Intracel S-200, Labmaster	

The Real Environment (DAOS)	Laboratory Software Associates	No	V.7-IBM PC or PC/AT, MS-DOS; VAX, VMS; V.6-LSI-11, RT-11	Real World or Data Translation (DT)	Development system
IPROC	C. Lingle	Yes	IBM PC or PC/AT, MS-DOS	None	Analysis only
ADCIN & related programs	J. Pumplin	Yes, at extra cost	IBM PC or PC/AT, MS-DOS	DT-2800 series	Stimulus and acquisition
AQ	see Robinson & Giles (1986)	Yes	IBM PC or PC/AT, MS-DOS	DT-2801A	Acquisition, display
Dempster Software	J. Dempster	Yes	IBM PC or PC/AT, MS-DOS	DT-2801, DT-2801a, DT-2821	Stimulus, Acquisition and Analysis

of channels (section 3.2.4.), and of the effect on channel function of the lipid environment (section 3.2.5.)

4. Studies of the mechanisms of regulation of channel activity and identification of the cellular components involved in this regulation (section 3.2.6.)
5. Identification and description of roles of channels in cellular processes and broader biological functions. Bilayer methods allow study of channels from subcellular membranes and, in particular, have given access to the complex intracellular membrane systems of muscle (section 3.2.7.). Patch-clamp methods allow mapping of the location of sites of transmitter release (section 3.2.8.) and of channels. Exciting ongoing work is revealing the part played by ion channels in the immune response and regulation of cell division (Cahalan et al., 1985; Chandy et al., 1985; Deutsch et al., 1986; Sidell et al., 1986), and the list is growing.

3.2. Examples of Applications

3.2.1. Analyses of Channel Gating Kinetics

To date, the most detailed studies of gating kinetics have focused on the sodium channel, the large Ca^{2+}-activated potassium channel, affectionately known to some as "the beautiful big one," and the nicotinic ACh receptor channel. In each case, there have been studies in native membrane patches and in bilayer systems. For sodium channels, extensive patch-clamp studies on unmodified channels in two laboratories have culminated in the recent papers of Aldrich and Stevens (1983, 1987), Aldrich et al. (1983), Horn and Vandenberg (1984), and Vandenberg and Horn (1984), whereas detailed work on gating in reconstituted systems has thus far been limited to batrachotoxin-activated channels (Keller et al., 1986). The large, Ca^{2+}-activated K^+ channel (Moczydlowski and Latorre, 1983; Magleby and Pallotta, 1983a, b) and the nicotinic ACh receptor channel (e.g., Labarca et al, 1985; Colquhoun and Sakmann, 1985; Sine and Steinbach, 1986a,b, 1987—*see* reference lists in these papers for more citations) have been the subject of both bilayer and patch clamp studies. Among these papers, a variety of different tactics has been adopted for the analysis, and each approach is worthy of scrutiny for that reason.

One generalization can be made—that one ekes out the details of gating reaction schemes and the transition rates between different, kinetically defined open and closed states only at considerable computational expense. It is worth giving some thought, therefore, as to how that effort can best be expanded. Many thousands of events must be recorded and analyzed in order to accurately estimate the parameters of a complex kinetic scheme (*see* section 2.5.2.). To some extent, however, one may have to make a practical choice between the complexity of the analysis and model-testing, and the number of events and protocols included in the analysis.

In their study of the calcium-activated K^+ channel, Magleby and Pallotta (1983a,b), under steady-steady conditions, collected long records containing very many events, in an effort to identify transition rates over a wide range. Multiple rate constants, determined from the open and closed time distributions at different agonist (Ca^{2+}) concentrations and voltages, were assembled into a minimal scheme for channel gating. In order to account for the distributions of open (double-exponential) and shut (triple-exponential) times at a fixed concentration of free Ca^{2+}, a scheme with at least two open and three closed states was necessary. Occasional extremely long closed times were observed, suggesting that the channel could also enter an additional, very long closed or inactivated state. The complexity of the model also had to be increased, by the addition of another open state, to account for the dependence of the open times on varying [Ca^{2+}]. Burst kinetics, examined in more detail in the second paper (Magleby and Pallotta, 1983b), were encompassed within the framework of the schemes developed from the complete open and closed time distributions, as verified by calculations following the general methods of Colquhoun and Hawkes (1977, 1981, 1982, 1983). Analyses such as these, endeavoring to simultaneously and accurately resolve kinetic components over a wide time scale, demand long periods of data collection at high resolution. Enumeration of the number of exponential components in the open and closed time distributions yields a lower limit to the number of open and closed states, whereas conditional probability analysis of single-channel records offers a direct way to identify the pathways connecting the states. For the calcium-activated K^+ channel, our picture of the gating kinetics is based almost completely on the analysis of single-channel data. In contrast, for the sodium channel, there is a long history of kinetic analysis of macroscopic ionic and gating currents. This both assists in framing questions to be answered by single-channel experiments, and provides complementary information (*see* French and Horn, 1983) that ultimately must be integrated into a complete kinetic model.

For the sodium channel, which inactivates spontaneously and rapidly after opening in response to a depolarizing voltage step, different analytical tactics have been employed. In general, analyses must be carried out on a much smaller number of events. Nonetheless, a discerning choice of voltage step protocols, together with the application of a conditional probability analysis, has allowed the single-channel experiments to dramatically change our ideas of the basis of sodium channel kinetics. Aldrich et al. (1983) and Aldrich and Stevens (1983, 1987) used a combined "fate and rate" analysis to demonstrate that the rate of decline of the sodium current was limited by the activation (opening) rate. This conclusion contradicted the widely held intuition that the rate of decline of the macroscopic sodium conductance during a maintained depolarization was determined by the rate of the inactivation process. Rather than trying exhaustively to enumerate the states, they chose to divide the states into three functional groups—resting, open, and inactivated. They then experimentally determined the time-independent probabilities of transition and the rates of transition between pairs of states during selected pulse protocols. The inactivated state could be treated as an absorbing state, from which the channel did not return during the depolarizing step. Their seemingly paradoxical conclusion arises because the activation rate is substantially slower than the rate of inactivation. Thus, although the channel open time is determined primarily by the inactivation rate over a wide range of voltage, the rate at which the macroscopic current builds up and declines is largely determined by the rate at which new channels are recruited from the resting state into the open state, i.e., the rate of activation. Their analysis provided separate estimates of the rates of inactivation from the resting and from the open state, and the rate of deactivation (returning to rest from the open state). The combined effect of the voltage-dependencies of the different transition rates out of the open state accounted for the observation that the open time varied little over a wide range of the voltages at which sodium channels normally activate.

Similar conclusions were obtained by Horn and Vandenberg (1984) and Vandenberg and Horn (1984) using quite a different analytical approach. Horn and Lange (1983) developed a maximum likelihood method for estimation of the parameters in a kinetic scheme for channel gating. The power of this approach lies in its adjustment of the model parameters to simultaneously fit a complete set of single-channel data, taking into account *every* observed transition, rather than separately examining individual open, closed, or burst time distributions. Therein also lies the cause for its substantial computational expense (though as faster and faster microcomputers become available at reasonable cost, this factor becomes less inhibiting). In principle, a defined time-homogeneous Markovian model of arbitrary complexity can be fit to data from an arbitrary number of channels. There are, of course, some practical limits resulting from both computational tedium as the complexity goes up and the fact that transition rates become too rapid to be accurately resolved if too many channels are present. However, in the culmination of these studies, Horn and Vandenberg (1984) were able to apply rigorous statistical criteria to compare the abilities of 25 different kinetic schemes to account for sodium channel gating. The model that ranked highest, using the Akaike information criterion, possessed three sequential closed ("resting") states plus a single open state and a single inactivated state, and had seven free parameters. Notably, 11 of the schemes that ranked less well possessed more free parameters. Furthermore, some strong qualitative conclusions emerged. It was possible to eliminate from consideration the Hodgkin-Huxley model, with inactivation proceeding independently from resting and open states, as well as any model demanding that a channel open before it inactivates.

3.2.2. An Anticonvulsant Appears to Enhance Slow Inactivation of Sodium Channels

In pharmacological studies, as for intrinsic gating processes, single-channel recording enables more stringent tests of mechanism than are possible with macroscopic measurements alone. Distinct from the fast inactivation, discussed above, there exist one or more processes of slower onset and much longer duration that temporarily inactivate sodium channels. Slow inactivation occurs on a time scale of hundreds of milliseconds to seconds. Although obviously critical to long-term modulation of nervous activity and information processing, slow inactivation of sodium channels is made difficult to study at the level of macroscopic currents in whole cells or axons by requirements of long-term stability of the preparation and the reduction in the amplitude of the signal as inactivation occurs. A study at the single-channel level has the advantage that the amplitude of the unitary event remains unchanged, whereas only the likelihood of channel openings changes with the degree of slow inactivation. By using event-driven data acquisition, the single-channel experiments allow precise analysis of low probability channel openings occurring over a long time span (seconds). Recent studies by Quandt (1987a) clearly revealed the kinetic basis of slow inactivation of whole cell currents in the bursting behavior of single channels (Fig. 8). Enzymatic removal of the fast inactivation process enabled the slow inactivation process to be studied separately in steady-state recordings. The time course of onset of slow inactivation was mirrored in the long duration components of the closed time distribution, in particular that which reflects the long closed periods between bursts of openings. Further, Quandt (1987b) showed that inhibition of Na^+ channel currents by diphenylhydantoin (DPH), an anticonvulsant, can occur in the absence of the fast inactivation, apparently by enhancement of slow inactivation. The action of DPH was complex, but several points are clear. The unitary conductance was unchanged. Reduction in sodium currents was caused by a reduction in the overall probability of the channel being open, rather than a decrease in the amplitude of the unitary current.

The data suggest directly that DPH can bind to either the open state, converting it to a nonconducting state, or to the slow-inactivated state, increasing the duration between bursts of openings. Thus, DPH either extends the duration of slow inactivation or induces a similar, but longer-lived, process.

3.2.3. Reconstitution of Purified Channels—in Patch and Bilayer

Reconstitution of biochemically purified channels into liposomes or bilayers from which electrical recordings of their activity can be made is essential to verify their identity and functional integrity. This has been achieved for nicotinic acetylcholine receptor channels (Nelson et al., 1980; Tank et al., 1982; for a comprehensive review, *see* Anholt et al., 1985) and for sodium channels from rat brain (Hartshorne et al., 1985), as well as from eel electroplax (Rosenberg et al., 1984; Recio-Pinto et al., 1987). In each case,

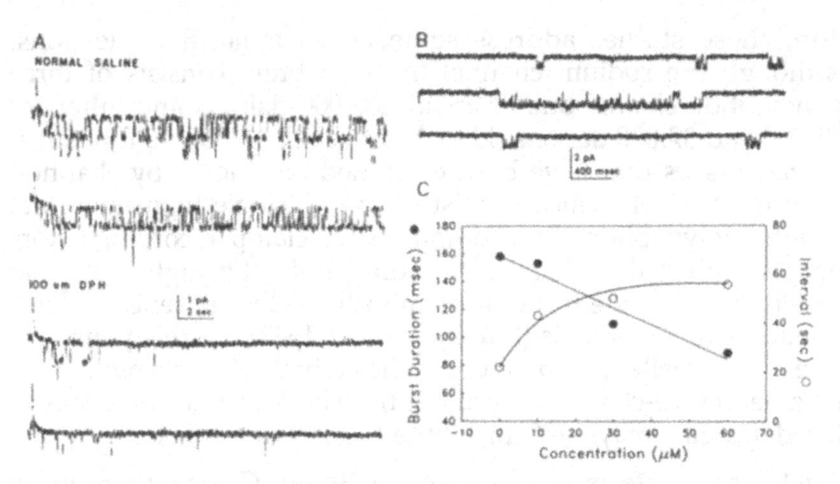

Fig. 8. Diphenylhydantoin (DPH) blocks reopening of Na^+ channels. (A) A patch of plasma membrane from a cultured neuroblastoma cell was isolated in the inside-out configuration. In the top pair of traces, Na^+ channels reopen (channel opening is downward) repeatedly throughout depolarizations to –10 mV from a holding potential of –90 mV following removal of fast inactivation with *N*-bromoacetamide. The onset of the 16-s depolarization is indicated by the capacitative transient. In the bottom pair of traces, exposure of the inside of the membrane to 100 μ*M* diphenylhydantoin (DPH) reversibly reduced the probability that channels reopen. CsF internal solution throughout, 9°C. (B) A continuous record (top to bottom) of membrane current at 0 mV. In the absence of fast inactivation, which was removed with the proteolytic enzyme papain, Na^+ channels burst during maintained depolarization. The interval between bursts reflects the rate of recovery from slow inactivation. (C) The effect of DPH on the burst activity of Na^+ channels for the experiment shown in B is plotted. The mean duration of bursts was reduced as the concentration of DPH was increased because of the ability of DPH to reduce reopening of Na^+ channels. The mean interval between bursts for a single channel was increased in the presence of DPH, because of a reduction in the rate of recovery from slow inactivation. Unpublished figure kindly provided by Dr. F. N. Quandt (*see also* Quandt, 1987b).

the properties of the channels have been compared in detail with characteristics observed in recordings from intact cells, excised cell membrane patches, or from channels incorporated into lipid bilayers from native membrane vesicles without any solubilization of the channel protein. Beyond demonstrating preservation of function, these studies address some broad molecular questions. Although the sodium channel from rat brain consists of three polypeptide chains, one of about 260,000 daltons and others of 39,000 and 37,000 daltons (Hartshorne et al., 1986), all functional characteristics that have been examined are shown by channels from electric eel, which consist of only a single high molecular weight polypeptide. In addition, patch-clamp recordings from oocytes injected with mRNA coding only the high molecular weight polypeptide from rat brain indicate the presence of functional sodium channels (Stühmer et al., 1987). At this point, the role of the smaller polypeptides in the rat brain channel remains an enigma. Single-channel recordings from both cellular and reconstituted systems may well contribute to resolving this issue.

3.2.4. Three Effects of Modifying a Single Carboxyl Group on Sodium Channels

By examining the modification of single sodium channels by trimethyloxonium (TMO), Worley et al. (1986) were able to conclude that modification of a single group on the protein simultaneously abolished the ability of the channels to bind saxitoxin (STX), reduced the unitary conductance by about 35%, and dramatically weakened the block of the channel by external Ca^{2+}. The three effects were always seen when the only channel in a membrane was modified. However, in experiments in which only one of two channels in the membrane was modified, it was always the channel with the modified (reduced) conductance that was immune to STX block, and STX always blocked channels whose conductance was unchanged. It is only in a single-channel experiment that this association of the three actions of TMO with the modification of a single group could have been made so clearly and directly. The use of the planar bilayer system facilitated the use of the labile TMO, which has a half-life of seconds at room temperature in aqueous solution. The open chamber allowed the direct addition of solid TMO to the chamber with stirring to bring the reagent into contact with the channels rapidly. Recording of current fluctuations resulting from an individual channel molecule was thus possible both before and after modification. Current–voltage relations showing the effect on single-channel conductance and the reduction in the channel-blocking action of external Ca^{2+} are presented in Fig. 9. Site-directed mutagenesis, in conjunction with m-RNA injection and patch-clamp recording,

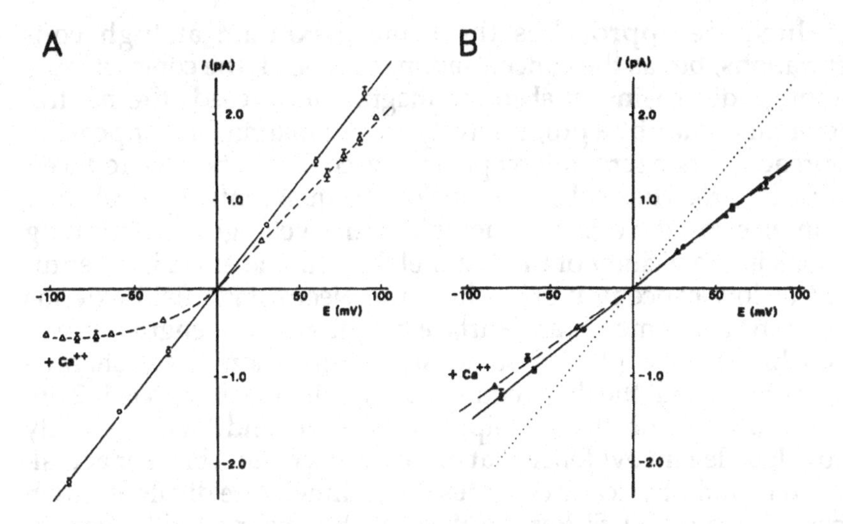

Fig. 9. Trimethyloxonium (TMO) modification both reduces the single-channel conductance of rat brain sodium channels and diminishes the potency of block by external Ca^{2+}. (A) Single-channel current plotted as a function of membrane potential before TMO modification. Open circles, symmetric 125 m*M* NaCl, no free divalents; open triangles, symmetric 125 m*M* NaCl, 10 m*M* external Ca^{2+}. (B) After TMO modification, ionic conditions as in A; the dotted line shows the control (zero Ca^{2+}) i-E relation obtained prior to modification (from A). TMO modification also eliminated saxitoxin block of the channel (not shown). Reproduced with permission from Krueger et al. (1986).

offers the potential to ask further questions about the functional role of specific parts of the channel molecule.

3.2.5. Lipid Effects—Effects of Lipid Surface Charge on Unit Conductance and Channel Block

It requires heroic efforts to change the lipid composition of the plasma membrane in an intact cell significantly (Steele et al., 1981), whereas by using a planar lipid bilayer, the composition may be selected at will with few restrictions. This possibility was astutely applied by Bell and Miller (1984) in a study of the effect of lipid surface charge on ion conduction through the K^+ channel from sarcoplasmic reticulum. In neutral membranes the single-channel conductance is an hyperbolic function of K^+ concentration, saturating at about 220 pS. With negatively charged membranes, the conductance approaches the same maximum at high concentrations, but as the concentration is reduced, the conductance, although decreasing in absolute magnitude, exceeds the neutral membrane value by a progressively greater margin, and appears to approach a nonzero intercept as the concentration is reduced towards zero. These observations can be quantitatively explained as an electrostatic effect of the lipid surface charge concentrating cations in the vicinity of the channel mouth. The effect is accentuated at low concentrations where the electrostatic field extends farther from the membrane surface (i.e., the Debye length is greater). This analysis offered some new information about channel geometry, in that the data require that the channel mouth is 1–2 nm removed from the charged lipid surface. Beyond this, the study provided elegant evidence that one class of channel blocker actually enters and physically occludes the channel. The divalent cation "bisQ11" causes a flickering block of the channel with discrete blocking events discernible in the single-channel record, enabling direct measurement of unidirectional binding and dissociation rates. The binding rate, but not the dissociation rate, was increased by making the membrane surface negative, exactly as if the blocking ions were concentrated in solution near the channel mouth, hence increasing their rate of entry into the channel. The effect of the surface charge on the binding rate was predicted quantitatively from the effect on ion conduction, consistent with the idea that conducting and blocking ions followed the same pathway into the channel. Also in accord with this picture is the independence of the dissociation rate from surface charge, given independent evidence that the binding site for bisQ11 is buried deep within the channel and perhaps placed almost symmetrically between the charged lipid surfaces.

3.2.6. Modulation of Membrane Excitability—Control of Channel Activity by Phosphorylation

Long-term modulation of activity of excitable cells has obvious physiological significance as a potential basis for hormonal effects on nerve and muscle activity, and for the imprinting of reflex responses or the learning of more complex behavior. Over the past decade or so, observations have rapidly accumulated of the modulation of electrical activity in a variety of cells by intracellular second messengers. Prominent among the mechanisms involved is the cAMP-dependent protein kinase system. Its specific effects vary from species to species and from cell type to cell type (for a recent review, *see* Levitan, 1985). The study of this particular modulator system in molluscan neurones provides a clear example of the complementary roles that patch recording from cellular membranes and single-channel recording from reconstituted systems can play in defining the in vivo role, and elucidating the site and mechanism of action of such an intracellular control system.

Biochemical studies on the cAMP-dependent protein kinase have provided several important tools for investigating the role of the kinase as a modulator of ion channel activity (Levitan et al., 1983; Levitan, 1985). The inactive holoenzyme consists of a tetramer of two catalytic (CS) and two regulatory (R) subunits. Dissociation of the R subunits in the presence of cAMP allows the free CS to catalyze ATP-dependent protein phosphorylation. Both CS and R subunits and an highly specific protein kinase inhibitor (PKI) have been purified to homogeneity. Introduction of the purified CS into neurons of an identified cluster in *Helix* enhances a Ca^{2+}-activated K^+ conductance. Patch–clamp experiments using inside-out excised patches have revealed a relatively low conductance channel (40–60 pS maximum in symmetric K^+ solutions) whose opening probability is enhanced by addition of CS in the presence of Mg^{2+} and ATP. Thus, the phosphorylation target must be a membrane-associated protein (Ewald et al., 1985). Similar experiments using channels incorporated into bilayers formed on the tips of patch pipets or large, painted bilayers, where the channels are effectively at infinite dilution in exogenous lipid, suggest that the kinase acts either on the channel protein itself or some very tightly associated protein (Ewald et al., 1985; Levitan, 1986; *see* Fig. 10, this chapter). These studies indicate that exogenous CS is sufficient to modulate channel activity in a manner that would account for cAMP-dependent modulation in intact cells. Complementary experiments introducing the R subunit, or highly specific PKI, into cells would, by inhibiting the upregulation of the channel, point directly to the physiological role of endogenous kinase. In this case, the modulation results from an increase in probability of opening of a single channel. Other potential means of regulation include changing the number of active channels (either by promoting production or destruction of channels, or by all-or-none effects on the open probability) or by changes in the single-channel current. The former appears to play a role in Ca^{2+} channel regulation (Tsien et al., 1986); the latter has not been observed to date.

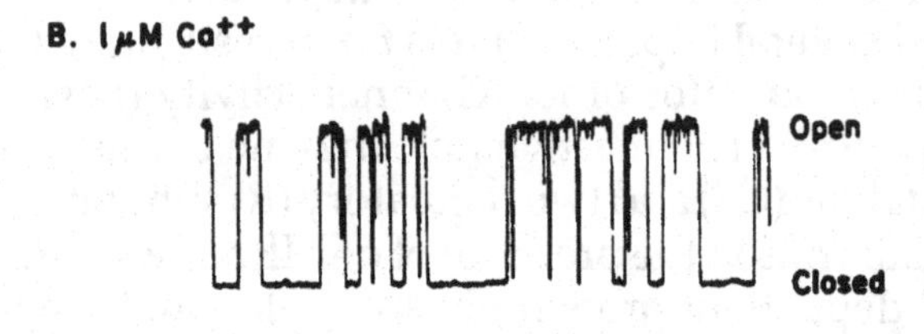

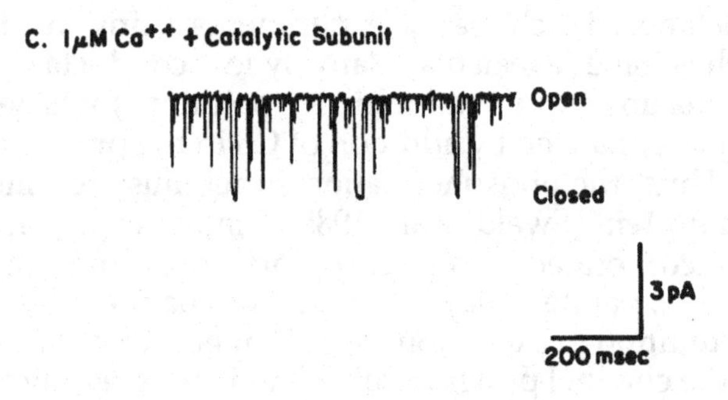

Fig. 10. Phosphorylation of a Ca^{2+}-dependent K^+ channel increases the probability of the channel being open. The channel was incorporated into a bilayer formed at the tip of a patch electrode from membrane vesicles prepared from an identified group of *Helix* neurons. (A) The channel was almost never open in 0.01-μM free Ca^{2+}. (B) In 1-μM free Ca^{2+}, the channel was open about 15–20% of the time. (C) After phosphorylation with catalytic subunit, the channel remains open almost all of the time with only brief closures. The voltage was +40 mV in each record; bandwidth was 1 kHz. Reproduced with permission from Levitan (1986), based on experiments of Ewald et al. (1985).

3.2.7. Channels from Subcellular Membranes—Making the Inaccessible Accessible

Planar bilayer methods have been successfully applied to study several channels from subcellular membranes that are not directly accessible to patch electrodes. The K^+ channel from sarcoplasmic reticulum has contributed much to our understanding of fundamental channel biophysics, thanks to the work of Christopher Miller and his collaborators (*see* Miller, 1983 for references). More recently, a Ca^{2+} channel likely to be responsible for Ca^{2+} release for activation of the contractile machinery of skeletal muscle has been successfully studied (Smith et al., 1985). The voltage-dependent anion channel from mitochondrial membrane—the bane of some attempted work on other channel types—has also proved a fruitful subject for study. Although the planar bilayer system has provided the only system for examination of these channels from subcellular organelles, this may not remain true in the future. Techniques for cellular and vesicular fusion (e.g., Zimmerman and Vienken, 1982) may make a variety of small membrane-bound structures amenable to patch-clamp recording in the future.

3.2.8. A Bioassay with Millisecond Time Resolution and Single-Molecule Sensitivity

An ingenious application of the patch-clamp method, reported in back-to-back papers by Hume et al. (1983) and Young and Poo (1983) uses the channels in an excised patch as a probe rather than the object of study. Pipets, bearing outside-out patches from muscle with a high density of ACh receptor channels, were brought near to growth cones of embryonic neurones in tissue culture. Single-channel currents recorded from the patch reflected the release of acetylcholine from the nearby neurone.

Hume et al. (1983) studied ACh release from chick ciliary ganglion neurons, finding no evidence of spontaneous release, but detecting release reliably when the neuron was stimulated with short trains of pulses, and occasionally following a stimulus that produced a single-action potential. Of interest is their observation of delays, up to several tens of milliseconds duration, in the appearance of the transmitter. Careful control experiments showed that this delay could not be accounted for by diffusion time required for the ACh to reach the pipet, even if the source were a distant part of the neuron. This raised the possibility that there might be differences in the mechanism of release from growth cones and from mature nerve terminals, for which the delay between stimulation and release is much shorter.

Using *Xenopus* embryonic neurones, Young and Poo (1983) did observe spontaneous release of ACh that appeared to be quantal in nauture, based on the bursts of activity induced in the excised patch. However, the mechanism by which release was triggered is not clear, since they did not record from the neurons, but only monitored the current from the excised patch used as a detector. Notably, a smaller channel type, which was not blocked by α-bungarotoxin, was also activated when the probing pipet approached a growth cone. This illustrates the possibility of a much broader application of the excised patch as a probe to monitor the release of channel-activating and channel-modulating factors, and the investigation of the role of these factors in developmental processes.

4. Concluding Remarks

The various methods of single-channel recording offer a direct and powerful approach toward an understanding of the roles of ion channels in cellular processes. However, although they are providing an unprecedented picture of some of the trees in the forest of biological function, their power will be fully realized only when the recordings are interpreted in conjunction with a larger view of the "thickets," provided by macroscopic recordings, and the detailed structure of the "leaves" gleaned from biochemical and molecular biological studies.

Acknowledgments

We thank the following for reading a draft of the manuscript and for comments and discussion: Drs. M. Delay, R. Horn, K. Magleby, F. Sachs, and J. C. Tanaka. We are most grateful to Dr. F. N. Quandt for providing Fig. 8, and for prepublication copies of manuscripts that were provided by Drs. R. J. Bauer, A. L. Blatz, B. F. Bowman, D. S. Duch, R. E. Furman, J. L. Kenyon, S. R. Levinson, K. L. Magleby, O. B. McManus, E. Recio-Pinto, F. J. Sigworth,

S. J. Sine, J. C. Tanaka, and B. W. Urban. Work in our own laboratory was supported by the Alberta Heritage Foundation for Medical Research and the Medical Research Council of Canada.

Appendix

Sources of interfacing hardware and software:

Address	Contact
Axon Instruments, Inc. 1101 Chess Drive Foster City, CA 94404	Phone: (415) 571-9400 Fax: (415) 571-9500
Laboratory Software Associates 1st Floor, 130 Mount St. Heidelberg, Victoria 3084 Australia	Phone: (03) 457-6566 Fax: (03) 457-6827 Telex: 38578 LABSOFT
Indec Systems, Inc. 1283A Mt. View—Alviso Road Sunnyvale, CA 94089	Phone: (408) 745-1842
INTRACEL Limited Broad Lane Cottenham Cambridge CB4 4SW ENGLAND	Phone: 0954-50957 Telex: 817674 Fax: 0954-51833
Dr. C. Lingle Florida State University Department of Biological Sciences Tallahassee, Florida 32306	
Dr. J. Pumplin Michigan State University Department of Physics East Lansing, Michigan 48824	Phone: (517) 355-9275
Dr. J. Dempster Dept. of Physiology and Pharmacology University of Strathclyde Glasgow, G1 1XW SCOTLAND	Phone: (041) 552-440, ext 2320

Source of Teflon film (inquire about availability of samples and minimum orders):

CHEMFAB
Chemical Fabrics Corporation
Toralon Division
5490 Dexter Way
Mangonia Park
West Palm Beach, FL 33407
Phone: 1-800-262-2990 (US calls only)
1-407-842-2990

References

Aldrich R. W. and Stevens, C. F. (1983) Inactivation of open and closed sodium channels determined separately. *Cold Spring Harbor Symp. on Quantitative Biology* **48,** 147–153.

Aldrich R. W. and Stevens C. F. (1987) Voltage-dependent gating of single sodium channels from mammalian neuroblastoma cells. *J. Neurosci.* **7,** 418–431.

Aldrich R. W. and Yellen G. (1983) The analysis of nonstationary channel kinetics, in *Single-Channel Recording* (Sakmann B. and Neher E., eds.), Plenum, New York, pp. 287–299.

Aldrich R. W., Corey D. P., and Stevens C. F. (1983) A reinterpretation of mammalian sodium channel gating based on single channel recording. *Nature* **306,** 436–441.

Alvarez, O. (1986) How to set up a bilayer system, in *Ion Channel Reconstitution* (Miller C., ed.), Plenum, New York, pp. 115–130.

Andersen O. S. (1983) Ion movement through gramicidin A channels. Single-channel measurements at very high potentials. *Biophys. J.* **41,** 119–133.

Anholt R., Lindstrom J., and Montal M. (1985) The molecular basis of neurotransmission: Structure and function of the nicotinic acetylcholine receptor, in *The Enzymes of Biological Membranes* Vol. 3 (Martonosi, A. N., ed.) Plenum, New York, pp. 335–401.

Armstrong C. M. and Chow R. H. (1987) Supercharging: A method for improving patch-clamp performance. *Biophys. J.* **52,** 133–136.

Auerbach A. and Sachs F. (1985) High-resolution patch-clamp techniques, in *Voltage and Patch Clamping with Microelectrodes* (Smith T. G., Jr., Lecar H., Redman S. J., and Gage P. W., eds.) American Physiol. Soc., Bethesda, MD, pp. 121–149.

Bauer R. J., Bowman B. F., and Kenyon J. L. (1987) Theory of the kinetic analysis of patch-clamp data. *Biophys. J.* **52,** 961–978.

Bean R. C., Shepherd W. C., Chan H., and Eichner J. (1969) Discrete conductance fluctuations in lipid bilayer protein membranes. *J. Gen. Physiol.* **53,** 741–757.

Bell J. E. and Miller C. (1984) Effects of phospholipid surface charge on ion conduction in the K^+ channel of sarcoplasmic reticulum. *Biophys. J.* **45,** 279–287.

Blatz A. L. and Magleby K. L. (1986a) Quantitative description of three modes of activity of fast chloride channels from rat skeletal muscle. *J. Physiol.* **378,** 141–174.

Blatz A. L. and Magleby K. L. (1986b) Correcting single channel data for missed events. *Biophys. J.* **49,** 967–980.

Brett R. S., Dilger J. P., Adams P. R., and Lancaster B. (1986) A method for the rapid exchange of solutions bathing excised membrane patches. *Biophys. J.* **50,** 987–992.

Cahalan M. D., Chandy K. G., DeCoursey T. E., and Gupta S. (1985) A voltage-gated potassium channel in human T-lymphocytes. *J. Physiol.* **358,** 197–237.

Chandy K. G., DeCoursey T. E., Cahalan M. D., and Gupta S. (1985) Electroimmunology: The physiologic role of ion channels in the immune system. *J. Immunol.* **135,** 787s–791s.

Clay, J. R. and DeFelice L. J. (1983) Relationship between membrane excitability and single channel open-close kinetics. *Biophys. J.* **42,** 151–157.

Cohen F. S. (1986) Fusion of liposomes to planar bilayers, in *Ion Channel Reconstitution* (Miller C., ed.), Plenum, New York, pp. 131–139.

Cohen F. S., Akabas M. H., and Finkelstein A. (1982) Osmotic swelling of phospholipid vesicles causes them to fuse with a planar phospholipid bilayer membrane. *Science* **217,** 458–460.

Cohen F. S., Zimmerberg J., and Finkelstein A. (1980) Fusion of phospholipid vesicles with planar phospholipid bilayer membranes. II. Incorporation of a vesicular membrane marker into the planar membrane. *J. Gen. Physiol.* **75,** 251–270.

Colquhoun D. and Hawkes, A. G. (1977) Relaxation and fluctuations of membrane currents that flow through drug-operated channels. *Proc. R. Soc. Lond. B* **199,** 231–262.

Colquhoun D. and Hawkes A. G. (1981) On the stochastic properties of single ion channels. *Proc. R. Soc. Lond. B* **211,** 205–235.

Colquhoun D. and Hawkes A. G. (1982) On the stochastic properties of bursts of single ion channel openings and of clusters of bursts. *Phil. Tran. Roy. Soc. Lond. B* **300,** 1–59.

Colquhoun D. and Hawkes A. G. (1983) The principles of the stochastic interpretation of ion-channel mechanisms, in *Single-Channel Recording* (Sakmann B. and Neher E., eds.), Plenum, New York, pp. 135–175.

Colquhoun D. and Hawkes A. G. (1987) A note on correlations in single ion channel records. *Proc. Roy. Soc. Lond. B* **230,** 15–52.

Colquhoun D. and Sakmann B. (1985) Fast events in single-channel currents activated by acetylcholine and its analogues at the frog muscle end-plate. *J. Physiol.* **369,** 501–557.

Colquhoun D. and Sigworth F. J. (1983) Fitting and statistical analysis of single-channel records, in *Single-Channel Recording* (Sakmann B. and Neher E., eds.) Plenum, New York, pp. 191–263.

Corey D. P. and Stevens C. F. (1983) Science and technology of patch-recording electrodes, in *Single-Channel Recording* (Sakmann B. and Neher E., eds.), Plenum, New York, pp. 53–68.

Coronado R. (1985) Effect of divalent cations on the assembly of neutral and charged phospholipid bilayers in patch-recording pipettes. *Biophys. J.* **47,** 851–857.

Coronado R. and Latorre R. (1983) Phospholipid bilayers made from monolayers on patch-clamp pipettes. *Biophys. J.* **43,** 231–236.

Coronado R. and Miller C. (1979) Voltage-dependent caesium blockade of a cation channel from fragmented sarcoplasmic reticulum. *Nature* **280,** 807–810.

Coronado R. and Miller C. (1980) Decamethonium and hexamethonium block K^+ channels of sarcoplasmic reticulum. *Nature* **288,** 495–497.

Cota G. and Armstrong C. M. (1988) Potassium channel "inactivation" induced by soft-glass patch pipettes. *Biophys. J.* **53,** 107–109.

Cull-Candy S. G. and Parker I. (1983) Experimental approaches used to examine single glutamate-receptor ion channels in locust muscle fibers, in *Single-Channel Recording* (Sakmann B. and Neher E., eds.), Plenum, New York, pp. 389–400.

DeFelice L. J. (1981) *Introduction to Membrane Noise* (Plenum, New York).

Dempster J. (1988) Computer analysis of electrophysiological signals, in *Microcomputers in Physiology* (Fraser P. J., ed.), IRL Press, Oxford, pp. 51–93.

Deutsch C., Krause D., and Lee S. C. (1986) Voltage-gated potassium conductance in human T-lymphocytes stimulated with phorbol ester. *J. Physiol. (Lond.)* **372,** 405–423.

Ehrenstein G., Lecar H., and Nossal R. (1970) The nature of the negative resistance in bimolecular lipid membranes containing excitability-inducing material. *J. Gen. Physiol.* **55,** 119–133.

Ehrenstein G., Blumenthal R., Latorre R., and Lecar H. (1974) Kinetics of the opening and closing of individual excitability-inducing material channels in a lipid lilayer. *J. Gen. Physiol.* **63,** 707–721.

Eisenman G., Latorre R., and Miller C. (1986) Multi-ion conduction and selectivity in the high-conductance Ca^{++}-activated K^+ channel from skeletal muscle. *Biophys. J.* **50,** 1025–1034.

Ewald D. A., Williams A., and Levitan I. B. (1985) Modulation of single Ca^{2+}-dependent K^+-channel activity by protein phosphorylation. *Nature* **315,** 503–506.

Fischmeister R. and Hartzell H. C. (1987) Cyclic guanosine 3',5'-monophosphate regulates the calcium current in single cells from frog ventricle. *J. Physiol. (Lond.)* **387,** 453–472.

Frauenfelder H., Parak F., and Young, R.D. (1988) Conformational substates in proteins. *Ann. Rev. Biophys. Biophys. Chem.* **17,** 451–479.

Fredkin D. R., Montal M., and Rice J. A. (1985) Identification of aggregated Markovian models: Application to the nicotinic acetylcholine receptor. *Proc. Berkeley Conf. Honor of Jerzy Neyman and Jack Kiefer* vol. 1, (Le Carn L. M. and Olshen R. A., eds.), Wadsworth, Belmont, CA pp. 269–289.

French A. S. and Stockbridge, L. L. (1988) Fractal and Markov behavior in ion channel kinetics. *Can. J. Pharmacol.* **66,** 967–970.

French R. J. and Horn R. (1983) Sodium channel gating: models, mimics and modifiers. *Ann. Rev. Biophys. Bioeng.* **12,** 319–356.

French R. J., Worley J. F. III, and Krueger B. K. (1984) Voltage-dependent block by saxitoxin of voltage-dependent sodium channels incorporated into planar lipid bilayers. *Biophys. J.* **45,** 301–310.

Furman R. E. and Tanaka J. C. (1988) Patch electrode glass composition affects ion channel currents. *Biophys. J.*, **53,** 287–292.

Garber S. S. and Miller C. (1987) Single Na^+ channels activated by veratridine and betrachotoxin. *J. Gen. Physiol.* **89,** 459–480.

Green W. N., Weiss L. B., and Andersen O. S. (1987) Batrachotoxin-modified sodium channels in planar lipid bilayers. Ion permeation and block. *J. Gen. Physiol.* **89,** 841–872.

Hamill O. P., Marty A., Neher E., Sakmann B., and Sigworth F. J. (1981) Improved patch-clamp techniques for high-resolution current recording from cells and cell-free membrane patches. *Pflügers Arch.* **391,** 85–100.

Hanke W, Methfessel C., Wilmsen H-U., Katz E., Jung G., and Boheim G. (1983) Melittin and a chemically modified trichotoxin form alamethicin-type multi-state pores. *Biochim. Biophys. Acta* **727,** 108–114.

Hartshorne R., Tamkun M., and Montal M. (1986) The reconstituted sodium channel from brain, in *Ion Channel Reconstitution* (Miller C., ed.), Plenum, New York, pp. 337–362.

Hartshorne R. P., Keller B. U., Talvenheimo J. A., Catterall W. A., and Montal M. (1985) Functional reconstitution of the purified brain sodium channel in planar lipid bilayers. *Proc. Natl. Acad. Sci. USA* 82, 240–244.

Hladky S. B. and Haydon D. A. (1970) Discreteness of conductance change in bimolecular lipid membranes in the presence of certain antibiotics. *Nature* **225,** 451–453.

Hladky S. B. and Haydon D. A. (1972) Ion transfer across lipid membranes in the presence of gramicidin A. I. Studies of the unit conductance channel. *Biochim. Biophys. Acta* **274,** 294–312.

Horn R. (1984) Gating of channels in nerve and muscle: A stochastic approach. *Curr. Topics in Membranes and Transport* **21,** 53–97.

Horn R. and Lange K. (1983) Estimating kinetic constants from single channel data. *Biophys. J.* **43,** 207–223.

Horn R. and Vandenberg C. A. (1984) Statistical properties of single sodium channels. *J. Gen. Physiol.* **84,** 505–534.

Hume R. I., Role L. W., and Fischbach G. D. (1983) Acetylcholine release from growth cones detected with patches of acetylcholine receptor-rich membranes. *Nature* **305,** 632–634.

Jackson M. B. (1985) Stochastic behavior of a many-channel membrane system. *Biophys. J.* **47,** 129–137.

Jackson M. B., Wong B. S., Morris C. E., Lecar H., and Christian C. N. (1983) Successive openings of the same acetylcholine receptor channel are correlated in open time. *Biophys. J.* **42,** 109–114.

Kakei M. and Ashcroft F. M. (1987) A microflow superfusion system for use with excised membrane patches. *Pflügers Arch.* **409,** 337–341.

Kameyama M., Hofmann F., and Trautwein W. (1985) On the mechanism of β-adrenergic regulation of the Ca channel in the guinea-pig heart. *Pflügers Arch.* **405,** 285–293.

Keller B. U., Hartshorne R. P., Talvenheimo J. A., Catterall W. A., and Montal M. (1986) Sodium channels in planar lipid bilayers. Channel gating kitetics of purified sodium channels modified by batrachotoxin. *J. Gen. Physiol.* **88,** 1–23.

Korn S. J. and Horn R. (1988) Statistical discrimination of fractal and Markov models of single-channel gating. *Biophys. J.* **54,** 871–877.

Krueger B. K. and Blaustein M. P. (1980) Sodium channels in presynaptic nerve terminals. Regulation by neurotoxins. *J. Gen. Physiol.* **76,** 287–313.

Krueger B. K., Worley J. F. III, and French R. J. (1986) Block of sodium channels in planar lipid bilayers by guanidinium toxins and calcium. *Ann. N.Y. Acad. Sci.* **479,** 257–268.

Labarca P., Coronado R., and Miller C. (1980) Thermodynamic and kinetic studies of the gating behavior of a K^+-selective Channel from the sarcoplasmic reticulum membrane. *J. Gen. Physiol.* **76,** 397–424.

Labarca P., Rice J. A., Fredkin D. R., and Montal M. (1985) Kinetic analysis of channel gating. Application to the cholinergic receptor channel and the chloride channel from *Torpedo californica. Biophys. J.* **47,** 469–478.

Latorre R. (1986) The large calcium-activated potassium channel, in *Ion Channel Reconstitution* (Miller, C., ed.), Plenum, New York, pp. 431–467.

Läuger P. (1985) Ionic channels with conformational substates. *Biophys. J.* **47,** 581–591.

Läuger P. (1988) Internal motions in proteins and gating kinetics of ionic channels. *Biophys. J.* **53,** 877–884.

Levi R. and DeFelice L. J. (1986) Sodium-conducting channels in cardiac membranes in low calcium. *Biophys. J.* **50,** 5–9.

Levitan I. B. (1985) Phosphorylation of ion channels. *J. Membrane Biol.* **87,** 177–190.

Levitan I. B. (1986) Phosphorylation of a reconstituted potassium channel, in *Ion Channel Reconstitution* (Miller C., ed.), Plenum, New York, pp. 523–532.

Levitan I. B., Lemos J. R., and Novak-Hofer I. (1983) Protein phosphorylation and the regulation of ion channels. *Trends in Neurosci.* **6,** 496–499.

Liebovitch L. S. and Sullivan J. M. (1987) Fractal analysis of a voltage-dependent potassium channel from cultured mouse hippocampal neurons. *Biophys. J.* **52,** 979–988.

Liebovitch L. S., Fischbarg J., Koniarek J. P., Todorova I., and Wang M. (1987) Fractal model of ion-channel kinetics. *Biochim. et Biophys. acta* **896,** 173–180.

Magleby K. L. and Pallotta B. S. (1983a) Calcium dependence of open and shut interval distributions from calcium-activated potassium channels in cultured rat muscle. *J. Physiol. (Lond.)* **344,** 585–604.

Magleby K. L. and Pallotta B. S. (1983b) Burst kinetics of single calcium-activated potassium channels in cultured rat muscle. *J. Physiol. (Lond.)* **344,** 605–623.

McManus O. B., Blatz A. L., and Magleby K. L. (1987) Sampling, log binning, fitting, and plotting durations of open and shut intervals from single channels and the effects of noise. *Pflügers Arch.* **410,** 530–553.

McManus O. B., Weiss D. S., Spivak C. E., Blatz A. L., and Magleby K. L. (1988) Fractal models are inadequate for the kinetics of four different ion channels. *Biophys. J.* **54,** 859–870.

Miller C. (1978) Voltage-gated cation conductance channel from fragmented sarcoplasmic reticulum: Steady-state electrical properties. *J. Membrane Biol.* **40,** 1–23.

Miller C. (1983) Integral membrane channels: Studies in model membranes. *Physiol. Rev.* **63,** 1209–1242.

Miller C. (ed.) (1986) *Ion Channel Reconstitution* (Plenum, New York).

Millhauser G. L., Salpeter E. E., and Oswald R. E. (1988) Diffusion models of ion-channel gating and the origin of power-law distributions from single-channel recording. *Proc. Natl. Acad. Sci. USA* **85,** 1503–1507.

Moczydlowski E. and Latorre R. (1983) Gating kinetics of Ca^{2+}-activated K^+ channels from rat muscle incorporated into planar lipid bilayers. Evidence for two voltage-dependent Ca^{2+} binding reactions. *J. Gen. Physiol.* **82,** 511–542.

Montal M. and Mueller P. (1972) Formation of bimolecular membranes from lipid monolayers and a study of their electrical properties. *Proc. Natl. Acad. Sci. USA* **69,** 3561–3566.

Mueller P. (1975) Membrane excitation through voltage-induced aggregation of channel precursors. *Ann. NY Acad. Sci.* **264,** 247–264.

Mueller P., Rudin D. O., Tien H. T., and Wescott W. C. (1962) Reconstitution of cell membrane structure *in vitro* and its transformation into an excitable system. *Nature* **194,** 979–980.

Neher E. and Sakmann B., (1976) Single-channel currents recorded from membrane of frog muscle fibers. *Nature* **260,** 799–802.

Nelson N., Anholt R., Lindstrom J., and Montal M. (1980) Reconstitution of purified acetylcholine receptors with functional ion channels in planar lipid bilayers. *Proc. Natl. Acad. Sci. USA* **77,** 3057–3061.

Niles W. D. and Cohen F. S. (1987) Video fluorescence microscopy studies of phospholipid vesicle fusion with a planar phospholipid membrane. *J. Gen Physiol.* **90,** 703–735.

Niles W. D., Levis R. A., and Cohen F. S. (1988) Planar bilayer membranes made from phospholipid monolayers form by a thinning process. *Biophys. J.* **53,** 327–335.

Offner F. F. and Clark B. (1985) An improved amplifier for patch-clamp recording. *Biophys. J.* **47,** 142a.

Parsegian A. (1969) Energy of an ion crossing a low dielectric membrane: Solutions to four relevant electrostatic problems. *Nature* **221,** 844–846.

Patlak J. and Horn R. (1982) Effect of *N*-bromoacetamide on single sodium channel currents in excised membrane patches. *J. Gen. Physiol.* **79,** 333–351.

Prakash J., Jensen D. N., Paulos J. J., Grant A. O., and Strauss H. C. (1987) An integrating patch-clamping amplifier with on-chip capacitors and reset switch. *Biophys. J.* **51,** 70a.

Quandt F. N. (1987a) Burst kinetics of sodium channels which lack fast inactivation in mouse neuroblastoma cells. *J. Physiol. (Lond.)* **392,** 563–585.

Quandt F. N. (1987b) Modification of slow inactivation of single sodium channels by diphenylhydantoin in neuroblastoma cells. *Molecular Pharmacol.* **34,** 557–565.

Rae J. L. and Levis R. A. (1984) Patch voltage clamp of lens epithelial cells: Theory and practice. *Molecular Physiol.* **6,** 115–162.

Rao C. R. (1973) *Linear Statistical Inference and Its Applications* (J. Wiley & Sons, New York).

Recio-Pinto E., Duch D. S., Levinson S. R., and Urban B. W. (1987) Purified and unpurified sodium channels from eel electroplax in planar lipid bilayers. *J. Gen. Physiol.* **90,** 375–395.

Robinson K. and Giles W. (1986) A data acquisition, display and plotting program for the IBM PC. *Computer Methods and Programs in Biomedicine* **23,** 319–327.

Rosenberg R. L., Tomiko S. A., and Agnew W. S. (1984) Single channel properties of the reconstituted voltage-regulated Na channel isolated from the electroplax of *Electrophorus electricus*. *Proc. Natl. Acad. Sci. USA* **81,** 5594–5598.

Roux B. and Sauve R. (1985) A general solution to the time interval omission problem applied to single channel analysis. *Biophys. J.* **48,** 149–158.

Sachs F. (1983) Automated analysis of single-channel records, in *Single-Channel Recording* (Sakmann, B. and Neher, E., eds.), Plenum, New York, pp. 265–285.

Sachs F., Neil J., and Barkakati N. (1982) The automated analysis of data from single ionic channels. *Pflügers Arch.* **395,** 331–340.

Sakmann B. and Neher E. (eds.) (1983a) *Single-Channel Recording* (Plenum, New York).

Sakmann B. and Neher E. (1983b) Geometric parameters of pipettes and membrane patches, in *Single-Channel Recording* (Sakmann, B. and Neher, E., eds.), Plenum, New York, pp. 37–51.

Schindler H. and Quast U. (1980) Functional acetylcholine receptor from Torpedo marmorata in planar membranes. *Proc. Natl. Acad. Sci. USA,* **77,** 3052–3056.

Schuerholz T. and Schindler H. (1983) Formation of lipid-protein bilayers by micropipette guided contact of two monolayers. *FEBS Lett.* **152,** 187–190.

Sidell N., Schlichter L. C., Wright S. C., Hagiwara S., and Golub S. H. (1986) Potassium channels in human NK cells are involved in discrete stages of the killing process. *J. Immunol.* **137,** 1650–1658.

Sigworth F. J. (1983a) Electronic design of the patch clamp, in *Single-Channel Recording* (Sakmann B. and Neher E., eds), Plenum, New York, pp. 3–35.

Sigworth F. J. (1983b) An example of analysis, in *Single-Channel Recording* (Sakmann B. and Neher E., eds.), Plenum, New York, pp. 301–321.

Sigworth F. J. and Neher E. (1980) Single Na^+ channel currents observed in cultured rat muscle cells. *Nature* **287,** 447–449.

Sigworth F. J. and Sine S. M. (1987) Data transformations for improved display and fitting of single-channel dwell time histograms. *Biophys. J.* **52,** 1047–1054.

Sine S. M. and Steinbach J. H. (1986a) Acetylcholine receptor activation by a site-selective ligand: Nature of brief open and closed states in BC3H-1 cells. *J. Physiol. (Lond.)* **370,** 357–379.

Sine S. M. and Steinbach J. H. (1986b) Activation of acetylcholine receptors on clonal mammalian BC3H-1 cells by low concentrations of agonist. *J. Physiol. (Lond.)* **373,** 129–162.

Sine S. M. and Steinbach J. H. (1987) Activation of acetylcholine receptors on clonal mammalian BC3H-1 cells by high concentrations of agonist. *J. Physiol. (Lond.)* **385,** 325–359.

Smith J. S., Coronado R., and Meissner G. (1985) Sarcoplasmic reticulum contains adenine nucleotide-activated calcium channels. *Nature* **316,** 446–449.

Soejima M. and Noma A. (1984) Mode of regulation of the ACh-sensitive K-channel by the muscarinic receptor in rabbit atrial cells. *Pflügers Arch.* **400,** 424–431.

Starmer C. F., Dietz M. A., and Grant A. O. (1986) Signal discretization: A source of error in histograms of ion channel events. *IEEE Trans. Biomed. Eng. BME* **33,** 70–73.

Steele J. A., Poznansky M. J., Eaton D. C., and Brodwick M. S. (1981) Lipid vesicle-mediated alterations of membrane cholesterol levels: Effects on Na^+ and K^+ currents in squid axon. *J. Membrane Biol.* **63,** 191–198.

Steinberg I. Z. (1987) Frequencies of paired open-closed durations of ion channels. Method of evaluation from single-channel recordings. *Biophys. J.* **52,** 47–55.

Stühmer W., Methfessel C., Sakmann B., Noda M., and Numa S. (1987) Patch clamp characterization of sodium channels expressed from rat brain cDNA. *Eur. Biophys. J.* **14,** 131–138.

Suarez-Isla B. A., Wan K., Lindstrom J., and Montal M. (1983) Single-channel recordings from purified acetylcholine receptors reconstituted in bilayers formed at the tip of patch pipets. *Biochemistry* **22,** 2319–2323.

Tank D. W. and Miller C. (1983) Patch-clamped liposomes: recording reconstituted ion channels, in *Single-Channel Recording* (Sakmann B. and Neher E., eds.), Plenum, New York, pp. 91–105.

Tank D. W., Miller C., and Webb W. W. (1982) Isolated-patch recording from liposomes containing functionally reconstituted chloride channels from Torpedo electroplax. *Proc. Natl. Acad. Sci. USA* **79,** 7749–7753.

Tsien R. W., Bean B. P., Hess P. Lansman J. B., Nilius B., and Nowycky M. C. (1986) Mechanisms of calcium channel modulation by β-adrenergic agents and dihydropyridine calcium agonists. *J. Mol. Cell Cardiol.* **18,** 691–710.

Vandenberg C. A. and Horn R. (1984) Inactivation viewed through single sodium channels. *J. Gen. Physiol.* **84,** 535–564.

Vivaudou M. B., Singer J. J., and Walsh J. V., Jr. (1986) An automated technique for analysis of current transitions in multilevel single-channel recordings. *Pflügers Arch.* **407,** 355–364.

White S. H. (1986) The physical nature of planar bilayer membranes, in *Ion Channel Reconstitution* (Miller C., ed.), Plenum, New York, pp. 3–35.

Worley J. F. III, French R. J., and Krueger B. K. (1986) Trimethyloxonium modification of single batrachotoxin-activated sodium channels in planar bilayers. Changes in unit conductance and in block by saxitoxin and calcium. *J. Gen. Physiol.* **87,** 327–349.

Yellen G. (1982) Single Ca^{2+}-activated nonselective cation channels in neuroblastoma. *Nature* **296,** 357–359.

Yellen G. (1984) Ion permeation and blockade in Ca^{2+}-activated K^+ channels by external cations. *J. Gen. Physiol.* **84,** 157–186.

Young S. H. and Poo M. (1983) Spontaneous release of transmitter from growth cones of embryonic neurones. *Nature* **305,** 634–637.

Zimmerberg J., Cohen F. S., and Finkelstein A. (1980a) Fusion of phospholipid vesicles with planar phospholipid bilayer membranes. I. Discharge of vesicular contents across the planar membrane. *J. Gen. Physiol.* **75,** 241–250.

Zimmerberg J., Cohen F. S., and Finkelstein A. (1980b) Micromolar Ca^{2+} stimulates fusion of lipid vesicles with planar bilayers containing a calcium-binding protein. *Science* **210**, 906–908.

Zimmermann U. and Vienken J. (1982) Electric field-induced cell-to-cell fusion. *J. Membrane Biol. 67,* 165–182.

Whole-Cell and Microelectrode Voltage Clamp

Stephen W. Jones

1. Introduction

1.1. *Why Voltage Clamp?*

As other chapters in this book demonstrate, intra- and extracellular recording provide considerable information about the electrical behavior of cells. However, as Hodgkin and Huxley (1952) taught us long ago, analysis of the mechanisms that generate electrical activity is far easier with voltage clamp. If the behavior of ion channels depends on voltage, the first step in studying such channels is to control the voltage. Under voltage clamp, the kinetic behavior of a channel (or of a macroscopic assemblage of channels) is at its simplest. Without voltage clamp, there is a complex interaction between channel gating and membrane potential. When channels open, current through them affects the membrane potential, which in turn makes channels open (or close), affecting the membrane potential. . . .

Conceptually, voltage clamp is simple. Measure the membrane potential, compare that measurement to the desired value, and if there is a difference (voltage error), pass current in the direction appropriate to drive the membrane potential to the desired (command) potential. If for any reason ion channels open in the clamped membrane, the voltage clamp must pass a current equal and opposite to the current through the channels in order to hold the membrane potential constant. So, ideally, the current passed by the voltage clamp amplifier is nothing more or less than (–1 ×) the current through whatever channels happen to be open in the membrane at a particular time. This ability to measure the current through open ion channels directly is the central advantage of voltage clamp.

Under "current clamp," an injection of current into a passive

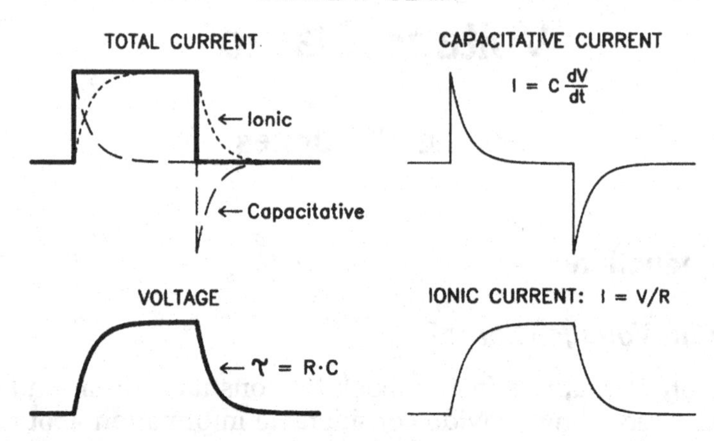

Fig. 1A. Current clamp. The illustration is for a "cell" with only passive leakage and capacitance. The heavy lines (left) show the applied current stimulus (top) and resulting voltage change (bottom). At the instant the current is applied, there is no driving force on the ion channels in the cell membrane, so no ionic current can flow through them. All of the initial current is capacitative; that is, all of the charge injected into the cell piles up on the inside of the membrane, thereby changing the membrane potential. Eventually, the current is entirely ionic, and the membrane potential no longer changes. When the current injection is suddenly terminated, equal and opposite ionic and capacitative currents flow. The charge moving through ion channels piles up on the outside of the membrane, returning the membrane potential to rest. The exponential time course of the component ionic and capacitative currents is shown with dashed lines at left and plotted separately for clarity at the right.

cell produces an exponential change in membrane potential, according to the time constant of the cell. Part of the current goes to charge the membrane, which changes the membrane potential. That is called the capacitative current. Once the membrane potential begins to change, ionic current is driven through ion channels (the physical basis of the membrane conductance). In general, under current clamp, both ionic and capacitative currents may be flowing at any given instant (Fig. 1A). However, when a cell is voltage clamped at a fixed potential, no capacitative current can flow, since that would change the membrane potential. Only when

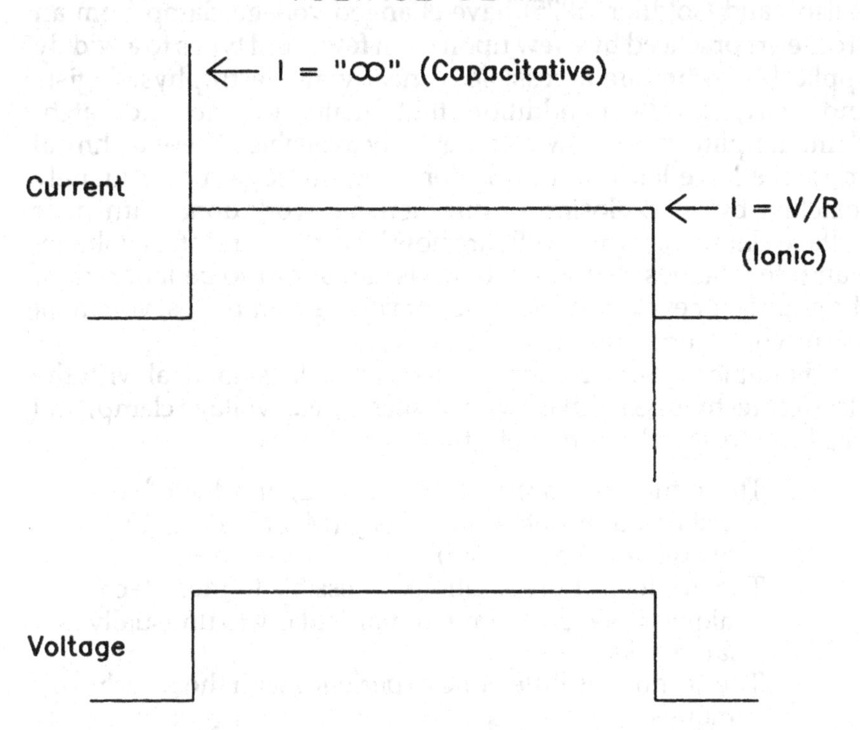

Fig. 1B. Voltage clamp. The illustration is for a "cell" with only passive leakage and capacitance. A sudden voltage step first causes a large, rapid capacitative current. Once the capacitative current is over, the cell is at a new constant voltage, where only ionic current flows. Here, only the passive current through the input resistance of the cell is shown; if any voltage-dependent currents were activated, they would simply add to the observed current measured at any time.

the voltage is commanded to change does a capacitative current flow (Fig. 1B). Ideally, the voltage change is instantaneous, but that would require an infinite current! In practice, the voltage change should be fast with respect to the kinetics of the ion channels being studied.

This chapter deals with the practical methods used for voltage clamp and their limitations. The intended audience is not only the producer of voltage-clamp data, but also the consumer. Recent developments, particularly whole-cell recording (Hamill et al.,

1981) and discontinuous single-electrode voltage clamp (dSEVC; Wilson and Goldner, 1975) have changed voltage clamp from an arcane art practiced by a few upon even fewer cell types to a widely applicable technique accessible to nearly all electrophysiologists and nearly all cells. In addition, high-quality voltage- and patch-clamp amplifiers are now commercially available. These technical advances have led to an explosion of knowledge, unfortunately matched by an explosion of questionable work done with poor voltage clamp. Since few cells are now beyond the reach of voltage-clamp techniques, few cell biologists can afford to be ignorant of the criteria for evaluating accuracy of voltage clamp. Yes, you must learn what "series resistance" is!

For at least three reasons, this chapter does not deal with the classical techniques of axial wire or sucrose gap voltage clamp, and has little to say about two-electrode voltage clamp.

1. These methods are well established, and have been evaluated in detail previously (Moore, 1971, 1985; Finkel and Gage, 1985)
2. The number of preparations accessible to those techniques is severely limited, particularly to unusually large cells
3. The author has little or no experience with those techniques.

1.2. Why Not Single-Channel Recording?

The patch-clamp technique described by Hamill et al. (1981) almost accidentally provides a nearly perfect voltage clamp, with noise levels so extraordinarily low that openings and closings of single-ion channels can be observed. It might be thought that this would make the rest of electrophysiology obsolete, but that is not entirely true. First, the molecularly tight seal between the electrode and cell membrane necessary for patch-clamp methods does not appear to be possible in most intact (in vivo) or semi-intact (e.g., brain slice) preparations. If anything like normal neuronal circuitry is necessary for the questions to be addressed, that is a problem. Second, the high resolution of single-channel recording is ideal for studies of trees (and even twigs), but not forests. Most cells contain many thousands of ion channels, of ten or more distinct types. To reconstruct the electrical behavior of a cell entirely from single-channel recording would be a heroic effort. Third, there may be differences between the behavior of ion channels in patches and in intact cells. Obviously, single-channel recording techniques, including bilayer methods (Wonderlin et al., 1989), complement but do not replace other approaches.

2. Voltage-Clamp Methods and Errors

There are two main considerations in voltage clamp: *point* clamp and *space* clamp. How accurately can the voltage be controlled, and how accurately can the membrane current be recorded, at the point in the cell where the electrode is? Are the voltages and currents well controlled in regions of the cell remote from the electrode(s)?

There are two main considerations to point clamp: clamp speed and steady-state accuracy. How rapidly does the voltage clamp respond to a change in current, resulting either from a voltage command or from an event, such as a synaptic potential? How close is the final membrane potential to the desired value? The severity of these errors vary with the voltage-clamp method and with the properties of the cell under study.

Neurons are unique in many ways, including their shapes. Perhaps no other cell type can be greater than 1 m long; a spinal motoneuron can extend from muscle fibers in a toe to its cell body in the spinal cord. Even the dendritic arbors of many neurons are large enough for the membrane potential to vary dramatically in different regions. These functionally important morphological characteristics of neurons present a problem for voltage-clamp methods.

2.1. Whole-Cell Recording and Series Resistance

The patch-clamp technology invented by Neher, Sakmann, and colleagues (Hamill et al., 1981) is most famous for its revelation that current across biological membranes really does flow through discrete ion channels. However, the ability to record from small, fragile cells in the whole-cell configuration (Fig. 2) may actually be its most significant contribution. Ten years ago, the number of cell types that could be accurately voltage clamped by existing methods was severely limited. Now nearly any cell, if it can be isolated or grown in cell culture, can be clamped. Electrophysiology is no

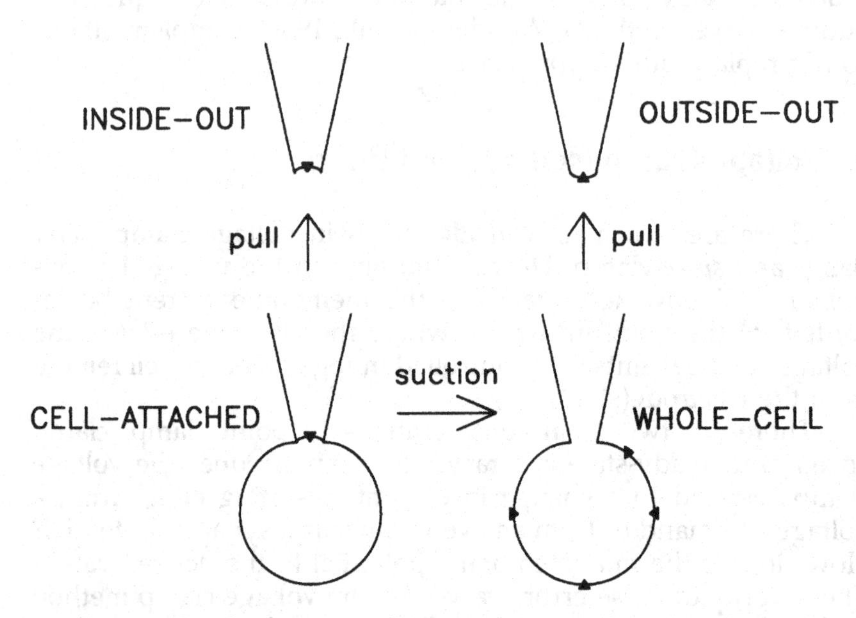

Fig. 2. Patch-clamp configurations. Initially, the electrode is attached by gentle suction to the surface of the cell (cell-attached; lower left). Here the electrode records the activity of any ion channels (Δ) active in the small patch of membrane under the electrode. Pulling away from the cell can give an excised, or inside-out patch, where the original patch of membrane has its cytoplasmic surface facing the outside world (upper left). Alternatively, brief application of stronger suction, or a large brief current pulse, can destroy the patch of membrane but leave the electrode attached to the cell. In that configuration (lower right), the electrode interior is directly connected to the cell interior, and the currents of the whole-cell are recorded. At this stage, an outside-out patch can sometimes be formed by pulling away; the cell membrane is stretched, breaks, and reseals.

longer a subdiscipline of neurobiology; that was merely an accident of the fact that some neurons are very large. All cells have ion channels, even the lowly red blood cell (Hamill, 1983), and channels are important for the function of even "nonexcitable" cells.

In whole-cell recording, an electrode with relatively low resistance (generally <10 MΩ) and relatively large diameter tip (>1 μm) is confluent with the interior of the cell (Fig. 2). Topologically, this is similar to the situation in intracellular recording. The blessing and curse of whole-cell recording is that small ions (and even proteins) can diffuse from pipet to cell (or the reverse). This allows experimental control of the intracellular milleu, but conversely requires that any soluble substance required for the normal electrical activity of the cell be provided by the experimenter. Some ion channels, notably calcium channels, are notoriously unstable, and require elaborate metabolic support systems for their long-term maintenance under whole-cell recording conditions. Others, notably sodium channels, are remarkably robust, and survive without obvious change for hours when only inorganic salts and a pH buffer are in the pipet.

The "perforated patch" method of Horn and Marty (1988) shows promise for recording metabolically labile currents. In that method, the pore-forming antibiotic nystatin is added to the patch pipet in the cell-attached configuration. Current flow through the nystatin channels provides electrical continuity with the inside of the cell, without allowing movement of large organic molecules and proteins.

The popularity of whole-cell recording has been greatly enhanced by the existence of excellent, commercially available patch-clamp amplifiers, particularly the EPC-7 (List-electronic, Darmstadt, West Germany) and the Axopatch (Axon Instruments, Burlingame, CA). The manuals for both are excellent, and are well supplemented by the book edited by Sakmann and Neher (1983). These amplifiers use a simple strategy for achieving voltage clamp: fundamentally, a current-follower circuit (Fig. 3). Since this method uses a single electrode to pass current and record voltage simultaneously, it will be called continuous single-electrode voltage clamp (cSEVC) here, to distinguish it from discontinuous SEVC (dSEVC) discussed below. The main practical limitation of cSEVC is that the voltage-clamp command is given not across the cell membrane only, but across the combination of the electrode and cell in series. This acts as a voltage divider. If the resistance of the cell is much greater than the resistance of the electrode, most of the voltage drop occurs across the cell, so that the cell is effectively voltage clamped. That may be true for a small cell, whose input resistance can be several GΩ, with a patch electrode with a resistance of a few MΩ. However, for large cells and large currents, the resistance of the cell may be comparable to that of the electrode, so that a significant fraction of the commanded voltage drops across

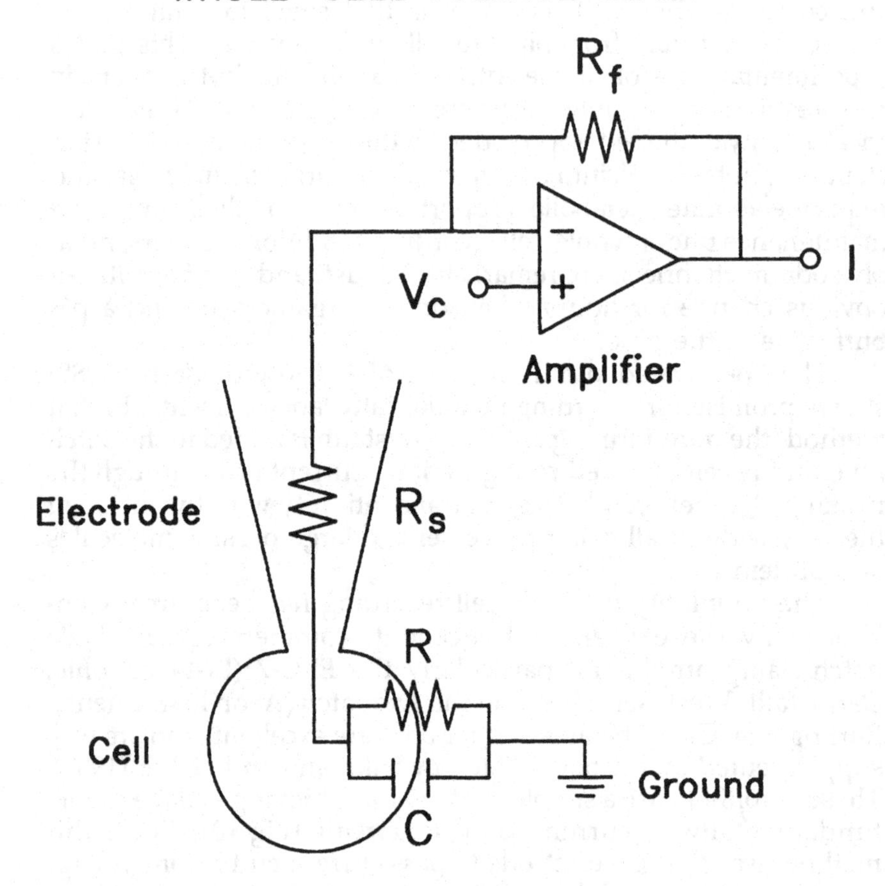

Fig. 3. A simplifed circuit for whole-cell voltage clamp. The amplifier acts to hold the membrane potential at the commanded level (V_C). The current is measured from the voltage drop across the feedback resistor in the amplifer (R_f). If the series resistance, R_s, is much lower than the membrane resistance R, the measured current is the current through the cell membrane, and if the membrane potential is held constant (so that capacitative currents do not flow through the cell membrane), the measured current is the ionic current through all ion channels open in the cell membrane at a given time.

the electrode and is not seen by the cell. This error is called the series resistance error, as it results from the electrode resistance, which is electrically in series with the cell membrane.

Series resistance limits both the speed and the steady-state accuracy of the clamp. The membrane potential will not respond to a voltage step instantaneously, but will approach the new voltage exponentially, with a time constant equal to the product of the series resistance and the cell capacitance. For example, with a 50-μm diameter cell of 100 pF capacitance, and a 10-MΩ electrode resistance, it would take 1 ms to reach 63% of the command potential. That would not give adequate clamp of fast currents. The steady-state voltage error is roughly the product of the series resistance and the ionic current. A 10-MΩ series resistance would give 1 mV of voltage error for each 100 pA of current; currents over 1 nA would be clamped very poorly. In other words, the voltage error resulting from series resistance is current-dependent. Also, it is not recorded by the electrode. The voltage signal that the amplifier puts out is not the recorded membrane potential, but simply a copy of the voltage command applied to the cell.

Series resistance can be compensated for, but only partially. With the Axopatch and List amplifiers, there are two stages to the procedure. First, the capacity transient is compensated for as well as possible. The capacity transient in whole-cell recording has at least two components, for charging the electrode and for charging the membrane. The time constant of the latter component is the product of the electrode (series) resistance and the cell capacitance; the dials used for compensation are in fact calibrated in those units (actually, series conductance on the List), so that compensation for the capacity transient provides a measure of the series resistance. It is important to note that the performance of the clamp has not yet been improved. The capacitance compensation is purely cosmetic, since it occurs at a stage after the membrane current has been recorded, and it does not affect the current or voltage at the cell. The second stage in the procedure is the actual series resistance compensation, where the measured values for series resistance and cell capacitance are used to shape the voltage command applied to the cell in an attempt to make the voltage change more rapid and accurate. Series resistance compensation involves positive feedback, so it is inherently unstable and can lead to disastrous oscillations if overdone. For it to be accurate and stable, the capacitance controls must be set carefully, which requires much practice.

A recommended strategy for setting the capacitance controls is to start with them all fully off. Observe the capacity transient on the oscilloscope at low gain, so that the entire transient is on scale. First, set the series resistance dial to 1.5–2 × the electrode resistance. Next, increase the whole-cell capacitance until the slow component is reduced. Go back and forth between the series resistance and capacitance dials until the slow component is as flat as possible, leaving only a fast component. Now, increase the fast capacitance compensation. Increase the gain on the oscilloscope, and "fine tune" the settings. This is best done with extremely light filtering (~20 kHz). After introducing series resistance compensation, adjust the fast (but not slow) capacitance controls to minimize oscillations.

The measured series resistance is invariably higher than the resistance of the electrode in the bath, and sometimes much higher. This may be because of partial occlusion of the electrode opening by the cellular contents. In general, low-resistance electrodes (<2 MΩ) will have series resistance roughly 1.5 × their original resistance, but higher resistance electrodes are likely to "clog up" to very high values. To make matters worse, it is more difficult to compensate for the series resistance when it is high. With 2 MΩ or lower series resistance, it is possible to routinely compensate for 80% of the series resistance with either the List or Axopatch, but at 10 MΩ, it is rare to be able to compensate at all.

How well does series resistance compensation really work in cSEVC? My experience with sodium and calcium currents in isolated frog sympathetic neurons suggests that the steady-state accuracy of the clamp is greatly improved, but the response time is slightly slower than expected (Jones and Marks, 1989). The delay may be the result of a small, slow component to the capacity transient, which cannot be adequately compensated, and leads to brief oscillations in the clamp. However, the quality of the clamp appears good even for ~10 nA currents, consistent with the expected voltage error of ~4 mV following compensation. Without series resistance compensation, currents are poorly clamped (Fig. 4).

Armstrong and Chow (1987) have recently described a "supercharging" method, where a large, brief current pulse is injected at the start and end of the voltage-clamp step. This can produce clamp settling times as fast as 5–15 μs. It is not yet clear whether this method will work with existing commercially available patch-clamp amplifiers.

In summary, the accuracy of point clamp with cSEVC is limited by series resistance. Low-resistance electrodes allow both faster and more accurate voltage clamp. Larger currents give more steady-state error; larger cells give slower clamp.

2.2. Discontinuous Single-Electrode Voltage Clamp

Series resistance considerations would suggest that it is hopeless to clamp a cell whose resistance is comparable to that of the electrode. That is the situation with (e.g.) a microelectrode in a cell in a brain slice, where both the cell and the electrode typically have resistances of tens of MΩ. However, a completely different strategy for voltage clamp can be used in such cases, and can even be useful for whole-cell clamp *(see below)*.

In a discontinuous voltage clamp (dSEVC), one electrode passes current and records voltage alternately (Fig. 5). For part of each cycle, typically 25–50%, the electrode passes current. That current inevitably causes a voltage change across the electrode, which interferes with accurate measurement of the membrane potential. For nearly all of the remainder of the cycle, the electrode does nothing. Meanwhile, the voltage across the electrode dissipates. If the electrode were an ideal resistor, that would happen instantly, but a real electrode has a certain amount of capacitance associated with it. The voltage across the electrode therefore settles according to its RC time constant (the product of the electrode's resistance and capacitance), which is generally a few microseconds. Then, the electrode can read the true membrane potential, in theory with no series resistance error. Next, the feedback circuit compares the recorded voltage to the voltage command, and determines the current to be applied to the cell during the next switching cycle. The usual strategy is to apply a fixed amount of current to the cell per mV of voltage error detected. (This is precisely the meaning of the "gain" knob on the Axoclamp.)

It is necessary to compensate for the electrode's capacitance as well as possible to produce rapid discharge of the voltage across the electrode. This is best done by observing the electrode voltage directly on an oscilloscope triggered once per switching cycle. The capacity compensation should be adjusted so that the electrode

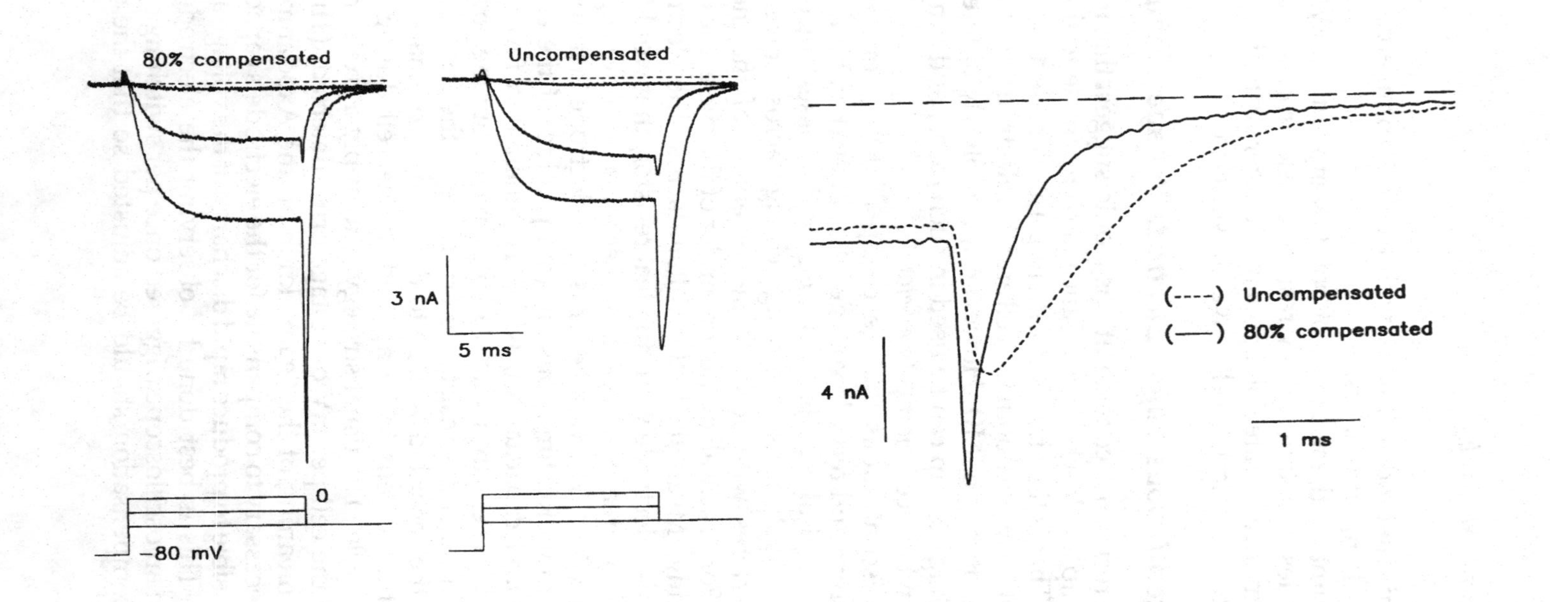
80% compensated
Uncompensated
3 nA
5 ms
0
−80 mV
4 nA
1 ms
(----) Uncompensated
(—) 80% compensated

Fig. 4. Effects of series resistance on calcium currents, with cSEVC under whole-cell clamp in a frog sympathetic neuron. (A) With sodium and potassium currents blocked (Jones and Marks, 1989), a cell was depolarized from –80 mV to activate calcium current, using Ba^{2+} instead of Ca^{2+} as the charge carrier. A few milliseconds later, the cell was partially repolarized, to –40 mV. That would increase the driving force on Ba^{2+}, and should cause an instantaneous increase in the current amplitude. The current would then decline, as calcium channels close (a "tail current"). Note that the currents with R_s compensation are noisier. Without compensation, the current at –20 mV appears to activate more slowly, as expected for R_s error (Fig. 9A). The voltage traces are illustrations of the protocol, not actual recorded membrane potential, and do not reflect R_s error. (B) Tail currents are greatly slowed without R_s compensation. This is an expanded view of data from A, upon repolarization from 0 to –40 mV. With R_s compensation, the current increased with a 10–90% rise time of 80 μs, of which 45 μs can be attributed to filtering at 8 kHz. Without R_s compensation, the rise time increased to 170 μs, and the time course of the tail current was greatly distorted. The measured series resistance in this cell was 2.0 MΩ, with 51 pF capacitance. The expected clamp time constant would be approximately 100 μs without and 20 μs with 80% R_s compensation, corresponding to 10–90% rise times of 220 and 44 μs, respectively.

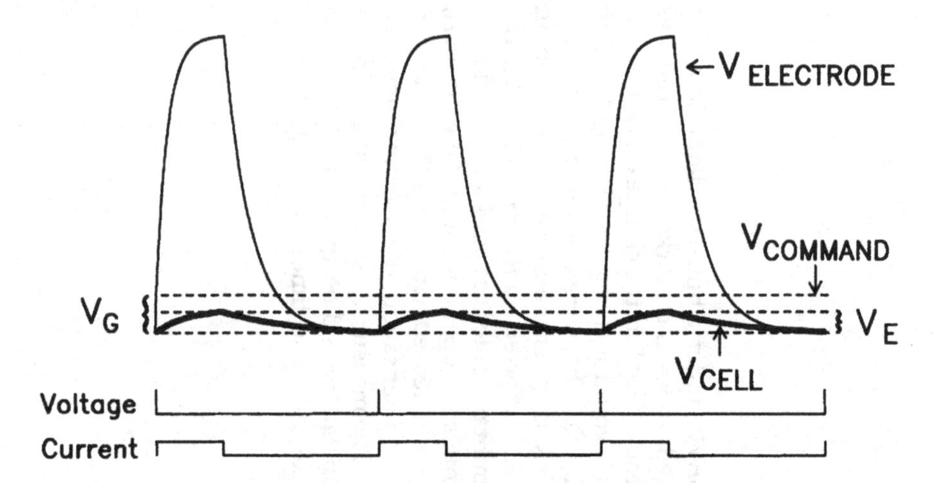

Fig. 5. A timing diagram for dSEVC. A switching cycle begins when the voltage is read, compared to the command potential, and the current to pass is computed from the measured voltage error (V_G) and the gain. That current then causes a voltage drop partially across the electrode (which charges rapidly) and partially across the cell (which charges slowly; heavy line). After the current is turned off, the electrode discharges rapidly; the cell slowly. At the end of the cycle, if the electrode has settled fully, the voltage reported by the electrode is the true cell membrane potential. During each cycle, the true cell membrane potential changes slightly (ripple); V_E is the amplitude of that fluctuation. Because of ripple, V_G overestimates the difference between the command potential and the average membrane potential, assuming the clamp is perfectly tuned.

voltage is flat at the end of each cycle, not still changing and not oscillating (Fig. 6). If that cannot be accomplished, the switching frequency must be reduced to allow more time for the electrode voltage to settle properly.

Once that is done, there are two main errors with dSEVC: low gain, and "ripple." Error resulting from low gain is reported accurately by the electrode; that resulting from ripple is not. In addition, for low switching frequency, clamp speed can be a problem; obviously, only events slow with respect to the switching frequency can be detected and controlled appropriately. Finkel and Redman (1984, 1985) have discussed these errors in detail; *see also* Jones (1987).

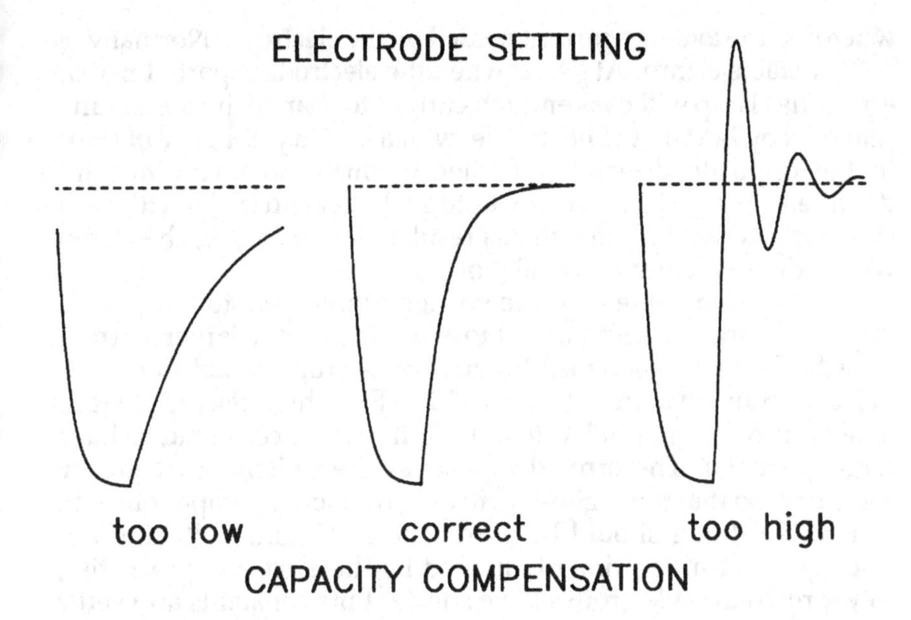

Fig. 6. Effect of capacitance compensation on electrode settling in dSEVC. This is a diagram of the raw electrode voltage during one switching cycle, which is customarily monitored on a separate oscilloscope. With the capacitance incorrectly set, the voltage across the electrode at the end of the switching cycle does not return to zero, so that the membrane potential reported by the electrode will be in error.

2.2.1. Low Gain

The gain (G) in dSEVC is defined as the amount of current passed for a given amount of voltage error detected by the electrode. The steady-state voltage error (V_G) measured by the electrode is therefore $V_G = I/G$, where I is the membrane current averaged over a switching cycle. At the critical gain (G_C):

$$G = G_C = FC \tag{1}$$

the clamp is critically damped, so that the cell reaches the command potential in one switching cycle (Finkel and Redman, 1984; Jones, 1987). (F = switching frequency; C = cell capacitance.) Defining $g = G/G_C$:

$$V_G = I/gFC \tag{2}$$

where I = the average current passed by the electrode. Normally, $g < 2$ for stable clamp. At $g = 2$, when the electrode reports 1 mV of error, the clamp will pass enough current to change the membrane potential by 2 mV in the next cycle, which would yield 1 mV of error in the opposite direction; in theory, this would continue indefinitely. For $g > 2$, the error would get larger and larger with each switching cycle, with disastrous results. For $1 < g < 2$, the clamp will undergo a damped oscillation.

For a given value of g, the voltage error owing to low gain is larger for larger currents, or more accurately for larger current density (A/cm^2), since membrane area is proportional to capacitance. In contrast to the situation with cSEVC, high cell capacitance is not a problem in dSEVC (except in that large cells tend to have large currents). The error decreases as the switching frequency increases, so that the highest switching frequency compatible with electrode settling should be used. The maximum switching frequency, in turn, is primarily limited by the electrode properties; lower resistance electrodes have shorter time constants and settle faster.

How large is this error in practice? For a 10-MΩ microelectrode, a switching frequency of 10 kHz can usually be reached. In a 100 pF cell, G_C = 0.1 nF × 10 kHz = 1 nA/mV. This performance is at least equal to that of cSEVC (for given electrode and cell properties), even with series resistance compensation. However, it is clear that the size of currents that can be accurately clamped is limited with dSEVC, to a few nA in the case illustrated.

As proposed originally by Sigworth (1983), dSEVC is also useful in whole-cell recording. With a 1-MΩ patch electrode, switching frequencies >50 kHz can be routinely achieved, giving clamp settling times <0.1 ms (Fig. 7) and adequate clamp of rapid currents up to 30 nA (Jones, 1987). Note that adequate clamp of such large currents requires high switching frequency to achieve sufficient gain. A 20 nA current in a 100 pF cell would give 20 mV of clamp error with a 10 kHz switching frequency, or 3 mV at 60 kHz (assuming critical gain).

There is a trick that appears to allow significantly increased gain. The basic problem with dSEVC is that the steady-state gain is limited by the response to a rapid transient (Eq. 2). If the gain is too high, too much current will be passed too quickly, and the clamp will oscillate. What is needed is some way to slow the response to a

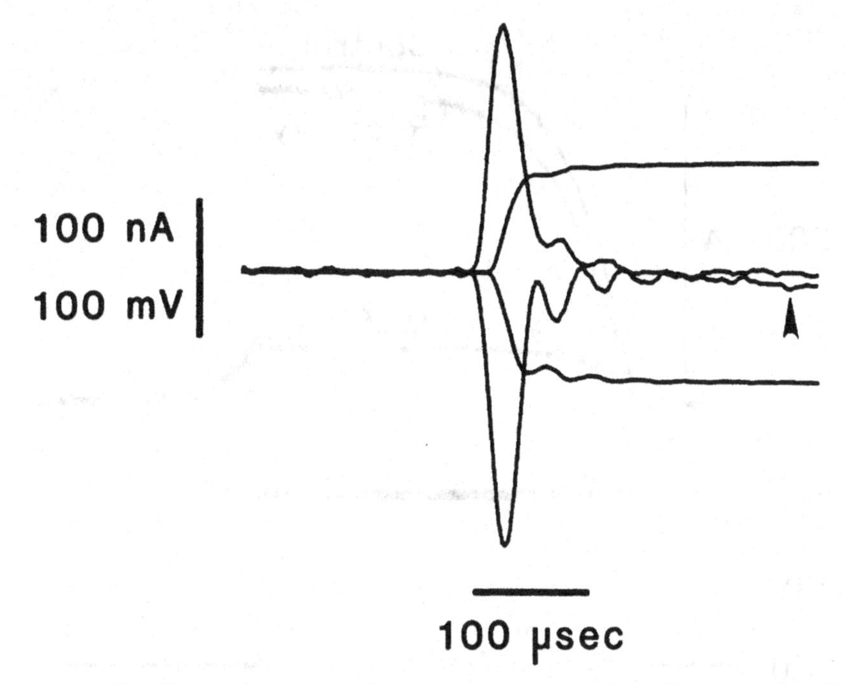

Fig. 7. Clamp settling and capacity transients in a dissociated bullfrog ganglion cell, using dSEVC with whole-cell recording. The patch pipet contained Cs^+, and external solutions included Mn^{2+} and TEA, to isolate sodium currents. Commands of ±80 mV from the holding potential of –80 mV produced large, rapid capacity transients that were 90% complete within 100 μs, despite slight oscillations. By 200 μs, a net inward Na current is visible for the depolarizing step (arrow). The records were filtered at 20 kHz and digitized at an effective rate of 400 kHz, from magnetic tape. The capacitative current is slowed somewhat by the filtering. The switching frequency was 50 kHz, with an electrode resistance of approximately 1 MΩ. The phase control was not used.

rapid transient, without lowering the steady-state gain. This seems to be possible with aggressive use of the "phase" control on the Axoclamp. Large phase lags (near –4) with long time constants (2–200 ms) slow the clamp, but clamp speed can be regained by increasing the gain. The gain and phase settings become extremely critical with this combination, and strange oscillations can be produced easily. Experiments with model cells suggest that the voltage reported by the electrode is accurate, so that can be used to

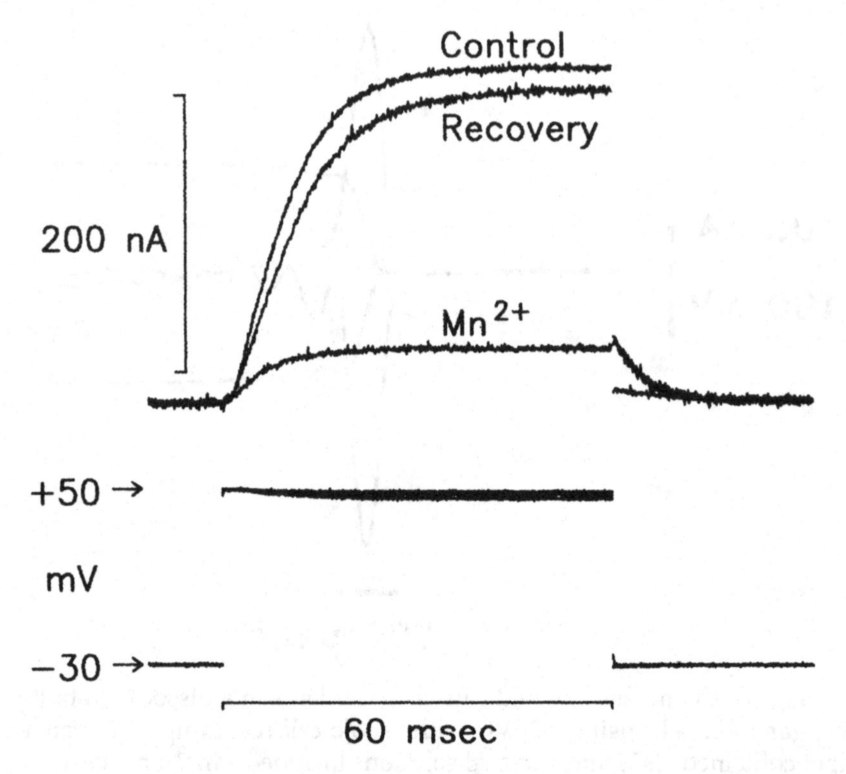

Fig. 8. Clamp of large outward currents with dSEVC in whole cell recording. The phase controls were used (*see* text). Three records are superimposed, taken from an isolated bullfrog sympathetic neuron before, during, and after replacement of Ca^{2+} with Mn^{2+}. By blocking calcium current, Mn^{2+} prevents the large Ca^{2+}-dependent potassium current; the current remaining in Mn^{2+} is primarily the delayed rectifier (*see* Adams et al., 1982a). This experiment was performed with Dr. Peter Pennefather.

determine the optimal settings. Since the error is current-dependent, it can be necessary to tune the clamp using a voltage step that activates a large current. Preliminary experiments indicate that currents > 100 nA can be clamped under whole-cell conditions with no obvious signs of bad clamp (Fig. 8). The phase control was designed for other purposes (Finkel and Gage, 1985; Finkel and Redman, 1985), so the theoretical basis for its use to increase the gain is unclear.

2.2.2. Ripple

"Ripple" is settling of the membrane potential toward the resting potential during each switching cycle. That is, while the electrode is waiting for the voltage across it to decay during a switching cycle, the cell membrane potential is also changing, at a rate determined by the membrane time constant. That means that the membrane potential will systematically vary during each switching cycle, and the voltage reported by the electrode is not the value averaged over the cycle (Fig. 5). The peak amplitude of this error (V_E) is:

$$V_E = (V_C - V_R)(1 - e^{-t/\tau}), \tag{3}$$

where V_C = command potential, V_R = resting potential, t = duration of the portion of the switching cycle where current is not being passed, and τ = cell time constant. (If voltage- or agonist-dependent currents are flowing, V_R is really the instantaneous zero-current voltage of the cell.) To a good approximation, for switching frequencies >> $1/\tau$, this reduces to:

$$V_E = (V_C - V_R)(t/\tau) = (V_C - V_R)d/FRC \tag{4}$$

where d = (1 – duty cycle) = the fraction of the switching cycle where current is not being passed. Since $V_C - V_R$ is the driving force, and driving force divided by resistance equals current:

$$V_E = I(d/FC) \tag{5}$$

(Jones, 1987). Error resulting from ripple therefore depends on many of the same factors as does error resulting from low gain. However, the error resulting from ripple is in the opposite direction. For most of the switching cycle, the membrane potential is closer to the command potential than at the moment when the electrode reads the voltage. Roughly, the error resulting from ripple averaged over one switching cycle is $V_E/2$. The errors resulting from low gain and ripple would cancel for $g = 2.9$ (if $d = 0.7$, as on the Axoclamp).

2.3. Classical Voltage-Clamp Techniques

The cSEVC and dSEVC methods described above attempt the difficult task of using a single electrode to do two things: pass current and record voltage. Earlier techniques for voltage clamp had used two separate electrodes, which has the advantage that the voltage-sensing electrode is not a series resistance and is not limited to recording voltage only part of the time. This results in improved clamp performance under most circumstances (Moore, 1985). However, most cell types are either too small or two inaccessible to allow penetration by two electrodes, so cSEVC and dSEVC are widely used.

The original axial wire voltage-clamp method, used for the giant axon of the squid, requires insertion of two electrodes parallel to the membrane. Gap methods use insulating materials, such as a concentrated sucrose solution or grease, to separate regions of the membrane of a long cell, with the ends cut to allow electrical access to the interior. These two methods can use large, macroelectrodes of very low resistance, which is electrically advantageous for both current passing and voltage recording. Clamp settling times <10 μs are possible with such methods. Alternatively, two microelectrodes (or two patch-clamp electrodes) can be used for a two-electrode voltage clamp. This technique is limited by the higher resistance of the electrodes, but clamp settling times under 100 μs are sometimes possible.

The electrodes are not a source of series resistance in these methods, but in many cases restricted extracellular spaces contribute significant series resistance. Voltage steps are effectively applied between the electrode and a ground placed in the bath outside the preparation, so that resistances either between the top of the electrode and the cell membrane, or between the cell membrane and ground, act as series resistances. If the cell to be clamped is tightly wrapped by connective tissue or glia, or if there are elaborate membrane infoldings, there may be resistances of the order of kΩ between the cell membrane and ground. If the currents under study are of the order of μA, as is often the case for the large cells appropriate for two electrode clamp, the voltage errors can be significant. As for cSEVC, series resistance compensation is possible.

Another practical problem with two microelectrode voltage clamp is capacitative coupling between the electrodes. This can lead to oscillations that limit the voltage-clamp gain. Elaborate shielding and grounding techniques are often necessary to reduce the coupling, in order to achieve fast clamp with two microelectrodes.

2.4. Simulation of Point Clamp Error

This review emphasizes the various sources of error in voltage clamp, but how important is all that, really? Should only biophysicists worry about a few millivolts of clamp error? Perhaps a mediocre clamp is better than none at all. As always in science, the ultimate concern is not quantitative error, but qualitative error: are the experimental limitations severe enough to make the conclusions drawn invalid?

One useful approach is to take an established model for voltage-dependent currents, such as the Hodgkin-Huxley model for the sodium and potassium currents of squid axon, and simulate the effect of clamp error on the true membrane potential and on the observed currents. Figure 9 does this for a hypothetical cell, with units scaled to values appropriate for whole-cell clamp. This specific model mimics series resistance error in cSEVC, but the fundamental limitations are the time constant of establishment of clamp (here, the series resistance times the cell capacitance) and the steady-state voltage error (here, the series resistance times the current). These records were produced by a modification of the AXOVACS program (Axon Instruments).

The middle row illustrates conditions where the clamp speed is adequate, but steady-state accuracy is poor (τ = 0.1 ms, 10 mV error/nA). The currents do not show obvious signs of poor clamp, despite an unacceptable voltage error of up to 20 mV for sodium currents (Fig. 9A) and 40 mV for potassium currents (Fig. 9B). (Remember that cSEVC does not measure the actual membrane potential, so that the experimenter would not know directly how serious the voltage error is.) A comparison with properly clamped currents (at the left) shows that the current amplitudes are reduced at most voltages, the delay in activation appears to be increased for sodium currents and decreased for potassium currents, and the sodium currents activate over too narrow a voltage range. Inactivation of sodium currents is also slowed. Clearly, quantitative measurements (time constants for activation and inactivation, voltage sensitivity) would be in error. Such errors might lead to qualitative mistakes in modeling; for example, delays in activation might be explained by nonexistent closed states between the resting and open states of the channel.

The row at the right illustrates both poor clamp speed and

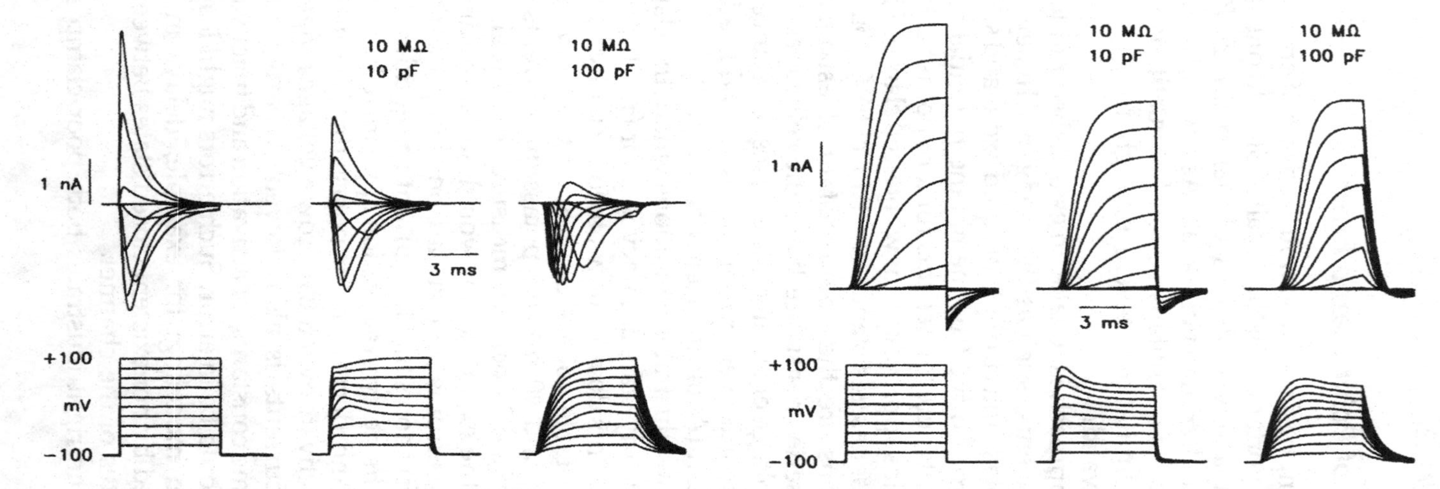

Fig. 9. Simulation of series resistance error using the Hodgkin-Huxley model. (A), sodium currents; (B), potassium currents. Leakage and capacitative currents are not shown. *See* text for further explanation.

poor accuracy (τ = 1 ms, 10 mV error/nA). Here poor clamp is obvious for sodium currents, which activate in a practically all-or-none manner, following a relatively long and variable latency. The currents are biphasic around the reversal potential, but in practice that might not be noted, since that occurs early in the pulse where the capacitative currents are still flowing, so clamp has obviously not been established. Potassium currents also activate with a latency, and tail currents are grossly distorted, but are not as dramatically altered as are sodium currents.

It is clear that inward currents are a severe test of a voltage clamp. If an inward current is unclamped, it can initiate a regenerative depolarization. That is the basis of the action potential. If an inward current is poorly clamped, an event closer to an action potential than to proper voltage clamp can result. As the gain of a voltage clamp is increased, there is a continuum between pure current clamp and perfect voltage clamp. For extremely poor voltage clamp, currents closely resemble upside-down action potentials, and the actual membrane potential is essentially an action potential, which bears little resemblence to the voltage command.

Poorly clamped currents like the sodium currents in the right column of Fig. 9A are not simply an oddity of a particular computer model. Similar currents abound in the literature, and are not absent from recent volumes of the Journal of Physiology.

2.5. Choice of Method

No voltage-clamp method is ideal for all preparations. Although the factors limiting speed and accuracy of clamp are similar for different methods, they are not identical. Factors to consider are

1. The size of the cell
2. The size of the current and
3. The speed of the current

cSEVC is the method of choice for small currents in small cells. Since the speed of the clamp is limited by the product of series resistance and cell capacitance, small cells can be clamped more rapidly than large ones. The low noise with cSEVC is also a major advantage. Another positive factor with cSEVC is that the low leakage current of the amplifier allows accurate determination of the resting potential even of extremely high resistance cells.

If two-electrode voltage clamp is not possible, dSEVC is preferable for large currents in large cells. With that method, clamp accuracy and speed can actually increase with cell size (Eq. 1, 2). With larger currents, the higher noise levels can usually be tolerated or reduced by averaging.

With cSEVC, the capacity transient is compensated, but not recorded directly. This can be an advantage or a disadvantage. It makes subtraction of the (residual) capacity transient easier, and allows the data to be recorded at a high gain without missing part of the transient. However, the information in the transient could be used to measure the cell capacitance, and to directly determine the speed of the clamp.

Work with dissociated bullfrog sympathetic neurons illustrates some of the factors involved. These are relatively large neurons, with diameters up to 50 μm and capacitances up to 100 pF. For comparable electrode properties, clamp speed is similar with dSEVC and cSEVC (Figs. 4 and 7). Without use of the phase control in dSEVC, steady-state accuracy is comparable to cSEVC, typically 1 mV error/2–5 nA current. With phase lag, dSEVC can clamp much larger currents when carefully tuned (Fig. 8). For small currents, such as the M-current, typically 1 nA in amplitude, the convenience and low noise levels of cSEVC are decisive factors. Series resistance compensation is rarely necessary, which simplifies the setup of the clamp and improves stability. For very large currents, such as the sodium current (~20 nA) and rapid potassium currents (~200 nA), dSEVC gives better results. Calcium currents, in the 1–10 nA range, can be studied with either method.

In some cases, specific amplifiers may introduce further limitations. For example, series resistance compensation on the List EPC7 cannot be set to values below 1 MΩ. The usual versions of the List and Axopatch are limited to whole-cell capacitances of <100 pF, and currents <20 nA. One strategy is to provide headstages with different properties, optimized for different purposes. A headstage that allows 10× higher currents and capacitances is available for the Axopatch. The Axoclamp also has different headstages available, which have different current-passing limits. In dSEVC, the usual x1 headstage is limited to 200–300 nA even with low-resistance electrodes. This can cause the amplifier to saturate during the capacity transient in a large cell, which slows the clamp and makes the shape of the capacity transient vary with the amplitude of the voltage command. This can be seen clearly using the model cell supplied with the Axoclamp.

2.6. Space Clamp

Regardless of the method used, a fundamental limitation of voltage clamp is that the membrane potential is not fully controlled in parts of the cell electrically far from the electrode(s). This can produce errors similar to those resulting from series resistance; current from electrically remote areas must pass though the cytoplasmic resistance between the remote channels and the electrode. For very remote areas, the membrane potential will be entirely unaffected by the clamp, and remote currents will not be recorded. For intermediate levels of error, the membrane potential and recorded current will be distorted. As for series resistance, the voltage errors owing to poor space clamp are not recorded by the electrode.

Since many of the most interesting cell types are electrically large, much theoretical effort has been devoted to the evaluation of space clamp (Jack et al., 1975; Johnston and Brown, 1983; Rall and Segev, 1985). Application of these results is limited by uncertainty in the actual electrical properties of the remote regions. In some cases, it is possible to simultaneously record from a distant region of the cell to evaluate the quality of clamp directly (Kramer, 1986).

A subtle problem is that the electrical size of the cell can change. Since the length constant of a cell increases with the membrane resistance, a conductance increase can worsen the clamp, and a conductance decrease can improve it. For example, blockade of potassium and other currents in hippocampal neurons reveals a chloride current, probably by making distal dendrites closer electrically (Madison et al., 1986).

3. Data Analysis

One of the joys of electrophysiology is that the immediate results of the experiment are instantly observable on the oscilloscope screen, but it is still necessary in most cases to transform the data in some way or other before final interpretation is possible. It is common for data analysis to take more time than the initial experiment, despite (or because of) the use of powerful computer-based analysis programs.

Probably the two most basic aims of voltage clamp are to determine what ionic currents are present and to determine the properties of those currents. The crucial point is that the current recorded under voltage clamp is a linear summation of current through all types of ion channels open in the membrane at a given instant. This allows direct measurement of the "macroscopic" kinetic behavior of an ensemble of ion channels, which is often enough to predict the "microscopic" kinetics of the ion channels themselves. Although voltage-dependent channels are emphasized here, the principles apply equally well to ligand-dependent channels.

3.1. Computer-Based Analysis

Although it is perfectly possible to conduct experiments and analyze voltage-clamp data without a computer, the convenience of modern computer systems for data collection and analysis make practical experiments and analyses that would be far too tedious otherwise. The ideal strategy, which is becoming increasingly common, is to use the computer at all stages in the process, from control of the experiment by generation of voltage commands, to preparation of publication-quality figures from the collected data.

The goal of this section is to summarize the principles and strategies involved in data analysis, with less attention to the details of specific computer programs. My experience has been with IBM AT-compatible microcomputers, with pCLAMP version 4 (Axon Instruments). Features of those and other systems have been reviewed recently (Wonderlin et al., 1989; Cachelin and Rice, 1987). The rapid rate of change in microcomputer hardware and software makes it extremely difficult to predict future directions. Historically, the relatively low cost of PC-compatible computers made it possible to include a computer routinely in each experimental setup. The entire computer system might cost less than the oscilloscope or stimulator alone. AT class computers also are powerful enough for nearly any desired experimental situation, with sampling rates >100 kHz possible simultaneously with analog and digital outputs for voltage commands, oscilloscope triggers, and so on. More powerful computers would add speed in analysis and the potential for data analysis on-line while the experiment is being conducted.

The primary advantage of using the computer to control the experiment is that, with proper software, it is possible to keep a full record of the experimental conditions along with the actual data. This allows the subsequent analysis programs to use that information directly in the analysis. It also allows the computer to substitute for an analog stimulator. For unusually complicated protocols, it can become necessary to combine computer- and stimulator-generated commands, often using the computer to trigger the stimulator.

The most straightforward strategy for data storage is to put the data directly into files on a hard disk. Since voltage-clamp data is generally command-driven, as opposed to single-channel data, which is usually event-driven, it is possible to alternately record data and write it to disk, rather than attempt to do both simultaneously. Large, fast hard disks, particularly the newer models with >40 MBytes and <30 ms access times, are a major advantage here. Use of analog or digital (VCR) tape to store data is a sensible alternative for single channel data, where long stretches of continuous data must be recorded at high rates. That is less satisfactory for command-driven data, since crucial timing information (beginning and end of voltage-clamp steps, for example) would be lost.

Since it is possible to generate several MBytes of data per hour of recording, it quickly becomes necessary to find some medium other than hard disks for archival storage of data. The standard method is to use digital tape drives, capable of storing data files directly. Sixty-MByte tape drives are now available for less than $1000. Their main disadvantage is speed; although the raw speed to read or write data is several MBytes per minute, it can take several minutes to find a file in the middle of such a tape. The new "worm" (write once-read mostly) optical disks, which are faster and can store several hundred MBytes on a single compact disk, appear to be an attractive alternative.

Although AT class computers are satisfactory for data collection, they are often undesirably slow for data analysis. One strategy is to provide separate, faster computers for off-line analysis. Computers based on the 80386 processor are coming down in price to the point where a 20-MHz system costs about $1000 more than a 10-MHz AT clone. Another advantage of using a separate analysis computer is that the entire experimental setup need not sit idle while data is being analyzed. It is most convenient if data can be shared among the computers using a local area network or a shared external tape drive.

Careful attention must be paid to sampling and filtering when using a computer to collect data. The considerations are detailed by Wonderlin et al. (1989). Briefly, the data should be sampled at several times the frequency of the fastest signal of interest, following analog filtering at no greater than half the sampling frequency. The safest strategy is to record data at as high a bandwidth as possible and use a digital filter later if necessary. A digital Gaussian filter is recommended. Note that, unlike an analog filter, a Gaussian filter averages data both backwards and forwards in time around each data point.

The amplification used is also important in data sampling. For the usual 12-bit A-D converter, the full range is ±2048 points. If set for ±10 V full scale, that is a resolution of ~5 mV. It is important to amplify the signal so that it is close to full scale and the resolution of the A-D converter does not affect the accuracy of the final analysis. That is straightforward with most patch-clamp amplifiers, which have a variable output gain, designed for precisely this purpose. It can be a problem with older designs, such as the Axoclamp, where an external amplifier may be necessary.

A final word of caution about computer analysis. It is necessary at every step to verify that the computer is maintaining the integrity of the data. The existence of an option in the analysis program does not mean that the option is either appropriate or valid for your data. The performance of the A-D converter, and the result of manipulation of the data by the analysis programs, should be checked for accuracy, linearity, and offsets. It is strongly recommended that the raw data be viewed on an old-fashioned analog oscilloscope as it is being collected. The temptation to analyze data automatically without first carefully inspecting for artifacts must be resisted.

3.2. Kinetic Analysis

One goal of voltage-clamp analysis is to describe the behavior of a class of ion channel. In essence, that is to ask what makes the channel open or close. This can be viewed either as an attempt to describe the molecular mechanisms involved in channel function or simply as an operational description of the behavior of the channel. Since such a description can be used to predict the response of the channel under different conditions, it aids in determining the role that a particular type of ion channel plays in the overall electrical activity of the cell. This section will assume that the experimental data reflect solely the behavior of one class of ion channel. Ways to determine whether that is in fact the case are discussed later.

3.2.1. Subtraction of Leakage and Capacitative Currents

Voltage-clamp steps elicit not only ionic current through voltage-dependent channels, but also capacitative and leakage currents. It is usually necessary to correct for the latter two types of current before analysis can proceed further. The standard technique is to compare voltage steps that activate voltage-dependent channels with steps that do not. Since most voltage-dependent channels open upon depolarization, a common practice is to use either hyperpolarizing steps or small depolarizing steps to elicit purely capacitative and leakage currents. This is illustrated in Fig. 10, for simulated data based on the Hodgkin-Huxley model. Simple addition of currents upon de- and hyperpolarizing voltage steps produces what should be a pure ionic current.

Figure 10 also illustrates one problem with leak subtraction. In the raw data, it is apparent that the membrane potential did not fully reach its steady-state value for a fraction of a millisecond. By definition, while a capacitative current is flowing, the membrane potential must be changing, so that the cell is not fully voltage clamped. The subtracted current looks much better than the raw data, but that is deceptive. In this example, the delay in activation of the current is largely artifactual, as is the delay in channel closing at the end of the voltage step. It is good practice to measure currents only after the capacitative current is effectively over, whether leak subtraction is used or not. In practice, that may not be a problem, since attempts to compensate for the capacitance and/or increase clamp gain often cause small oscillations that interfere with subtraction of capacitative currents. (Remember that with cSEVC the actual capacitative current is not recorded, so it may be difficult to determine when voltage-clamp control was actually established.) It is common practice to "blank" a brief period at the start and end of each voltage command to avoid showing imperfectly subtracted capacitative currents. That practice is fair if it is

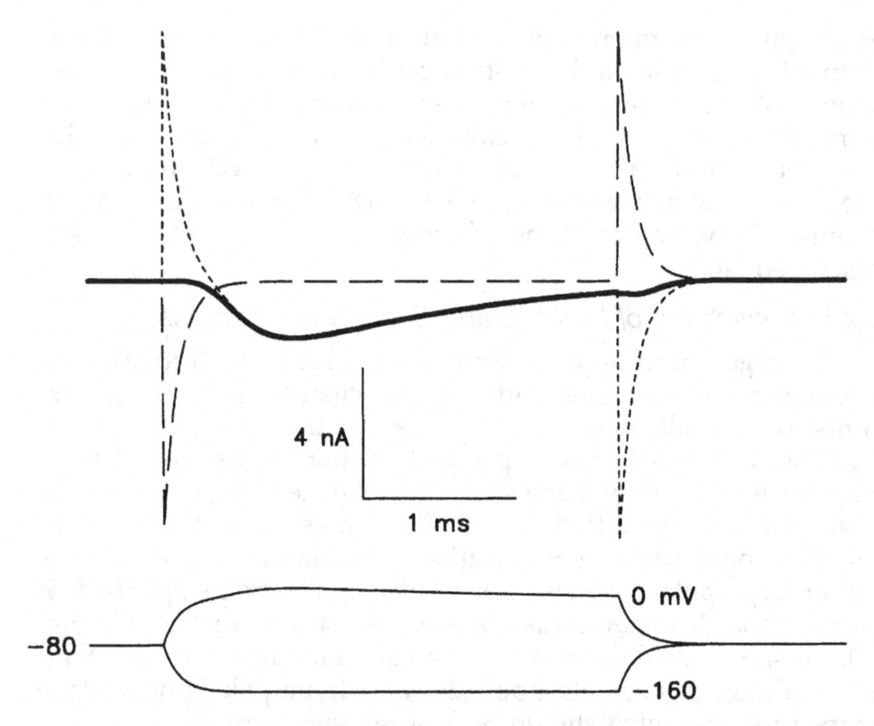

Fig. 10. Subtraction of leakage and capacitative currents. Currents were calculated from the Hodgkin-Huxley model for sodium current, with simulated whole-cell recording (R_s = 10 MΩ, C = 10 pF). The long dashed lines are the response to a hyperpolarizing voltage command, which generates only leak and capacity currents. A depolarizing command also elicits a sodium current (short dashed lines). Addition of the two current records yields the sodium current (solid line), here distorted by series resistance.

clearly stated that it has been done, and if the blanked period is short with respect with the time necessary to open or close the channels being studied.

In most cases, it is necessary to use rather small voltage steps for leak subtraction, and to scale them up before subtraction. That may be necessary either to avoid activation of active currents, or to avoid driving the cell to extreme hyperpolarized potentials, which might destroy the membrane. If so, it may be necessary to average several "leak" steps to prevent introduction of noise. When this is done, it is particularly important to check that the leakage and capacitative currents are stable, since any errors in their measurement would be magnified.

A fundamental assumption of leak subtraction is that there is a region of membrane potentials where no voltage-dependent currents exist. That may or may not be true. Many cells have "anomalous rectifier" currents that activate upon hyperpolarization, which prevents the use of hyperpolarizing steps to measure leak. Some cells may have no region where the membrane currents are truly passive. For example, even though hippocampal pyramidal cells have a linear current–voltage (I–V) around rest, that may result from a coincidental addition of voltage-dependent currents that are active near rest (Halliwell and Adams, 1982). At the least, it is necessary to establish that, under the experimental conditions used, the leakage currents are linear and not time-dependent before leak subtraction is attempted.

An alternative way to correct for leak and capacity currents is to subtract currents in the presence and absence of a pharmacological blocking agent. For example, calcium currents are often studied as difference currents $\pm Co^{2+}$ or another ion that blocks calcium channels. This obviously assumes that the blocker has no "side effects" on other ion channels.

3.2.2. Current–Voltage Curves

Sometimes it appears that the main function of current–voltage (I–V) curves is to make electrophysiology papers uninterpretable to biochemists. At least, that was my experience as a biochemistry graduate student. However, I–V curves are perhaps the most useful way to summarize the basic properties of a voltage-dependent current.

Two factors determine the amount of current measured at a particular voltage: the conductance and the driving force. The conductance (G) depends on how many channels are open and on how easily current can flow through a single open channel. The driving force is the difference between the membrane potential (V) and the reversal potential of the channel (V_R). Thus:

$$I = G(V - V_R) \qquad (6)$$

To a first approximation, an open channel is simply a resistor and follows Ohm's law. That is, G is constant with voltage. The current through the channel will be zero at the reversal potential. For more positive voltages, the current will be positive (outward

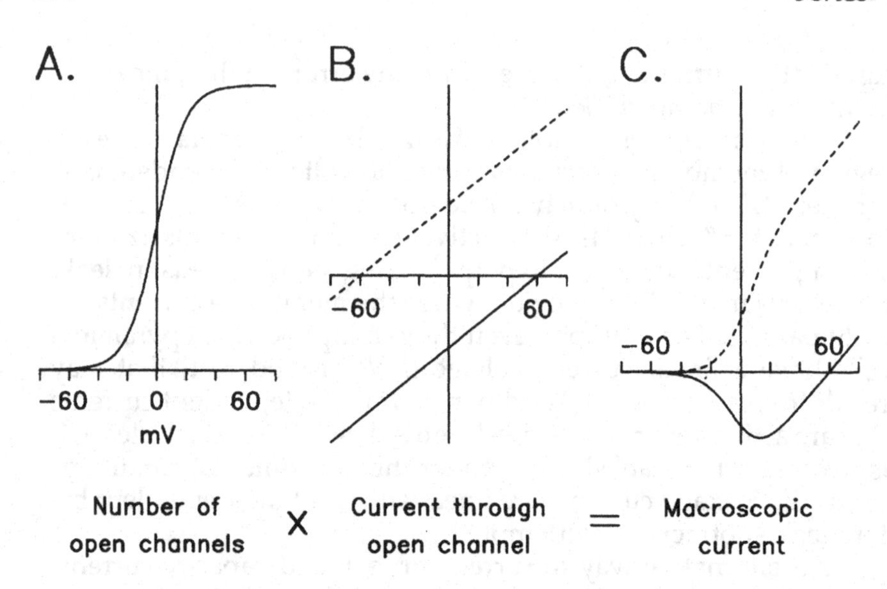

Fig. 11. Current–voltage curves for hypothetical voltage-dependent currents. The vertical scales are arbitrary. (A) Activation curve for a voltage-dependent current, assuming that the channel open probability changes e-fold for 5 mV, with half of the channels open at 0 mV. (B) Current–voltage relations for channels that obey Ohm's law, with reversal potentials of –60 mV (dashed line) or +60 mV (solid line.) (C) Current–voltage relations for currents with gating as in A and ion selectivity as in B.

current is defined as positive); negative to reversal, current will flow into the cell. This is illustrated in Fig. 11B, for channels with reversal potentials of –60 mV and +60 mV (approximations to potassium and sodium channels, respectively). The slope of the I–V curve is the conductance; a constant conductance produces a linear I–V curve.

For a voltage-dependent channel, the probability that the channel is open depends on voltage. Figure 11A shows this for a hypothetical channel that opens upon depolarization, with the channel open half the time at 0 mV. To calculate the current at any voltage from a cell with many thousands of channels, simply multiply the current per channel (at each voltage) times the number of channels open (at that voltage). The result is Fig. 11C. Because the potassium channel is nearly always closed around its reversal potential, the current through it is usually outward. The I–V curve approaches a straight line at extreme positive potentials,

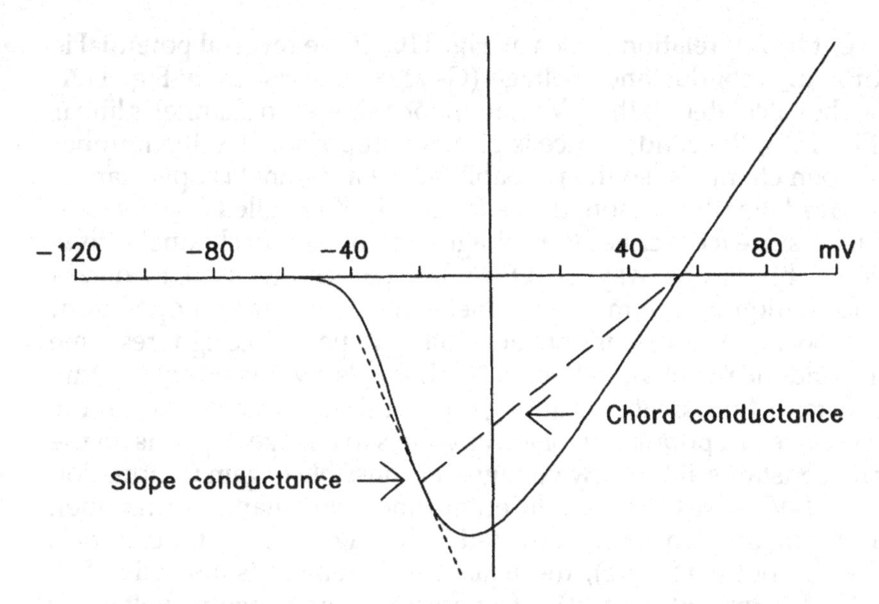

Fig. 12. Chord and slope conductance. Current–voltage curves for a hypothetical current, with the current units (vertical axis) arbitrary.

where all of the channels are open. The I–V curve for sodium current looks quite different; as the cell is depolarized, the amount of inward current first increases (as more and more channels open) and then decreases (as the driving force decreases). Around the reversal potential, nearly all of the channels are open, so the I–V curve is linear.

From Eq. 6, conductance is current divided by driving force. Graphically, it is the slope of the chord drawn between a point on the I–V curve and the reversal potential, so it is sometimes called the chord conductance (Fig. 12). Another measure, called the slope conductance, is sometimes encountered. It is the slope of the line tangent to the I–V curve. Slope conductance is not a direct measure of ion permeability, since it reflects also the voltage dependence of channel activation. The chord conductance is always positive (or zero), but the slope conductance can be either positive or negative. In some cases where the reversal potential is difficult or impossible to measure, slope conductance can be a useful measure if the conductance is not strongly voltage dependent.

The slope conductance reflects the dependence of channel opening on voltage. The result of a voltage-clamp experiment might be I–V relations, such as Fig. 11C. If the reversal potential is known, a conductance-voltage (G–V) relation (such as Fig. 11A) can be calculated. If the I–V relation for a single ion channel is linear (Fig. 11B), the conductance is directly proportional to the number of open channels, so the probability that a channel is open can be inferred. For this reason, the G–V curve is often called an activation curve, since it describes the voltage-dependence of channel activation. This is one way in which I–V curves are used to obtain information on the molecular mechanisms of channel operation.

So far, the discussion of voltage dependence ignores time dependence. Voltage affects ion channels by changing the rate constants for transitions among open and closed states, so that the response of a population of ion channels to voltage depends on the rate constants at the new voltage. It is possible to gain information from I–V curves taken at different times. For channels that open and remain open when activated by voltage (that is, for channels that do not inactivate), the usual measurement is a steady-state I–V, determined when the response to the change in voltage is complete. For channels that inactivate with time, such as voltage-dependent sodium channels, there is little or no steady-state current, so an I–V measured at the point of peak current is a better indication of the voltage dependence of activation.

Voltage-dependent currents can differ in rates of activation, inactivation, and deactivation. Deactivation is an inelegant term for decay of a current because of reversal of the activation process, rather than because of a separate inactivation mechanism. Currents during deactivation are often called tail currents. Fig. 4B illustrates calcium current tails, observed as the channels close upon repolarization of the cell from 0 to –40 mV.

Another important subclass of I–V curve is the instantaneous I–V. It consists of measurements of current as soon as possible following voltage steps. For the measurement to be effectively instantaneous, the capacitative current must be over before channels begin to open or close. If that is true, the change in current with voltage reflects the conductance of ion channels open when the voltage step is given. This provides an alternative method for determining the voltage dependence of an ionic current. This is illustrated for Hodgkin-Huxley potassium currents in Fig. 13A. The initial amplitude of the tail current following steps to different voltages is an assay of the amount of potassium conductance activated during the step. Since the tail currents are measured at

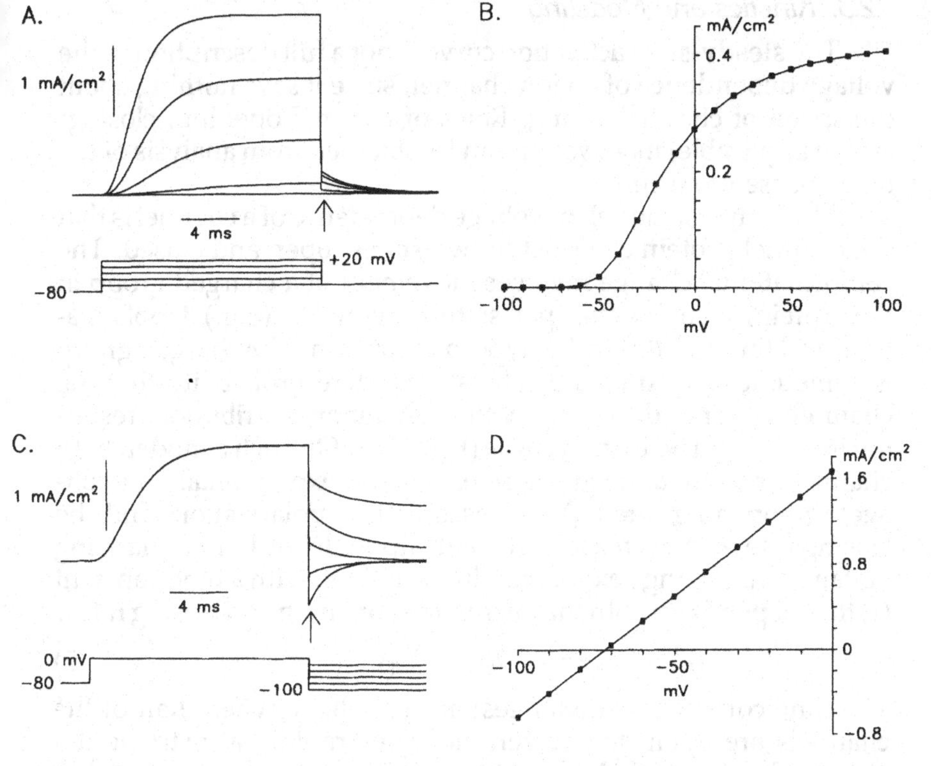

Fig. 13. Current–voltage relations for tail currents, calculated from the Hodgkin-Huxley model for potassium current. (A) Currents evoked by depolarizing steps from –80 mV, with tail currents measured at –60 mV following each step. Initial amplitudes of tail currents (arrow) are plotted vs the voltage during the depolarizing step in B. (C) Tail currents measured at different voltages, following a depolarizing step to 0 mV. Initial amplitudes of tail currents (arrow) are plotted in D. The tail current amplitude would be zero at –72 mV, the reversal potential for the current.

the same voltage, the activation curve calculated by this method (Fig. 13B) does not depend on the assumption that ion channels obey Ohm's law. A different protocol (Fig. 13C–D) uses the instantaneous I–V to determine the reversal potential of the potassium current, and to test the assumption that the channels are ohmic. Most channels do show deviations from Ohm's law; for a clear explanation of how this results from the mechanisms of ion permeation through channels, *see* Hille (1984).

3.2.3. Kinetics and Modeling

The steady-state activation curve is not a full description of the voltage dependence of an ion channel, since it says nothing about the speed of channel gating. Rates of channel opening, closing, and (if applicable) inactivation can be obtained from analysis of the time course of currents.

The simplest model for voltage dependence of a channel is that the channel protein can exist in two states, open and closed. The conformational change involves movement of a charged group in the protein, such that the open state is favored by (e.g.) depolarization, and the closed state by hyperpolarization. The charged group is sometimes called a gate. The steady-state probability that the channel is open is described by the Boltzmann distribution, resulting in a sigmoid activation curve (Fig. 11A). On such a model, both channel opening and closing rates depend exponentially on voltage; the opening rate (k_1) increases with depolarization, and the closing rate (k_{-1}) decreases. Currents upon de- or hyperpolarizing voltage steps change exponentially with time, with a time constant (τ) that depends on both the microscopic opening and closing rates:

$$\tau = 1/(k_1 + k_{-1}). \quad (7)$$

The time constant is the longest at the voltage where half of the channels are open; the current activates rapidly at extreme depolarized voltages and deactivates rapidly at extreme hyperpolarized voltages.

Few if any currents obey such a simple model. Some currents activate with a sigmoid time course, rather than exponentially (Fig. 13A). This is generally explained either by multiple closed states, which the channel must pass through before opening, or, as on the Hodgkin-Huxley model, multiple gates, all of which must be open for the channel to be open. Channels that inactivate require at least a three-state model, which complicates the kinetics. Often the rate constants for channel opening and closing do not depend symmetrically on voltage. The general principles are

1. Channels are voltage-dependent because of charged groups, which can exist in different conformational states
2. The time course of voltage-dependent currents contains information about the microscopic kinetic processes involved in channel opening and closing and

3. The dependence of current kinetics on membrane potential can be explained by the action of voltage on charged groups within the channel protein.

As always in kinetics, it is possible to describe a given set of data by physically distinct models. That is, the microscopic kinetic processes are not uniquely determined by the macroscopic data. Other information, such as the kinetic behavior of single channels, can be helpful in distinguishing models. The original paper by Hodgkin and Huxley (1952) is still a good introduction to how plausible models can be obtained from voltage-clamp data.

3.3. Separation of Currents

A major stage in the analysis of voltage-clamp data is to decide whether the observed currents result from the activity of a single class of ion channels or from several different channel types. It is not unusual for a single cell to contain a dozen voltage- and neurotransmitter-dependent ion channels. Under physiological conditions, voltage steps elicit a complex mixture of inward and outward currents that are initially all but uninterpretable (Fig. 14).

There are three general classes of tools that can be used to help isolate a single current type from this mess:

1. Ion dependence
2. Kinetics and
3. Pharmacology.

The first two are the most fundamental, and are well illustrated in the original work of Hodgkin and Huxley (1952).

3.3.1. Ion Dependence

The most basic criterion for identification of a current is the identity of the ion(s) that pass through the channel. This is best established by varying the concentrations of various ions, and observing shifts in reversal potential. If a channel is permeable solely to Na^+ ions, changes in the external Na^+ concentration should shift the reversal potential as expected from the Nernst equation for Na^+. This was how Hodgkin and Huxley determined that the early transient current in squid axon was carried by Na^+.

One mistake that is often made is to completely replace one ion with another and look for elimination of the current. This fails to distinguish a change in the current because of lack of permeant

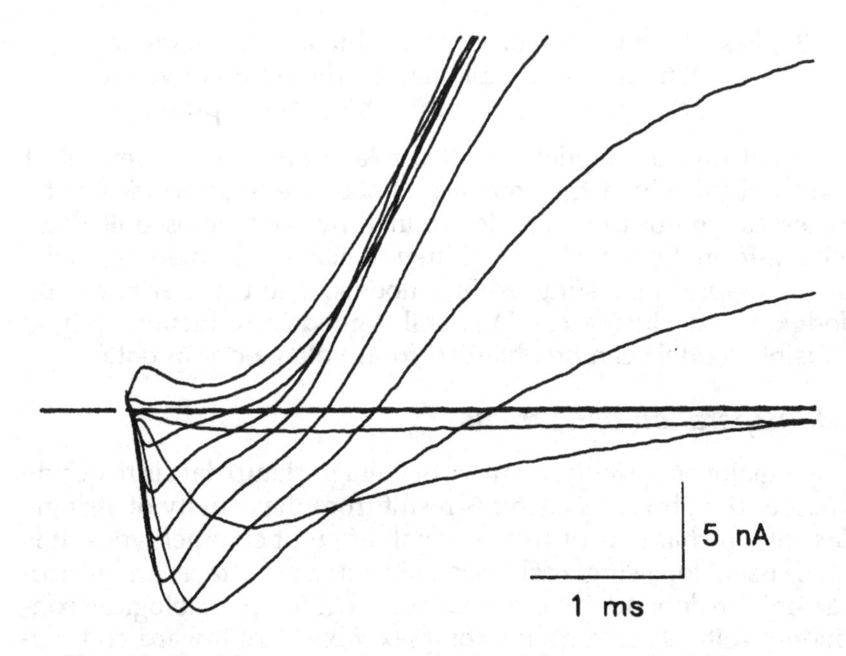

Fig. 14. Rapid currents in an isolated frog sympathetic neuron. cSEVC was used with whole-cell recording, with essentially normal intra- and extracellular solutions. Voltage steps were in 10-mV increments from the holding potential of –60 mV up to +60 mV. Records are leak subtracted, and 70 μs was blanked at the start of the step because of imperfect cancellation of capacity transients. Note that the potassium currents at large depolarizations cross over; this is because of the nonmonotonic voltage dependence of I_C (*see* Fig. 17). The largest outward currents are not fully shown, since the headstage used was limited to 20 nA.

ions from pharmacological blockade of the current by the ion used for substitution. The advantage of the reversal potential is that it is a null point measurement, so that it is relatively insensitive to the mechanism of ion permeation. That is absolutely true if the channel is permeable to only one ion; then the reversal potential equals the equilibrium potential, and is unaffected by factors other than ion concentrations.

Such terms as "sodium channel" are best thought of as names for a particular channel type, rather than as a statement that all current through the channel is carried by Na^+ ions. No real channel is perfectly selective for one and only one ion. Furthermore, it is useful to distinguish equilibrium potentials, defined for an ion by the Nernst equation, from reversal potentials, defined for a channel. These semantic problems are particularly acute for the calcium channel, often studied using Ba^{2+} rather than Ca^{2+} as the charge carrier. Under some circumstances, the calcium channel is highly permeable to monovalent cations, such as Na^{+}, but it still makes sense to retain the name "calcium channel" to reflect its role under physiological conditions.

3.3.2. Kinetics

Currents can often be distinguished by their kinetic properties. These properties include voltage-dependence (both of rate constants and of steady-state behavior), and the presence or absence of particular states (such as inactivated states or multiple closed states). Kinetic criteria are particularly useful under voltage clamp, where the true time course of channel opening and closing are directly reflected in the time course of the observed currents.

In the squid axon, sodium and potassium currents can be separated fairly well simply by measuring the current at different times. At ~1 ms, nearly all of the current is sodium current, since the potassium current activates more slowly, but later (~10 ms) nearly all of the current is potassium current, since most of the sodium current has inactivated.

Many channels appear to have only one open state, which means that tail currents decay exponentially with time. This can make tail currents easier to interpret than currents activated during a depolarizing voltage step. Complex behavior of tail currents is often taken as evidence for existence for multiple currents. For example, a tail current that is fit by the sum of two exponentials might indicate that each component results from the activity of one class of channel. Of course, other explanations are possible.

3.3.3. Pharmacology

Pharmacology is perhaps the most popular criterion for identification of currents, but it is also the most treacherous. It must be recalled that Hodgkin and Huxley rigorously separated the sodium and potassium currents of squid axon without resort to tetrodotoxin (TTX) or tetraethylammonium (TEA). The defining criterion for a sodium current is that it is carried (predominantly, under normal conditions) by Na^{+} ions, not that it is sensitive to

TTX. The First Law of Pharmacology ("All drugs have two effects; the one you know and the one you don't know") must be kept in mind.

This is not to deny that pharmacological differences exist among otherwise similar ion channels and that those differences are valuable in distinguishing currents. However, many useful drugs have rather complex actions, which must be taken into account. TTX is particularly valuable since it appears simply to prevent current through sodium channels, with no effect on the kinetics of the current. That is not true for many other widely used pharmacological tools. For example, the "blockade" of calcium channels by multivalent cations or dihydropyridine "antagonists" is voltage- and state-dependent, so that the time course of calcium current is affected. This complicates interpretation. Suppose that a drug blocks a macroscopic current only partially and the remaining current has different kinetics. That could be because of full blockade of one current type, revealing a second type of current in isolation, but it could also be because of a partial effect on a single preexisting current type, if the drug's action is time- and voltage-dependent.

3.3.4. Bullfrog Sympathetic Neurons

The principles discussed above for isolation of ionic currents are well illustrated in frog sympathetic neurons (Adams et al., 1982a, 1986; Pennefather et al., 1985; Jones, 1987). These cells have at least eight distinct voltage-dependent currents and respond to a comparable number of neurotransmitters. Some of the currents have unexpected properties, which illustrate the point that a combination of criteria is necessary to establish firmly the nature of a current. Rapid depolarizations activate inward and outward currents that appear similar to those of squid axon and other well-known preparations (Fig. 14), but closer analysis reveals the presence of at least three inward and five outward currents in bullfrog sympathetic neurons.

3.3.4.1. INWARD CURRENTS. It was obvious even before the use of voltage-clamp methods that these cells had both sodium and calcium currents. However, they also have a current with some of the properties of each.

Using conditions designed to isolate sodium currents (extracellular TEA and intracellular Cs^+ to block potassium currents and extracellular Mn^{2+} to block calcium currents), a rapid inward

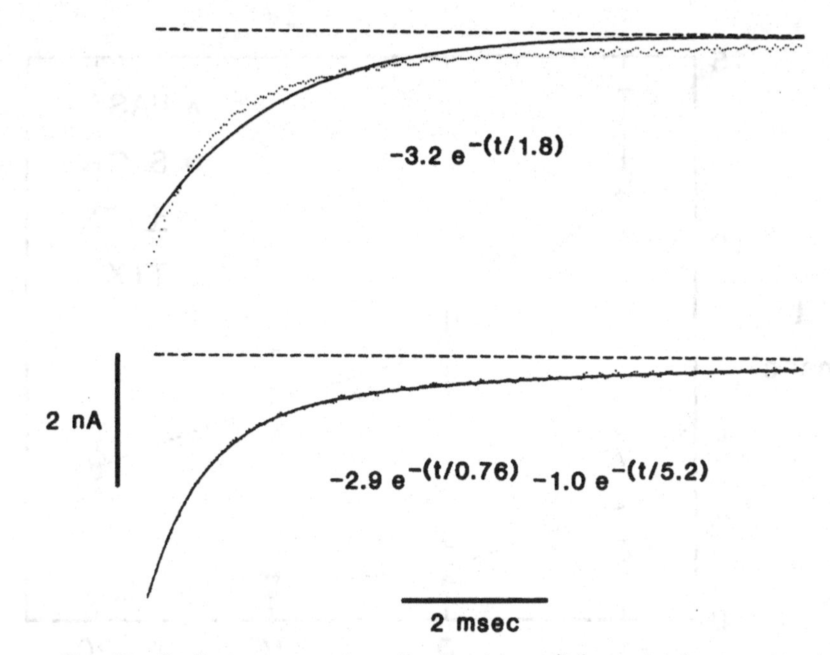

Fig. 15. Sodium current inactivation is not fit by a single exponential process in bullfrog sympathetic neurons. The points are the actual time-course of sodium current inactivation at 10 mV; the solid lines are best single (above) and double (below) exponential fits (Jones, 1987) to the data. The equations give initial amplitudes in nA and time constants in ms.

current is activated that looks much like the sodium current of squid axon (Jones, 1987). The current reverses to an outward current near the expected value for the equilibrium potential for Na^+, and more importantly, the reversal potential changes appropriately when extracellular Na^+ is decreased. The first hint that this is actually a combination of two distinct ionic currents was that inactivation was not well fit by a single exponential process (Fig. 15). That of course is not in itself conclusive evidence for multiple current types; it could be explained equally well by a complex inactivation process for a single current.

Pharmacological evidence strengthened the case for existence of two distinct currents (Jones, 1987). The classical sodium channel blocker TTX did inhibit most of the current, but even at high concentrations approximately 25% of the total inward current remained. Furthermore, that current inactivated more slowly than

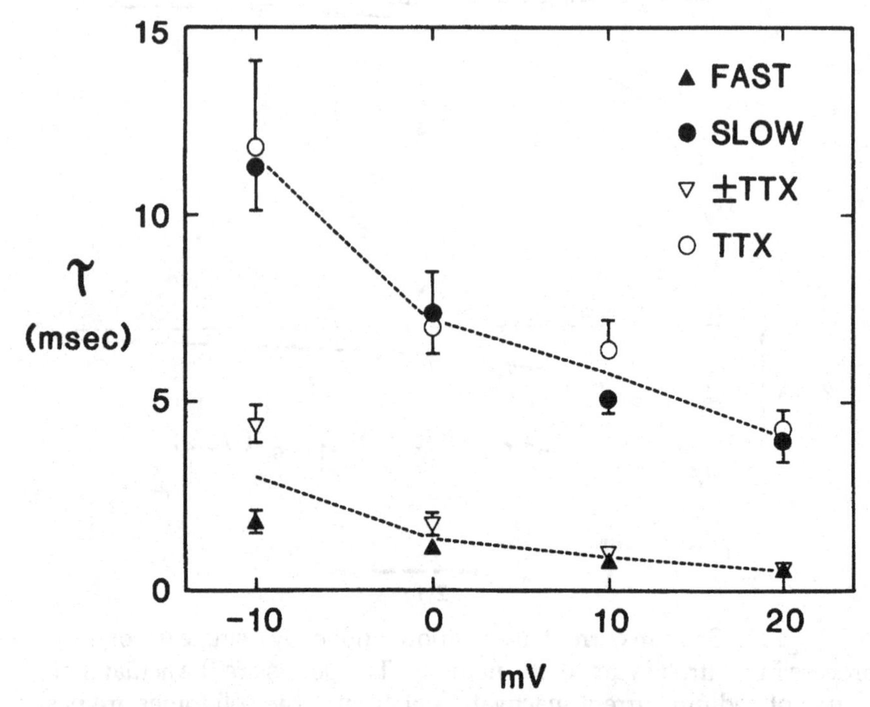

Fig. 16. Time constants for inactivation at different voltages. Solid symbols are from double-exponential fits, such as those of Fig. 15. Open symbols are single-exponential fits to the current remaining in TTX (o) or to the difference current ±1 μ*M* TTX (∇). Values are shown with error bars representing the standard error of the mean when larger than the size of the symbol. There is reasonable agreement between the kinetic and pharmacological means of separating the two sodium currents. It should be noted that double-exponential fitting routines can be inaccurate if the time constants differ by less than fivefold.

the total current. The difference current, that is the total current minus that in the presence of TTX, inactivated rapidly. Inactivation of the TTX-sensitive and TTX-resistant currents could each be fit by a single exponential process, with inactivation of the resistant current approximately 4 × slower. So, both kinetic evidence and pharmacological evidence were consistent with the interpretation that the total current consisted of a mixture of two currents, one TTX-sensitive and rapidly inactivating and the other TTX-resistant and slowly inactivating (Fig. 16). Still, this conclusion depends strongly on the action of one drug, which is dangerous. For example, if the blocking action of TTX were voltage-dependent, it might appear to affect the kinetics of inactivation. The ideal situation would be to have a second drug that selectively blocks the slowly inactivating current. Fortunately, Cd^{2+} ions do just that. In the presence of Cd^{2+}, the current shows only a rapidly inactivating component, and TTX and Cd^{2+} combined block >95% of the total current.

However, at this point, it became necessary to ask whether the TTX-resistant current really is a sodium current. After all, a TTX-resistant, Cd^{2+}-sensitive inward current with relatively slow inactivation sounds a lot like a calcium current! The crucial experiment, of course, is ion replacement. The reversal potential of the TTX-resistant current changed with the Na^+ concentration, and not with divalent ion concentrations (Jones, 1987). Both sodium currents appeared to be blocked by saxitoxin (STX).

Similar TTX-resistant, Cd^{2+}-sensitive sodium currents exist in sensory neurons (Kostyuk et al., 1981; Ikeda et al., 1986) and in cardiac muscle (DiFrancesco et al., 1985). Such currents can be easily confused with rapidly inactivating calcium currents, which are present in some of the same cells. Remember that what makes a sodium current a sodium current is not its kinetics or its pharmacology, but its selectivity for Na^+ ions.

When sodium currents are prevented by replacement of external and internal Na^+ ions by an impermeant cation such an *N*-methyl-D-glucamine, and Ca^{2+} or Ba^{2+} is present extracellularly, well-isolated calcium currents can be observed (Jones and Marks, 1989). The calcium current is kinetically different from the sodium currents in that it activates more slowly (with $\tau < 3$ ms, depending on voltage), and inactivates slowly and incompletely ($\tau > 100$ ms). The calcium current is highly sensitive to Cd^{2+}, even more sensitive than is the TTX-resistant sodium current. Surprisingly, the calcium current was fully blocked by commercial preparations of STX, although rather high concentrations were necessary. In fact, the apparent potency of STX varied from batch to batch against calcium (but not sodium) current, so that the action on calcium current might be the result of an impurity. This is another example of why it is necessary to use caution in interpreting pharmacological actions on ion channels.

There is good evidence from single-channel recording that two classes of calcium channel exist on frog sympathetic neurons, similar to the "L" and "N" channels of sensory neurons (Lipscombe et al., 1988). However, those currents are difficult to separate under whole-cell recording conditions, so their quantitative contribution to the macroscopic calcium current remains unclear.

3.3.4.2. OUTWARD CURRENTS. The first potassium current to be studied in these cells had several novel properties (Adams et al., 1982a).

1. The current was activated in the region around the resting potential, whereas other outward currents required much stronger depolarization to be activated. This property made it quite easy to study in isolation, since it was the only voltage-dependent current visible at –30 mV or more hyperpolarized, so that complex cocktails of blocking agents were not needed
2. The current had very slow kinetics, with time constants ~100 ms. It did not inactivate, even after many minutes. Activation and deactivation were fit reasonably well as single exponential processes
3. The current was inhibited strongly by activation of muscarinic receptors. This property gave it the name "M-current," or I_M
4. It was resistant to the classical potassium channel blocker TEA.

I_M is obvious in these cells upon small, slow depolarizations. However, the more traditional voltage-clamp protocol of large, rapid depolarizations elicits different potassium currents (Fig. 14); those currents are quite large, so the presence of I_M could be missed entirely. Perhaps the most striking feature of current-voltage relations from rapid depolarizing steps is that the amount of outward current first increases, and then decreases upon depolarization (Fig. 17). This "N-shape" is not observed in the presence of calcium channel blockers, or when intracellular Ca^{2+} is strongly buffered with BAPTA. That fact and other evidence suggest that much of the outward current is the result of a Ca^{2+}-dependent potassium channel, sometimes called I_C, which is activated by an increase in intracellular Ca^{2+} (Adams et al., 1982b). I_C is voltage- as well as Ca^{2+}-dependent; it deactivates rapidly upon hyperpolarization, much faster than can be explained by removal of intracellular Ca^{2+}. Most of the outward current remaining following blockade of

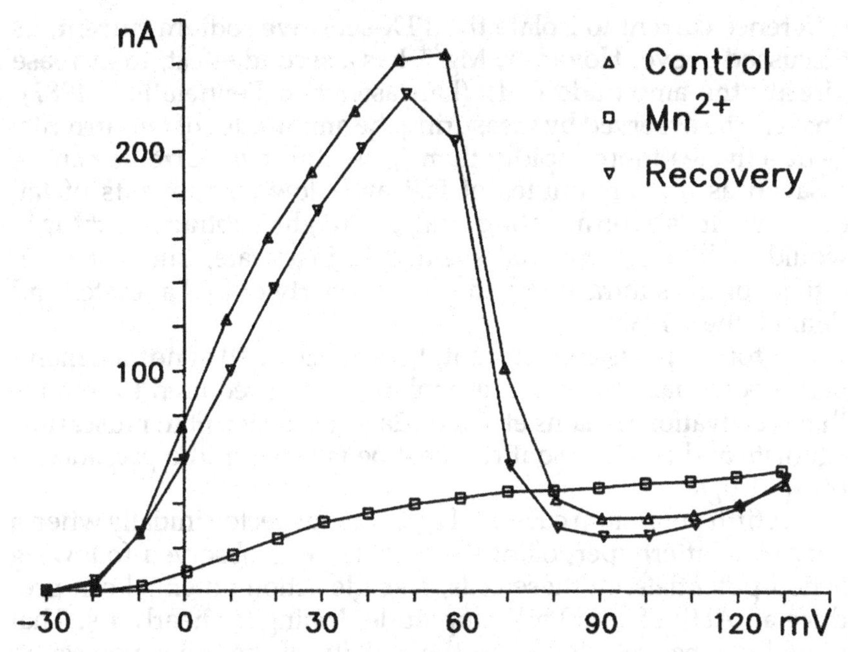

Fig. 17. The "N-shaped" current–voltage relationship in a frog sympathetic neuron and blockade by Mn^{2+}. Data are from the same experiment as Fig. 8. Currents were measured at the end of 60-ms depolarizations. Most of the outward current between −20 and +60 mV is I_C, and is blocked by perfusion with Mn^{2+}. The enhancement of the current above +80 mV may be the result of potentiation of I_K by Mn^{2+}. Direct measurement of calcium currents indicates that peak calcium entry occurs near 0 mV, with very little inward current above +50 mV. The continued increase in outward current between 0 and +50 mV, and the sharp cutoff above +60 mV, may indicate that little calcium is necessary to activate I_C at strongly depolarized voltages. However, the sharp cutoff would also be expected if there were a significant voltage error (Marty and Neher, 1985).

calcium channels is attributed to a "delayed rectifier" potassium current, I_K, similar to that of axons, but with slower activation (Adams et al., 1982a).

No selective blockers of I_K are yet known; TEA blocks both I_K and I_C equally well. One approach to studying I_C in isolation would be to look at the difference current $\pm Mg^{2+}$ or some other calcium channel blocker. That would be analogous to the use of the ± TTX difference current to isolate the TTX-sensitive sodium current, as discussed above. However, Mg^{2+} has a second effect, to increase directly the amplitude of I_K (Lancaster and Pennefather, 1987). That can be observed by measuring the amplitudes of tail currents; I_C deactivates more rapidly than I_K, so the two currents can be assayed as the amplitudes of fast and slow components of tail currents. It is worth noting that a straight subtraction $\pm Mg^{2+}$ would yield a current that seemed to inactivate, but that is an artifact of the subtraction and not a property of I_C (Lancaster and Pennefather, 1987).

A fourth potassium current, I_A, is inactivated at normal membrane potentials. Strong hyperpolarization is required to remove that inactivation (Adams et al., 1982a). I_A is resistant to muscarinic agonists and to TEA, so it can best be isolated in the presence of those drugs.

A fifth potassium current, I_{AHP}, was suspected initially when a very slow afterhyperpolarization (AHP) was observed following action potentials. In these cells, a single action potential can produce an AHP of 20–30 mV amplitude, lasting for nearly 1 s. That would not be expected from the activity of the other potassium currents. Voltage-clamp studies confirmed the existence of a rapidly activating outward current, which decayed with a time constant of >100 ms (Pennefather et al., 1985). That current was distinct from the other potassium currents. I_{AHP} is not voltage dependent, is much less sensitive to TEA, and is blocked by apamin and curare. Note that once again a combination of pharmacological and kinetic distinctions were necessary to establish the existence of a new current. Also, curare, classically a neuromuscular blocking agent, has an unrelated "side effect" on I_{AHP}.

Recent evidence suggests the existence of a sixth potassium current, as yet unnamed, whose role is to set the resting potential (Jones, 1989). It is often assumed, but rarely proven, that the resting potential of a cell is set by a voltage-insensitive potassium channel that is open at rest. There is such a resting current in bullfrog sympathetic neurons, present at negative membrane potentials where I_M and the other voltage-dependent currents are deactivated. Preliminary evidence indicates that that current is:

1. Potassium selective
2. Not Ca-dependent
3. Not strongly voltage-dependent

4. Not blocked by muscarinic agonists and
5. Blocked by Ba^{2+}.

Although other potassium currents, notably I_M, are also blocked by Ba^{2+}, overall these properties seem to indicate the existence of a new current type. No selective blocker is yet known for this current.

3.3.5. Summary

This discussion illustrates how the seemingly simple criteria of ion dependence, kinetics, and pharmacology can be expanded to separate the large number of ionic currents present in a single cell type. It only hints at the next questions: why are there so many currents, and what are their functions? Those questions are beginning to be addressed (Adams et al., 1986). Briefly, voltage-dependent currents do more than generate the action potential. They actively participate in the integration of synaptic information, and influence the cell's "decision" to fire an action potential or not. In frog sympathetic ganglia, I_M and I_{AHP} appear to play a particularly important role in regulation of excitability.

4. Conclusions

The goal of this chapter has been to summarize the goals and limitations of whole-cell patch clamp and other voltage-clamp techniques. With proper precautions, it is possible to approach the molecular basis of electrical excitability from analysis of ionic currents. Voltage clamp thus supplements both the more detailed picture provided by single-channel analysis and the more global picture of electrical activity that results from other electrophysiological techniques.

Many of the sections of this review could be (and have been) the subject of entire books. Patch- and voltage-clamp methods have been described in detail elsewhere (Sakmann and Neher, 1983; Smith et al., 1985; Standen et al., 1987), and the conceptual basis of electrical excitability is elegantly presented by Hille (1984).

Acknowledgment

Work in the author's laboratory is supported by NIH grant NS 24471.

References

Adams P. R., Brown D. A., and Constanti A. (1982a) M-currents and other potassium currents in bullfrog sympathetic neurones. *J. Physiol.* **330,** 537–572.

Adams P. R., Constanti A., Brown D. A., and Clark R. B. (1982b) Intracellular Ca^{2+} activates a fast voltage-sensitive K^+ current in vertebrate sympathetic neurones. *Nature* (Lond). **296,** 746–749.

Adams P. R., Jones S. W., Pennefather P., Brown D. A., Koch C., and Lancaster B. (1986) Slow synaptic transmission in frog sympathetic ganglia. *J. Exp. Biol.* **124,** 259–285.

Armstrong C. M. and Chow R. H. (1987) Supercharging: A method for improving patch-clamp performance. *Biophys. J.* **52,** 133–136.

Cachelin A. B. and Rice P. D. (1987) Microcomputers in the laboratory, in *Microelectrode Techniques. The Plymouth Workshop Handbook* (Standen N. B., Gray P. T. A., and Whitaker M. J., eds.), pp. 229–248, The Company of Biologists, Cambridge, England.

DiFrancesco D., Ferroni A., Visentin S., and Zaza A. (1985) Cadmium-induced blockade of the cardiac fast Na channels in calf Purkinje fibres. *Proc. Roy. Soc. B* **223,** 475–484.

Finkel A. S. and Gage P. W. (1985) Conventional voltage clamping with two intracellular microelectrodes, in *Voltage and Patch Clamping with Microelectrodes* (Smith T. G. Jr., Lecar H., Redman S. J., and Gage P. W., eds.), American Physiological Society, Bethesda, Maryland, pp. 47–94.

Finkel A. S. and Redman S. (1984) Theory and operation of a single microelectrode voltage clamp. *J. Neurosci. Meth.* **11,** 101–127.

Finkel A. S. and Redman S. J. (1985) Optimal voltage clamping with single microelectrode, in *Voltage and Patch Clamping with Microelectrodes* (Smith T. G., Jr., Lecar H., Redman S. J., and Gage P. W., eds.), American Physiological Society, Bethesda, Maryland, pp. 95–120.

Halliwell J. V. and Adams P. R. (1982) Voltage–clamp analysis of muscarinic excitation in hippocampal neurons. *Brain Res.* **250,** 71–92.

Hamill O. P. (1983) Potassium and chloride channels in red blood cells, in *Single Channel Recording* (Sakmann B. and Neher E., eds.), Plenum, New York, pp. 451–471.

Hamill O. P., Marty A., Neher E., Sakmann B., and Sigworth F. J. (1981) Improved patch–clamp techniques for high-resolution current recording from cells and cell-free membrane patches. *Pflugers Arch.* **391,** 85–100.

Hille B. (1984) *Ionic Channels of Excitable Membranes* (Sinauer Associates, Sunderland, Massachusetts).

Hodgkin A. L. and Huxley A. F. (1952) A quantitative description of membrane current and its application to conduction and excitation in nerve. *J. Physiol.* **117,** 500–544.

Horn R. and Marty A. (1988) Muscarinic activation of ionic currents measured by a new whole-cell recording method. *J. Gen. Physiol.* **92,** 145–159.

Ikeda S. R., Schofield G. G., and Weight F. F. (1986) Na^{+} and Ca^{2+} currents of acutely isolated adult rat nodose ganglion cells. *J. Neurophysiol.* **55,** 527–539.

Jack J. J. B., Noble D., and Tsien R. W. (1975) *Electric Current Flow in Excitable Cells* (Oxford University Press, Oxford, England).

Johnston D. and Brown T. H. (1983) Interpretation of voltage–clamp measurements in hippocampal neurons. *J. Neurophysiol.* **50,** 464–486.

Jones S. W. (1987) Sodium currents in dissociated bull-frog sympathetic neurones. *J. Physiol.* **389,** 605–627.

Jones S. W. (1989) On the resting potential of isolated frog sympathetic neurons. *Neuron* **3,** 153–161.

Jones S. W. and Marks T. N. (1989) Calcium currents in bullfrog sympathetic neurons. I. Activation kinetics and pharmacology. *J. Gen. Physiol.* **94,** 151–167.

Kostyuk P. G., Veselovsky N. S., and Tsyndrenko A. Y. (1981) Ionic currents in the somatic membrane of rat dorsal root ganglion neurons—I. Sodium currents. *Neuroscience* **6,** 2423–2430.

Kramer R. H. (1986) Axonal contribution to subthreshold currents in *Aplysia* busting pacemaker neurons. *Cell. Molec. Neurobiol.* **6,** 239–253.

Lancaster B. and Pennefather P. (1987) Potassium currents evoked by brief depolarizations in bull-frog sympathetic ganglion cells. *J. Physiol.* **387,** 519–548.

Lipscombe D., Madison D. V., Poenie M., Reuter H., Tsien R. Y., and Tsien R. W. (1988) Spatial distribution of calcium channels and cytosolic calcium transients in growth cones and cell bodies of sympathetic neurons. *Proc. Natl. Acad Sci. USA* **85,** 2398–2402.

Madison D. V., Malenka R. C., and Nicoll R. A. (1986) Phorbol esters block a voltage-sensitive chloride current in hippocampal pyramidal cells. *Nature* **321,** 695–697.

Marty A. and Neher E. (1985) Potassium channels in cultured bovine adrenal chromaffin cells. *J. Physiol.* **367,** 117–141.

Moore J. W. (1971) Voltage clamp methods, in *Biophysics and Physiology of Excitable Membranes* (Adelman W. J., ed.), Van Nostrand, New York, pp. 143–167.

Moore J. W. (1985) Comparison of voltage clamps with microelectrode and sucrose-gap techniques, in *Voltage and Patch Clamping with Microelectrodes* (Smith T. G., Jr., Lecar H., Redman S. J., and Gage P. W., eds.), American Physiological Society, Bethesda, Maryland, pp. 217–230.

Pennefather P., Lancaster B., Adams P. R., and Nicoll R. A. (1985) Two distinct Ca-dependent K currents in bullfrog sympathetic ganglion cells. *Proc. Natl. Acad. Sci. USA* **82**, 3040–3044.

Rall W. and Segev I. (1985) Space-clamp problems when voltage clamping branched neurons with intracellular microelectrodes, in *Voltage and Patch Clamping with Microelectrodes* (Smith T. G., Jr., Lecar H., Redman S. J., and Gage P. W., eds.), American Physiological Society, Bethesda, Maryland, pp. 191–215.

Sakmann B. and Neher E. (eds.) (1983) *Single-Channel Recording*. (Plenum, New York).

Sigworth F. J. (1983) Electronic design of the patch clamp, in *Single-Channel Recording* (Sakmann B. and Neher E., eds.), Plenum, New York, pp. 3–35.

Smith T. G., Jr., Lecar H., Redman S. J., and Gage P. W. (eds.) (1985) *Voltage and Patch Clamping with Microelectrodes* (American Physiological Society, Bethesda, Maryland).

Standen N. B., Gray P. T. A., and Whitaker M. J. (eds.) (1987) *Microelectrode Techniques. The Plymouth Workshop Handbook* (The Company of Biologists, Cambridge, England).

Wilson W. A., and Goldner M. M. (1975) Voltage clamping with a single microelectrode. *J. Neurobiol.* **6**, 411–422.

Wonderlin W. F., French R. J., and Arispe N. J. (1989) Recording and analysis of currents from single ion channels, in *Neuromethods*, Vol. 14 (Boulton A. A., Baker G. B., and Vanderwolf C. H., eds.), Humana Press, Clifton, New Jersey, in press.

Multisite Optical Measurement of Membrane Potential

Hans-Peter Höpp, Jian-Young Wu, Chun X. Falk, Marc G. Rioult, Jill A. London, Dejan Zecevic, and Lawrence B. Cohen

1. Introduction

An optical measurement of membrane potential using a molecular probe might be beneficial in a variety of circumstances. "Such a probe could, we believe, provide a powerful new technique for measuring membrane potential in systems where, for reasons of scale, topology, or complexity, the use of electrodes is inconvenient or impossible" (B. M. Salzberg, personal sentence). The possibility of using optical methods was first suggested in 1968 by the discovery of potential-dependent changes in intrinsic optical properties of squid giant axons (Cohen et al., 1968). Shortly thereafter, Tasaki et al. (1968) found stimulus-dependent changes in fluorescence of stained axons, and in 1971 a search was begun (Cohen et al., 1971) for dyes that would give signals large enough to be useful for monitoring membrane potential. By now more than 1000 dyes have been tested for their ability to act as molecular transducers of changes in membrane potential into changes in three types of optical signals: absorption, birefringence, and fluorescence. This screening effort has resulted in the discovery of dyes with a signal-to-noise ratio 100 times larger than was available from any signal in 1971. Several of these dyes (*see*, e.g., Fig. 1) have been used to monitor changes in potential in a variety of preparations. For reviews, *see* Cohen and Salzberg (1978), Waggoner (1979), Salzberg (1983), Grinvald et al (1988), and Dasheiff (1988). An earlier discussion of methods was published (Cohen and Lesher, 1986).

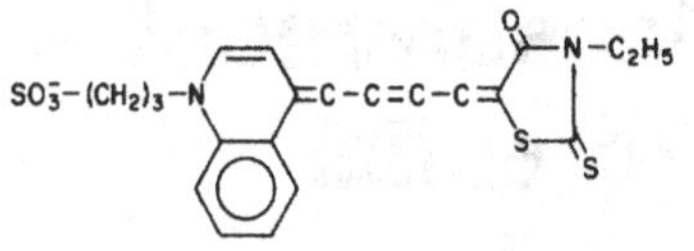

XVII, Merocyanine, Absorption, Birefringence

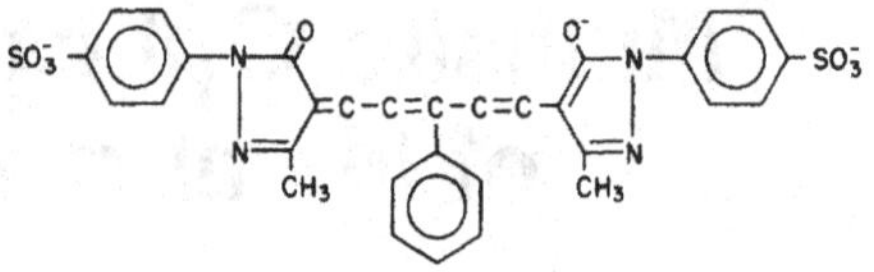

RH155, Oxonol, Absorption

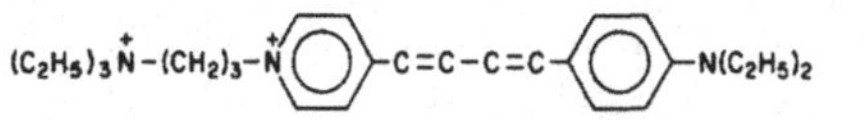

RH414, Styryl, Fluorescence

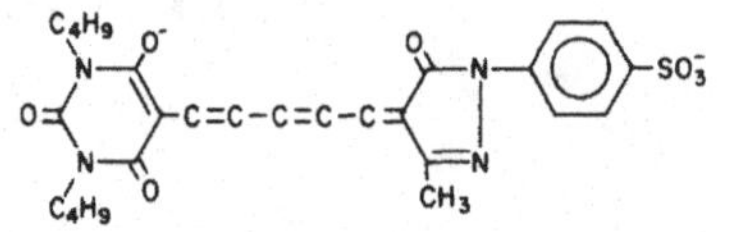

XXV, Oxonol, Fluorescence, Absorption

Fig. 1. Structures of several dyes that have been used to monitor membrane potential. The merocyanine (XVII) was the dye used in the experiments illustrated in Figs. 3 and 4. Dye XVII and the oxonol XXV are available from Dr. A. S. Waggoner, Center for Fluorescence, Carnegie Mellon University, 4400 Fifth Ave., Pittsburgh, PA, as WW375 and WW781. Dye XVII is available commercially as NK 2495 from Nippon Kankoh-Shikiso Kenkyusho Co. Ltd. The oxonol, RH155, and styryl, RH414, are available from Amiram Grinvald, Department of Neurobiology, Weizmann Institute, Rehovot, Israel. RH414 is available commercially as dye 1112 from Molecular Probes, Junction City, OR. RH155 is available as NK3041.

We begin with the evidence that has been used to show that optical signals are potential-dependent. Then we discuss the selection of signal type, dye, light source, photodetectors, optics, and computer hardware and software. This concern about apparatus arises because the signal-to-noise ratios in optical measurements are often smaller than one would like; attention to detail is required in order to maximize signal size. Some of the discussion of methods is most relevant to our own apparatus; other aspects of the paper are more general and would apply to any multisite optical measurement.

All of the optical signals described in this paper are fast signals, as defined earlier (Cohen and Salzberg, 1978). These signals are presumed to arise from membrane-bound dye; they follow changes in membrane potential with time courses that are rapid compared to the rise time of an action potential.

2. Some Optical Signals Are Potential-Dependent

The squid giant axon has provided a useful preparation for distinguishing among possible origins for optical signals and for screening new dyes. A simple filter spectrofluorimeter and spectrophotometer have been used to measure changes in dye-related optical properties (Cohen et al., 1974). In this apparatus and in other preparations where light scattering is not large, the light intensity reaching the photodiode in an absorption measurement was about $10^3 \times$ larger than in fluorescence. In birefringence measurements, the intensity was intermediate.

Figure 2 shows the results of a measurement of light absorption during an action potential in a squid axon stained with a merocyanine dye (XVII; Roman numerals refer to dyes in Fig. 1 or in Table I of Gupta et al. [1981]). The dotted trace is the light intensity transmitted through the axon at 750 nm; the smooth curve is the potential measured between internal and external electrodes. Because the two measurements had very similar time courses, it seemed likely that the absorption signal was related to the changes in membrane potential and not to the changes in membrane permeability that occur during the action potential. More direct evidence for potential dependence can be obtained from voltage-clamp experiments, such as the one illustrated in Fig. 3. The top trace is the absorption signal; clearly, it has a time course similar to that of the membrane potential (middle trace), and distinctly different from that of the permeability changes or the ionic currents (bottom trace). This kind of result has been obtained for many dyes, including all four illustrated in Fig. 1.

Inspection of Fig. 3 might suggest that signals with a time course similar to the currents or permeability are smaller than 5% of the total signal. In fact, a conclusion this strong is unwarranted, because the result in Fig. 3 was obtained with a somewhat arbitrary amount of compensation for the resistance in series with the axon membrane. Although the compensation used implied a series resistance within the range of previously reported values, the series resistance was not measured independently in this experiment. This ambiguity has been resolved by Salzberg and Bezanilla (1983).

In a voltage clamp experiment with four potential steps, the absorption change was linearly related to membrane potential over the range ± 100 mV from the resting potential. In additional ex-

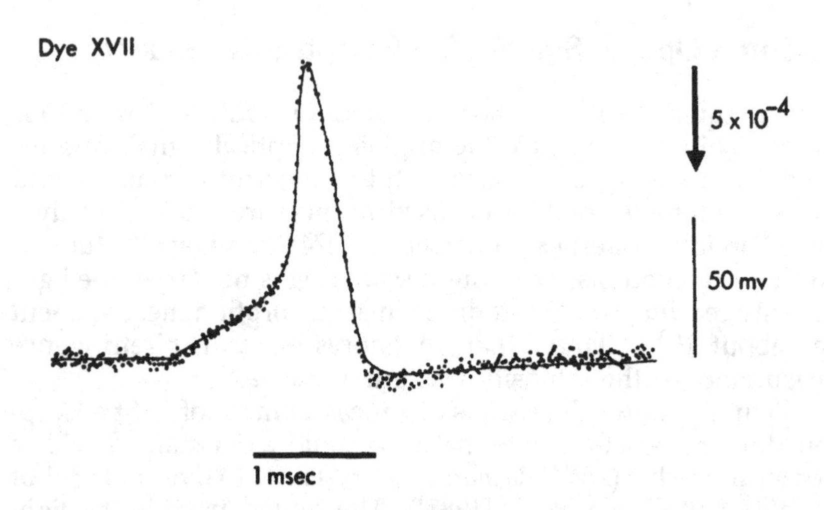

Fig. 2. The change in absorption (dots) of a giant axon stained with dye XVII during a membrane action potential (smooth trace) simultaneously recorded with an internal electrode. The change in absorption and the action potential had the same time course. In this figure and in Fig. 4, the direction of the arrow adjacent to the optical trace indicates the direction of an increase in absorption; the size of the arrow represents the stated value of a change in absorption, ΔI, in a single sweep divided by the resting absorption resulting from the dye, I. Incident light of 750 nm was used; 32 sweeps were averaged. The response time constant of the light measuring system was 5 μs. Redrawn from Ross et al. (1977).

periments with larger steps carried out with dyes XVII and XXII, the absorption signals were found to be linearly related to membrane potential over the range –130–+200 mV from the resting potential (Gupta et al., 1981). A linear relationship between optical signal and membrane potential has been obtained with absorption, fluorescence, and birefringence signals with many dyes. Thus, in many instances, there is strong evidence that the signals obtained with millisecond potential steps in squid axons depend in some manner on changes in the transmembrane potential (Cohen et al., 1970; Conti and Tasaki, 1970; Patrick et al., 1971; Davila et al., 1974; Conti, 1975; Ross et al., 1977; Gupta et al., 1981), although there was some earlier disagreement about this conclusion (Conti et al., 1971; Tasaki et al., 1972).

However, it is certain that nonpotential dependent signals can also be found. Russell et al. (1979) reported slow ion-dependent

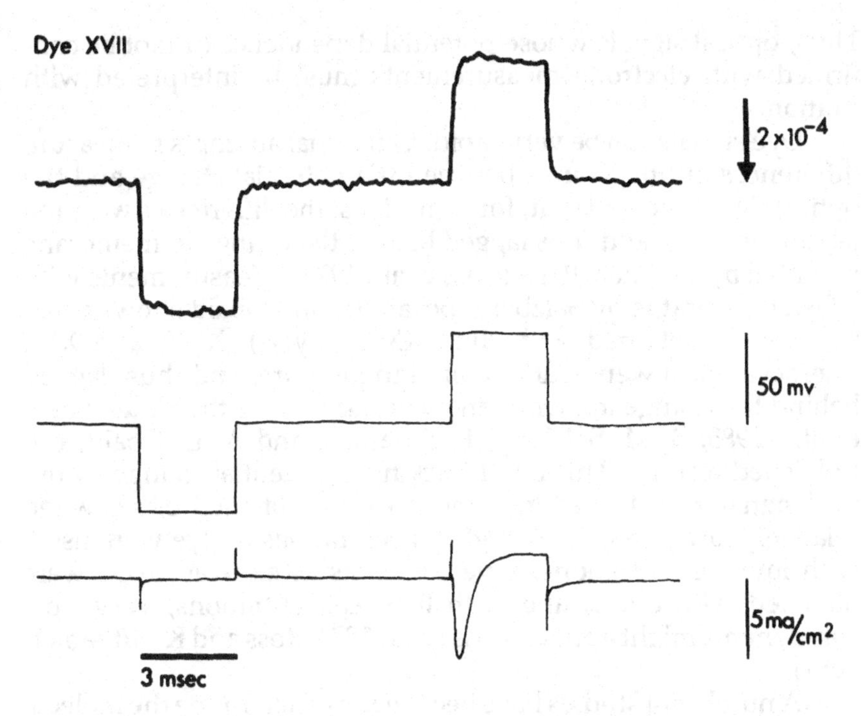

Fig. 3. Changes in absorption of a giant axon stained with dye XVII (top trace) during hyperpolarizing and depolarizing steps (middle trace). The bottom trace is the current density. The absorption changes had the same shape as the potential changes, and were insensitive to the large currents and conductance changes that occurred during the depolarizing step. The holding potential was the resting potential, and hyperpolarization is represented downward; inward currents are downward. Incident light of 750 nm was used; 128 sweeps were averaged; the time constant of the light-measuring system was 20 μs. Redrawn from Ross et al. (1977).

optical changes from suspensions of sarcoplasmic reticulum. Recently, Irena Klodos and Biff Forbush (personal communication) have found fluorescence changes related to the conformation of the Na-K ATPase with styryl dyes. Several kinds of optical signals are quite slow (100–1000 ms) (e.g., Orbach and Cohen, 1983; Orbach et al., 1985; Kauer et al., 1987; Lev-Ram and Grinvald, 1986; Blasdel and Salama, 1986; Grinvald et al., 1986). Clearly, it would be useful to have independent evidence from electrode recordings that these optical signals represent a change in membrane potential, but it is not easy to think of practical experiments to obtain this evidence.

Thus, optical signals whose potential dependence cannot be confirmed with electrode measurements must be interpreted with caution.

Dye signals can be very rapid. Our initial attempts to measure differences in time course between the potential change and the optical signal showed that, for some dyes, the differences were too fast to measure, and thus lagged behind the change in membrane potential by less than 10 μs (Ross et al., 1974). Measurements with a faster apparatus by Salzberg, Bezanilla, and Obaid showed that the signals obtained with dyes XXVI (styryl), XVII, and XXII (merocyanines) were still too rapid to measure, and thus, lagged behind the change in membrane potential by less than 2 μs (Loew et al,., 1985; B. M. Salzberg, F. Bezanilla, and A. L. Obaid, unpublished results). This rapid tracking of potential change by optical signal was, however, sometimes not obtained either when relatively low or relatively high concentrations of dye were used; with low concentrations, time constants as slow as 70 μs were obtained with dye I, and with high concentrations, very slow components might appear (Ross et al., 1974; Ross and Krauthamer, 1984).

A number of studies have been made to determine the molecular mechanisms that result in potential-dependent optical properties. This subject is discussed in the reviews cited above.

3. Choice of Absorption, Birefringence, or Fluorescence

Sometimes it is possible to decide in advance which kind of optical signal will give the best signal-to-noise ratio, but in other situations an experimental comparison is necessary. The choice of signal type may depend on the optical characteristics of the preparation. Birefringence signals are relatively large in preparations that, like giant axons, have a cylindrical shape and radial optic axis. However, in preparations with spherical symmetry (e.g., molluscan cell soma), the birefringence signals in adjacent quadrants will cancel, and thus in these preparations, birefringence can be measured only with a detection system with high spatial resolution. Achieving this spatial resolution leads to degradation of the signal-to-noise ratio (Boyle and Cohen, 1980). Because birefringence can be measured at wavelengths outside the absorption band of the dye, eliminating photodynamic damage and dye bleaching, birefringence would be preferable to absorption or fluorescence in measurements of propagation along axons. In one experiment where birefringence should have been tested, it was not (Shrager et al., 1987).

An instance where the preparation dictated the choice of signal was in measurements from mammalian cortex. Here transmitted light measurements are not feasible (without subcortical implantation of a light guide), and the small size of absorption signals that can be detected in reflected light (Ross et al., 1974; Orbach and Cohen, 1983) meant that fluorescence would probably be optimal (Orbach et al., 1985). Blasdel and Salama (1986) suggest that dye-related reflectance changes can be measured from cortex; this result has been indirectly questioned (Grinvald et al., 1986).

An additional factor that affects the choice of absorption or fluorescence is that the signal-to-noise ratio in fluorescence is relatively sensitive to the amount of dye bound to extraneous material. Figure 4 illustrates a spherical cell surrounded by extraneous material. In Fig. 4A, we assume that dye binds only to the cell; in Fig. 4B, we assume that there is 10 × as much dye bound to extraneous material. To calculate the transmitted intensity, we assume that there is one dye molecule for every 2.5 phospholipid molecules (a large concentration if one would also expect to maintain physiological function) and an extinction coefficient of 10^5 (some of the best available dyes have extinction coefficients this large). The amount of light transmitted by the cell is still 0.99 of the incident light. Thus, even if this dye were to disappear completely as a result of a change in potential, the fractional change, I/I, in transmission would be only 1% (10^{-2}). The amount of light reaching the photodetector in fluorescence will be much lower, say 0.0001 I. Several factors account for the lower intensity. First, only 0.01 I is absorbed by the dye. Second, we assumed a fluorescence efficiency (photons emitted/photons absorbed) of 0.1. Third, if we assume a light-collecting system of 0.8 NA (numerical aperture), only 0.1 of the emitted light reaches the photodetector. However, even though the light reaching the fluorescence detector is small, disappearance of dye would result in a 100% decrease in fluorescence—a fractional change of 10°. Thus, the fractional change in fluorescence can be much larger than the fractional change in transmission in situations where dye is bound only to the cell membrane and there is only one cell in the light path. However, the relative advantage of fluorescence is reduced if dye binds to extraneous material. When

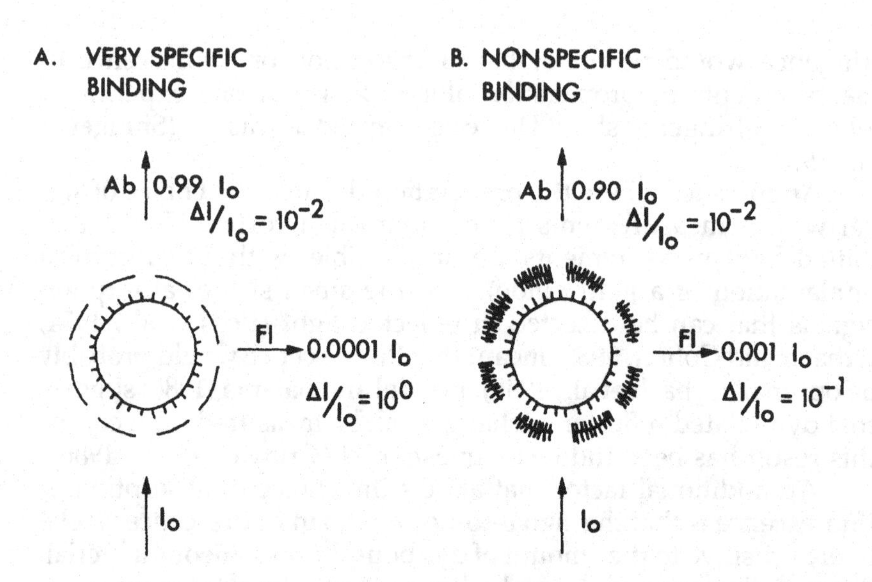

Fig. 4. (A) The light transmission and fluorescence intensity when only a neuron binds dye and (B) when both the neuron and extraneous material binds dye. In (A), assuming that 1 dye molecule is bound/2.5 phospholipid molecules, 0.99 of the incident light is transmitted. If a change in membrane potential causes the dye to disappear, the fractional change in transmission is 1%, but in fluorescence it is 100%. In (B), nine times as much dye is bound to extraneous material. Now the transmitted intensity is reduced to 0.9, but the fractional change is still 1%. The fluorescence intensity is increased tenfold, and, therefore, the fractional change is reduced by the same factor. Thus, extraneously bound dye degrades fluorescence fractional changes and signal-to-noise ratios more rapidly. Redrawn from Cohen and Lesher (1986).

10 × as much dye is bound to the extraneous material as was bound to the cell membrane (Fig. 4B), the transmitted intensity is reduced to approximately 0.9. If a potential change again causes the cell-bound dye to disappear, the fractional change in transmission is nearly unaffected. In contrast, the resting fluorescence intensity is now higher by a factor of 10, so the fractional fluorescence change is reduced by the same factor. Thus, the fluorescence fractional change is more severely affected. It does not matter whether the extraneous material happens to be connective tissue, glial membrane, or neighboring neuronal membranes. In Fig. 4B, the fractional change in fluorescence was still larger than in transmission.

However, the light intensity in fluorescence was about 10^3 smaller, and this reduces the signal-to-noise ratio in fluorescence *(see below)*. Partly because of the signal degradation resulting from extraneous dye, fluorescence signals have been most often used in monitoring activity from tissue-cultured neurons, whereas absorption has been preferred in measurements from ganglia and brain slices. Recently, very large signals have been obtained in the absorption mode from *Aplysia* neurons in tissue culture (Parsons et al., 1989a,b). In ganglia and brain slices, the fractional changes in both transmission and fluorescence are small; they range between 10^{-4}–10^{-2} for a 100-mV potential change.

4. Dyes

The choice of dye is a very important factor in maximizing the signal-to-noise ratio in an optical measurement. In only a few instances, where there is a large density of synchronously active membrane, has it been possible to obtain large signals without testing a number of dyes. In addition, in some preparations, photodynamic damage resulting from illumination of the dye in the presence of oxygen may also affect the choice of dye. Pharmacological effects and dye bleaching are also considered in this section.

4.1. Screening

Using squid giant axons, more than 1000 dyes have been tested for signal size in response to changes in membrane potential. This screening was made possible by synthetic efforts of three laboratories. Alan Waggoner, Jeff Wang, and Ravender Gupta of Amherst College; Rina Hildesheim and Amiram Grinvald at the Weizmann Institute; and Joe Wuskell and Leslie Loew at the University of Connecticut Health Center have each synthesized a large number of dyes. Included in these syntheses were about 100 analogs of each of the four dyes illustrated in Fig. 1. In each of these four groups, there were 10–20 dyes that gave similarly large signals (within a factor of 2 of the dye illustrated) on squid axons.

However, dyes that gave nearly identical signals on squid axons gave very different responses on other preparations, and thus many dyes had to be tested to maximize the signal. Examples of preparations where a number of dyes had to be screened are the

Navanax and *Aplysia* buccal ganglia (London et al., 1987; Zecevic et al., 1989), rat cortex (Orbach et al., 1985), and tissue-cultured neurons (Ross and Reichardt, 1979; Grinvald et al., 1981a). Some of the dyes were unable to penetrate through connective tissue or along intercellular spaces to the membrane of interest. Others appeared to have a relatively low affinity for neuronal vs non-neuronal (connective) tissue. However, in some instances, the dye penetrated well and the staining appeared to be specific, but nonetheless the signals were small. Ross and Krauthamer (1984) have reported a case where supraesophageal ganglia from different species of the same genus *(Balanus)* had qualitatively different signals.

4.2. Pharmacologic Effects

In many preparations, high concentrations of dye had pharmacologic effects. However, in many instances, the dye concentration needed to obtain the maximum signal size was lower than the concentration at which pharmacologic effects were detected. These include the squid giant axon (Cohen et al., 1974; Gupta et al., 1981), neuroblastoma cells in tissue culture (Grinvald et al., 1982a), the barnacle supraesophageal ganglion (Salzberg et al., 1977; Grinvald et al., 1981b), embryonic semilunar ganglion (Sakai et al., 1985), and the *Navanax* buccal ganglion (London et al., 1987). In the *Navanax* experiments, the buccal ganglion was stained in a minimally dissected preparation, and feeding behavior was measured with and without staining. Here complex synaptic interactions are probably required to generate the correct behavior. Thus, in many instances, pharmacologic effects are not known to be a major difficulty. However, in the optical experiments on salamander olfactory bulb, frog optic tectum, and mammalian cortex, the ability to detect pharmacologic effects is limited. In these preparations, one can only say that pharmacologic effects were not disastrous.

4.3. Photodynamic Damage and Dye Bleaching

In certain experiments—for example, on neuroblastoma neurons using the styryl dye RH414 (Fig. 1)—photodynamic damage (resulting from the interaction of light, dye, and oxygen) limited the duration of the experiments (Grinvald et al., 1982a). In others—for example, on *Navanax* buccal ganglia using the oxonol dye RH155 (Fig. 1)—it was difficult to detect photodynamic damage (London et al., 1987). Similarly, dye bleaching has caused difficulties in some preparations, but not in others. Grinvald et al. (1982a) reported a 3% bleaching from 350 ms of illumination using a styryl dye, whereas we found bleaching difficult to detect after 10 min of illumination using the oxonol (London et al., 1987). This difference in severity is, in part, the result of the difference in dyes that were used; however, in addition, higher light intensities were used in the experiments where damage and bleaching were severe. Thus, advantages of increased intensities in terms of signal-to-noise ratio *(see below)* may be counterbalanced by increased damage and bleaching. Since both effects are dye-dependent (Cohen et al., 1974; Ross et al., 1977; Gupta et al., 1981), additional dye screening may be necessary in preparations where they cause difficulty. Although bleaching and photodynamic damage are sometimes correlated (Gupta et al., 1981), in other instances bleaching can occur without detectable damage (Ross and Krauthamer, 1984).

5. Measuring Technology

The limit of accuracy with which light can be measured is set by the shot noise that arises from the statistical nature of photon emission and detection. Fluctuations in the number of photons emitted per unit time will occur, and if an ideal light source (tungsten filament) emits an average of 10^{14} photons/ms, the root-mean-square (RMS) deviation in the number emitted is the square root of this number or 10^{7} photons/ms. In this shot noise limited (ideal) case, the signal-to-noise ratio is proportional to the square root of the number of measured photons and the square root of the bandwidth of the photodetection system (Braddick, 1960; Malmstadt et al., 1974). The dependence on the number of measured photons is illustrated in Fig. 5. In A, on the top, is plotted the result of using a random number table to distribute 20 photons into 20 time windows. In B, on the bottom, the same procedure was used to distribute 200 photons into the same 20 bins. One can see by inspection that, relative to the average light level, there is more noise in the top trace with 20 photons than in the bottom trace with 200 photons. On the right side of Fig. 5, the signal-to-noise ratios

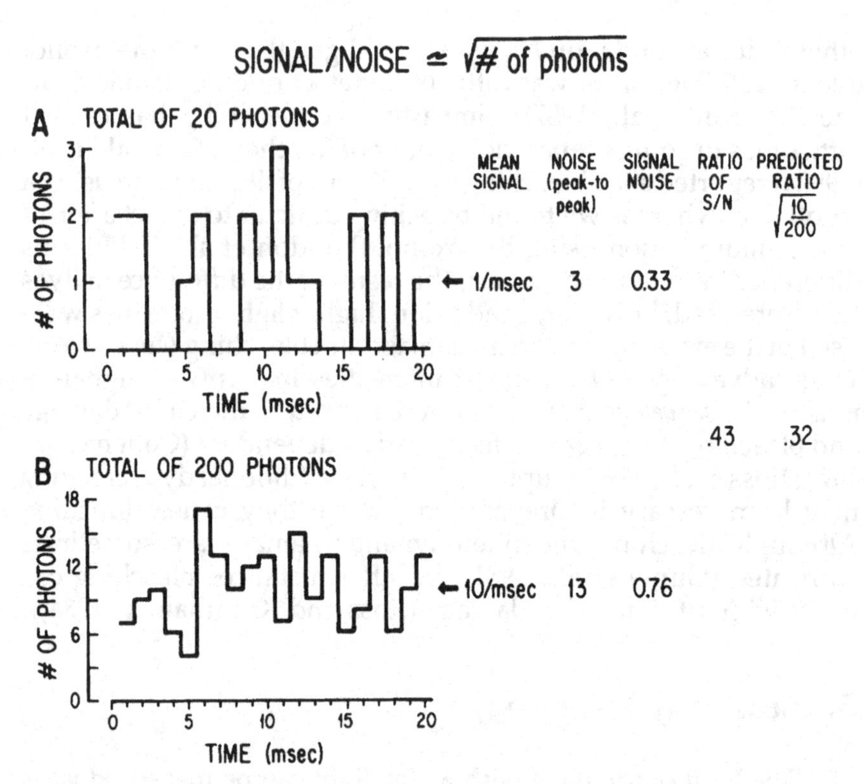

Fig. 5. Plots of the results of using a table of random numbers to distribute 20 photons (top, A) or 200 photons (bottom, B) into 20 time bins. The result illustrates the fact that, when more photons are measured, the signal-to-noise ratio is improved. On the right, the signal-to-noise ratio is measured for the two results. The ratio of the two signal-to-noise ratios was 0.43, which is close to the ratio predicted by the relationship that the signal-to-noise ratio is proportional to the square root of the measured intensity.

are measured and indeed the improvement in signal-to-noise ratio is similar to that expected from the square-root relationship.

This ideal result, of signal-to-noise ratio proportional to the square root of measured intensity, is indicated by the dotted line in Fig. 6. In a shot-noise limited measurement, improvements in the signal-to-noise ratio can only be obtained by increasing the illumination intensity, improving the light-gathering efficiency of the measuring system, or reducing the amplifier bandwidth. Only a small fraction of the 10^{14} photons/ms emitted by a 3300°F tungsten

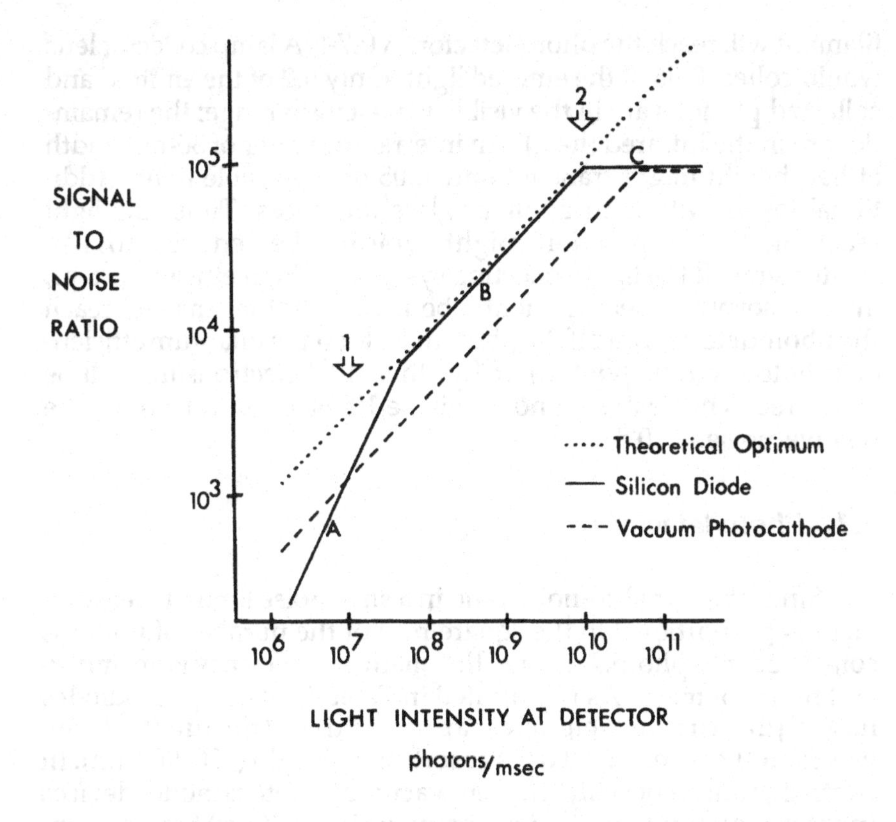

Fig. 6. Signal-to-noise ratio as a function of light intensity in photons per millisecond. The approximate light intensity per detector in fluorescence measurements from ganglia or vertebrate cortex using a tungsten filament bulb is indicated by arrow 1. The approximate intensity in absorption measurements in ganglia or brain slices is indicated by arrow 2. The theoretical optimum signal-to-noise ratio (dotted line) is the shot-noise limit. The signal-to-noise ratio expected with a silicon diode detector is indicated by the solid line. The silicon diode signal-to-noise ratio approaches the theoretical maximum at intermediate light intensities (segment B), but falls off at low intensities (segment A) because of dark noise and falls off at high intensities (segment C) because of extraneous noise. The expected signal-to-noise ratio for a vacuum photocathode detector is indicated by the dashed line. At low intensities, the vacuum photocathode is better than a silicon diode because it has less dark noise. At intermediate intensities, it is not as good because of its lower quantum efficiency. Redrawn from Cohen and Lesher (1986).

filament will reach the photodetector. A 0.7-NA lamp collector lens would collect 0.06 of the emitted light. Only 0.2 of the emitted and collected photons are in the visible wavelength range; the remainder are in the infrared (heat). An interference filter of 30-nm width at half-height might transmit only 0.05 of the visible light. Additional losses will occur at all air-glass interfaces. Thus, the light reaching the preparation might typically be reduced to 10^{10} photons/ms. If the light-collecting system has high efficiency, e.g., in an absorption measurement, about 10^{10} photons/ms will reach the photodetector, and if the photodetector has a quantum efficiency (photoelectrons/photon) of 1.0, then 10^{10} electrons/ms will be measured. The RMS shot noise will be 10^5 electrons/ms; thus, the relative noise is 10^{-5}.

5.1. Photodetectors

Since the signal-to-noise ratio in a shot-noise limited measurement is proportional to the square root of the number of photons converted into photoelectrons, the quantum efficiency is an important figure of merit. As is indicated in Table 1, silicon photodiodes have quantum efficiencies approaching the ideal at the wavelengths where most dyes absorb or emit light (500–900 nm). In contrast, only specially chosen vacuum photocathode devices (phototubes, photomultipliers, or image intensifiers) have a quantum efficiency as high as 0.15. Thus, in a shot-noise limited situation a silicon diode will have a signal-to-noise ratio that is more than 2.5 × larger. This advantage of silicon diode over vacuum photocathode is indicated in Fig. 6 by the fact that the diode curve (solid line) is higher than the vacuum photocathode curve (dashed line) over much of the intensity range (segment B).

There are three types of noise that can degrade the signal-to-noise ratio from the theoretical limit. The first is dark noise, the system's noise in the absence of light. The dark noise is generally far larger in a silicon diode system than in a vacuum photocathode system (Table 1). Thus, at low light levels ($<10^7$ photons/ms), a vacuum photocathode device will provide a larger signal-to-noise ratio. When the light level is reduced so that the shot noise is less than the dark noise of a silicon diode (about 10^8 photons/ms), the signal-to-noise ratio of the diode decreases linearly with light in-

Table 1
Detector Comparison

	Silicon diode	Vacuum photocathode
Quantum efficiency	0.9	0.15
Dark noise equivalent power	$\sim 10^7$ photons/s	$<10^5$ photons/s
1/f noise	Some diodes	No

tensity (segment A, Fig. 6). The crossover on signal-to-noise ratio between the silicon diode and vacuum photocathode device occurs at about 10^7 photons/ms (arrow 1, Fig. 6). This crossover occurs near the intensities obtained in fluorescence measurements from ganglia and intact cortex, and thus, the optimal choice of photodetector at these intensities requires a careful comparison.

Some silicon diodes may have an additional light-dependent noise. David Kleinfeld (personal communication) found this second type of noise, called excess noise or 1/f noise, in PIN 6D and PIN 10D diodes (United Detector Technology) and in a Hewlett-Packard 5082-4203 diode. One indication of the presence of excess noise was that the noise current was directly proportional to the photocurrent rather than having the square root proportionality of shot noise. We measured the relationship between noise and intensity in three diodes—a United Detector Technology PIN 5D, EG&G PV 444, and one element of a 12 × 12 Centronic array (Fig. 7). For both the PV 444 and the Centronic diode, the noise was proportional to intensity to the 0.55 power at the bandwidth we tested (10–100 Hz) (W. N. Ross, J. A. London, D. Zecevic, and L. B. Cohen, unpublished results), close to the expected relationship for shot noise. However, with the PIN 5D, the noise was proportional to intensity to the 0.65 power. This increased deviation from 0.5 suggests the presence of 1/f noise in this kind of diode, in agreement with the results obtained by Kleinfeld. Optical measurements on vertebrate preparations are made using frequencies lower than those we tested. Since measurements at low frequencies are more likely to suffer from interference from 1/f noise, additional testing would be informative.

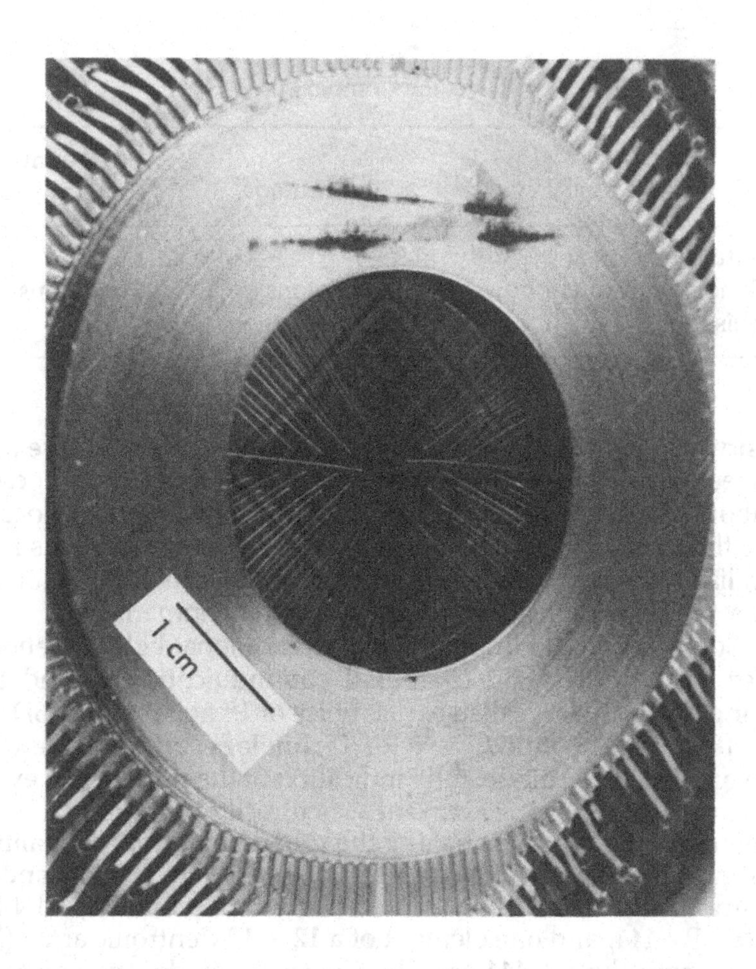

Fig. 7. A 10 × 10 array of silicon photodiodes manufactured by Centronic (New Addington, Croydon, England). The dark squares (1.4 × 1.4 mm) are the photodiodes. The insulating regions between the diodes are 0.1 mm thick. The output of each diode is carried from the array to a current to voltage converter (Fig. 9) via a wire. At the present time, a number of laboratories are using a 12 × 12 array that is identical to the one shown except for the additional 44 detectors.

5.2. Extraneous Noise

A third type of noise, termed extraneous noise, is more apparent at higher light intensities where the sensitivity of the measurement is high because the fractional shot noise is low. One type of extraneous noise, resulting from fluctuations in the output of the light source, will be discussed in the next section. Two other sources of extraneous noise are vibrations in the light path and movement of the preparation. A number of precautions for reducing vibrational noise are described in Salzberg et al. (1977). Embedding ganglia in 1–3% agar reduced vibrational noise (London et al., 1987). We recently found that the pneumatic isolation mounts on two vibration isolation tables that we used were providing only minimal isolation in the frequency range 20–60 Hz. By replacing the pneumatic mounts with air-filled soft rubber tubes, we have further reduced the vibration noise. (D. Zecevic, J. A. London, and L. B. Cohen, unpublished results). Nevertheless, it has been difficult to reduce vibrational noise to less than 10^{-5} of the total light. With this amount of vibrational noise, increases in measured intensity beyond 10^{10} photons/ms would not improve the signal-to-noise ratio (segment C of Fig. 6).

Noise resulting from movement is a major problem in measurement on in vivo preparations. Methods for reducing the movements or the resulting artifacts in molluscan and mammalian experiments have been described (London et al., 1987; Orbach et al., 1985).

5.3. Light Sources

Three kinds of sources have been used. Tungsten filament lamps are a stable source, but their intensity is relatively low, particularly at wavelengths less than 550 nm. Arc lamps and lasers are less stable, but can provide more intense illumination.

5.3.1. Tungsten Filament Lamps

It is not difficult to provide a power supply that is stable enough for the output of the bulb to fluctuate by less than 1 part in 10^5. In absorption measurements, where the fractional changes in intensity are relatively small, only tungsten filament sources have been used. On the other hand, fluorescence measurements often have larger fractional changes that will better tolerate light sources with systematic noise, and the measured intensities are low, which makes possible improvements in signal-to-noise ratio from more intense sources attractive. Hence, arc lamps or laser sources have sometimes been used in fluorescence measurements.

5.3.2. Arc Lamps

Both mercury and mercury-xenon arc lamps have been used to obtain higher intensities. However, comparison of the excitation intensity obtainable from tungsten filament and arc sources is not simple. Grinvald et al. (1982a) reported that the intensity from a mercury arc lamp was 50–100 × higher than that from a tungsten filament lamp using a 540-nm filter with a width at half-height of 18 nm. However, the advantage implied by such a comparison may be misleading. Because the excitation "action" spectrum of some dyes (i.e., the styryl, RH414) is quite broad, it is preferable to use a filter with a width at half-height of 90 nm, which will substantially increase the incident intensity from a tungsten filament source. Because the mercury lamp has a distinct emission line at 546 nm, using a wider filter adds little intensity with this source. Furthermore, the tungsten filament bulb can be overrun to increase its color temperature, increasing its output intensity by about 75%. Finally, the intensity will depend on the area of the object that is illuminated. Using critical illumination, the arc lamp, which approximates a point source, will be relatively preferred for smaller objects. Thus, the increase in intensity obtained by using an arc lamp in lieu of tungsten will often not be as great as a factor of 50–100 (P. Saggau, L. B. Cohen, and A. Grinvald, unpublished results).

The main difficulty with arc lamps is output intensity fluctuations (in part resulting from arc wander). Using a high-speed constant current power supply (KEPCO JQE 75-15[M] HS), the peak output fluctuations were 4×10^{-4} of the total intensity over the bandwidth of 0.5 Hz–1 kHz (Davila et al., 1974). J. Pine (personal communication) has utilized a feedback circuit onto the lamp power supply to further reduce output fluctuations. With such a circuit, the output noise was 10^{-4} over the bandwidth of 10 Hz–10 KHz. At present, it is difficult to make an *a priori* decision between a tungsten filament and an arc source for fluorescence measurements. The decision will depend on the fractional change, the predominant source of noise, the level of difficulty associated with dye bleaching or photodynamic damage, and possible improvements in quieting an arc lamp.

5.3.3. Lasers

It has been possible to take advantage of two useful characteristics of laser sources. First, the laser output can be focused onto a small spot in preparations with no scattering, allowing measurement of membrane potential from small processes in tissue-cultured neurons (Grinvald and Farber, 1981). Second, the laser beam can be positioned flexibly and rapidly using acoustooptical deflectors (Dillon and Morad, 1981; Hill and Courtney, 1985). However, lasers have thus far only been used in situations where the fractional change in intensity was large. There have been two difficulties: first, the amplitude of the laser light output is not stable, and second, there appears to be excess noise that may be the result of laser speckle (Dainty, 1984). Commercially available (Uniphase) helium-neon lasers with modified power supplies have intensity fluctuations of less than 2×10^{-5} of the total intensity (B. M. Salzberg, personal communication). Argon or krypton lasers have intensity fluctuations of 5×10^{-3} of the total intensity. However, the noise at the photodetector can be surprisingly large when the noise in the laser source appears to be small. In fluorescence measurements on invertebrate ganglia, we found that the fractional noise on each detector was substantially larger than the fractional noise present in the incident light (B. M. Salzberg, D. Senseman, L. B. Cohen, and A. Grinvald, unpublished results). This excess noise may be the result of laser speckle and interference from reflected light in the fluorescence measurement. It has been possible to partially eliminate this noise by introducing high-frequency mode scrambling into the laser beam (G. Ellis and B. M. Salzberg, personal communication).

Amplitude fluctuations in currently available solid-state lasers (Liconix, Sunnyvale, CA) are apparently very small, less than 10^{-4} of the total intensity (B. M. Salzberg, personal communication). They have not yet been tested in optical measurements.

5.4. Image-Recording Devices

The major motivation for developing optical methods for monitoring membrane potential was the possibility of making simultaneous multisite measurements of activity. A large number of factors must be considered in choosing an imaging system. Perhaps the most important considerations are the requirements for spatial and temporal resolution. Since, in a shot-noise limited measurement, the signal-to-noise ratio that can be achieved is proportional to the number of measured photons, increases in temporal resolution will reduce that number and reduce the signal-to-noise ratio; also, increases in spatial resolution will reduce the number of photons per pixel and degrade the signal-to-noise ratio. For example, increasing resolution by replacing a 10 × 10 array with a 100 × 100 array will reduce the photons per pixel by a factor of 100 and, thus, reduce the signal-to-noise ratio by a factor of 10. Several types of imaging devices will be considered.

5.4.1. Film

One type of imager that has outstanding spatial and temporal resolution is movie film. However, because it is difficult to obtain quantum efficiencies of even 1% with film (Shaw, 1979), there would be an automatic factor of 10 degradation in signal-to-noise ratio in comparison with a silicon diode. This and other difficulties, including frame-to-frame and within-frame emulsion nonuniformity, have discouraged attempts to use film.

5.4.2. Silicon Diode Arrays

Arrays of silicon diodes are attractive, because they have nearly ideal quantum efficiencies (close to 1.0). An array that has been used in several laboratories is the silicon diode array illustrated in Fig. 7. The dark squares (1.4 × 1.4 mm) in this 10 × 10 array (Centronic, Ltd.) are the individual detectors (pixels). The 10 × 10 and 12 × 12 arrays (with parallel readout, *see below*) have been used to monitor activity in invertebrate ganglia (Grinvald et al., 1981b), tissue cultured neurons (Grinvald et al., 1981a), brain slices (Grinvald et al., 1982b), nerve terminals in the pituitary (Salzberg et al., 1983), intact vertebrate CNS (Orbach and Cohen, 1983), and in vivo mammalian cortex (Orbach et al., 1985). A new array, with 464 pixels, is now available from Centronic (David Senseman, personal communication). D. Y. Ts'o, R. D. Frostig, E. E. Lieke, and A. Grinvald (personal communication) have recently used a charged coupled device (CCD) with 128,000 pixels to measure slow intrinsic cortical signals.

5.4.3. Vacuum Photocathode Cameras

Although the lower quantum efficiencies of vacuum photocathodes is a disadvantage, these devices are widely available and have high spatial resolution. With this high resolution, the number of photons per detector would be low and in the intensity region where the low dark noise of vacuum-photocathode devices would be an advantage (Fig. 6). One of these, a Radechon (Kazan and Knoll, 1968), has an output proportional to the changes in the input. Cameras have been used in recordings from mammalian cortex (Blasdel and Salama, 1986) and salamander olfactory bulb (Kauer, 1988).

In all of the systems now in use, the image recorder has been placed in the objective image plane of a microscope. However, Tank and Ahmed (1985) suggested a scheme by which a hexagonal close-packed array of optical fibers is positioned in the image plane and individual photodiodes are connected to the other end of the optical fibers.

5.4.4. Parallel Vs Serial Recording

The output of each detector in the array shown in Fig. 7 is followed by its own amplifier. With this parallel recording, the high-frequency cutoff of each amplifier is determined by the highest frequencies that are present in the signal of interest. If, for example, molluscan action potentials have frequency components up to 200 Hz, and one wants to know the time course accurately, then the amplifiers should have high-frequency cutoffs of 200 Hz, and the output of each amplifier should be recorded every millisecond.

An alternative, serial readout mechanism, which would save the construction and maintenance cost of the array of amplifiers, would be to follow the diode array with a multiplexer and then to use a single amplifier. However, this single amplifier must operate on 100 signals every millisecond, and thus, its high-frequency cutoff would have to be in the 100 kHz range—many times higher than in the case of a parallel readout. Thus, a serial recording will have a relatively large dark noise.

A second advantage of a parallel readout scheme is that lower recording accuracy is required. In a serial readout, determining the changes in intensity, which are a very small fraction of the total intensity, can be done only by subtracting two relatively large numbers: the total intensity at time t_2 minus the total intensity at an earlier time, t_1. To measure the total intensity with the accuracy of one part in 10^5 that would be necessary for some absorption signals, an analog-to-digital (A-D) conversion accurate to 17 bits would be needed. This kind of accuracy is not inexpensively achieved at the required data rates. However, in a parallel readout, either capacity coupling or DC subtraction *(see below)* can be used, so that the analog-to-digital converter needs to have only enough resolution to measure the change in intensity. In this situation, 8-bit resolution can be adequate.

5.5. Optics

The need to maximize the number of measured photons has also dominated the choice of optical components. The number of photons that are collected by an objective lens in forming the image is proportional to the square of the numerical aperture (NA) of the lens. In epifluorescence, the excitation light and the emitted light pass through the objective, and therefore the intensity reaching the photodiodes is proportional to the fourth power of its numerical aperture. Accordingly, objectives (and condensers) with high numerical apertures have been employed.

Direct comparison of the intensity reaching the image plane indicated that objectives may vary in their efficiency. With a Leitz 25X 0.4 NA lens, we obtained 100% more intensity than with a Nikon 20X 0.4 NA lens after correction for differences in magnification. Two Olympus lenses appeared to be intermediate in efficiency (J.-Y. Wu, H.-P. Höpp, C. X. Falk, and L. B. Cohen, unpublished results).

5.5.1. Depth of Focus

Salzberg et al. (1977) determined the effective depth of focus for a 0.4-NA objective lens on an ordinary microscope by recording an optical signal from a neuron when it was in focus and then moving the neuron out of focus by various distances. They found that the neuron had to be moved 300 µm out of focus to cause a 50% reduction in signal size. (This result will be obtained only when the diameter of the neuron image and the diameter of the detector are similar.) Using 0.6-NA optics, the effective depth of focus was reduced by about 50% (A. Grinvald, unpublished results). This large, effective depth of focus can be advantageous in some circumstances. If, for instance, one would like to record from all the neurons in a 500-µm-thick invertebrate ganglion, then, using 0.4-NA optics one could focus at the middle of the ganglion and the signals from neurons on the top and bottom of the ganglion would be reduced in size by less than 50%. (This situation will, of course, result in the superposition of signals from two or more neurons on some detectors; subsequent sorting out will be required.) On the other hand, the large effective depth of field makes it difficult to determine the depth in a preparation at which a signal originates.

5.5.2. Effects of Light Scattering

Light scattering can limit the spatial resolution of an optical measurement. London et al. (1987) measured the scattering of 705 nm light in *Navanax* buccal ganglia. They found that insertion of a ganglion in the light path caused light from a 30-μm spot to spread, such that the diameter of the circle of light that included intensities greater than 50% of the total was now about 50 μm. The spread was greater, to about 100 μm, with light of 510 nm. Since the blurring is not large compared to the average cell diameter at the wavelengths we used (705 nm), it does not lead to a large overestimate of cell size, but it does degrade the signal-to-noise ratio. Figure 8 illustrates the results of similar experiments that were carried out on the salamander olfactory bulb (Orbach and Cohen, 1983). The top section indicates that, when no tissue is present, essentially all of the light (750 nm) from the small spot falls on one detector. The bottom section illustrates the result when a 500-μm-thick slice of olfactory bulb is present. The light from the small spot is spread to about 200 μm. Certainly, mammalian cortex does not appear to scatter less than the olfactory bulb. Thus, light scattering will cause considerable blurring of signals in intact vertebrate preparations.

A second possible source of blurring is signal from regions that are out of focus. For example, if the active region is a cylinder (a column) that is perpendicular to the plane of focus, and the objective is focused at the middle of the cylinder, then the light from the middle will have the correct diameter at the image plane, but the light from the regions above and below are out of focus and will have a diameter that is too large. The middle section of Fig. 8 illustrates the effect of moving the small spot of light 500 μm out of focus. The light from the small spot is spread to about 200 μm. Thus, in preparations with considerable scattering or with out-of-focus signals the actual spatial resolution that can be achieved may be limited by the preparation and not by the number of pixels in the imaging device.

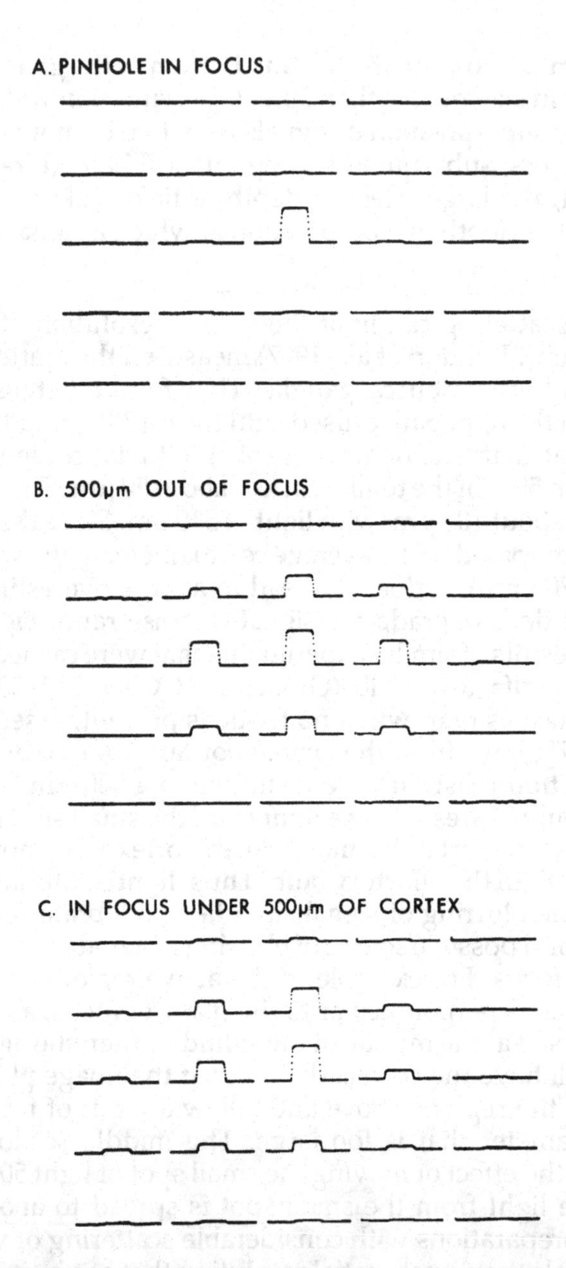

Fig. 8. Effects of focus and scattering on the distribution of light from a point source onto the array. (A) A 40-µm pinhole in aluminum foil covered with saline was illuminated with light at 750 nm. The pinhole was

5.5.3. Confocal Microscope

Petran and Hadravsky (1966) patented a modification of the microscope that substantially reduces both the scattered and out-of-focus light that contributes to the image. Egger and Petran (1967) showed that it provided much clearer images of dorsal root ganglia neurons. The Petran invention is based on the following idea. A small spot of the object is illuminated. If a pinhole is positioned at the corresponding spot in the image plane and only light passing through the pinhole is used to form the image, then most of the scattered light and the light from out-of-focus regions will be rejected. As is illustrated in Fig. 8, both the scattered and out-of-focus light reach a relatively large area of the image plane, and most of this light is rejected in the confocal microscope. To form a complete image, the spot and pinhole are moved in parallel to each pixel in the field of view. The implementation of this idea by Petran and Hadravsky involved symmetrical incident light and imaging pathways with a rotating disk of very carefully positioned pinholes. The pinholes in the image path corresponded exactly to the pinholes in the incident path.

One of the difficulties of Petran's design was that only 10^{-7} of the illuminating light is available for forming the image (Boyde et al., 1983), and thus, very bright sources were required; Egger and Petran (1967) used sunlight. However, if a scanning laser spot (Davidovitz and Egger, 1969) were used to illuminate the preparation, then the light loss in the illuminating path would be eliminated. Thus, by using a confocal microscope with a laser light source, one might be able to obtain signals from intact vertebrate pre-

---

in focus. More than 90% of the light fell on one detector. (B) The stage was moved downward by 500 μm. Light from the out-of-focus pinhole was now seen on several detectors. (C) The pinhole was in focus, but covered by a 500-μm slice of salamander cortex. Again the light from the pinhole was spread over several detectors. A 10 × 0.4 NA objective was used. Kohler illumination was used before the pinhole was placed in the object plane. The recording gains were adjusted so that the largest signal in each of the three trials would be approximately the same size in the figure. Redrawn from Orbach and Cohen (1983).

parations with much better spatial resolution than was achieved with ordinary microscopy. P. Saggau (personal communications) is attempting to construct a laser-based confocal microscope with frame rates in the millisecond range.

5.6. Amplifiers

In our apparatus, amplifier noise does not make a substantial contribution to either the dark noise or the total noise. The dark noise usually did not decrease when a National Semiconductor, LF 356N, (Fig. 9) was replaced with a Burr-Brown OPA 111AM, an amplifier with substantially lower noise specifications (suggested by J. Meyer). However, certain relatively noisy LF356N amplifiers did add to the dark noise, and thus, this amplifier appears to define the limit at which amplifier noise will add to the dark noise.

There are two difficulties with the use of AC-coupled amplifiers. First, the AC coupling will restrict the ability to measure slow signals (low frequencies), and second, the optical recording cannot begin the instant the light is turned on, because one must wait about 10 time constants for the input to the second amplifier to settle. This can be disadvantageous if there is significant photodynamic damage and the recording period is short compared to the settling time. A useful feature of the AC recording was that a second amplifier could have a high gain, so that a relatively large signal could be fed to the multiplexer and A-D converter. The advantage of using a high-gain, second-stage amplifier can also be obtained along with a DC recording, if the DC light level is measured when the light is first turned on and then this value is subtracted from all subsequent measurements. This can be implemented either with sample-and-hold amplifiers (Nakajima and Gilai, 1980) or digitally (Senseman and Salzberg, 1980).

5.7. Computer

The discussion in this section is limited to the hardware and software currently in use in our laboratory. The main concern arises from the large amounts of data that are generated by optical recordings. Recording from a 464-element array at 1 kHz generates about 1 Mbyte of datas. A discussion of the previously used PDP-11 system can be found in Cohen and Lesher (1986). The PDP-11's limited directly addressable memory space made pro-

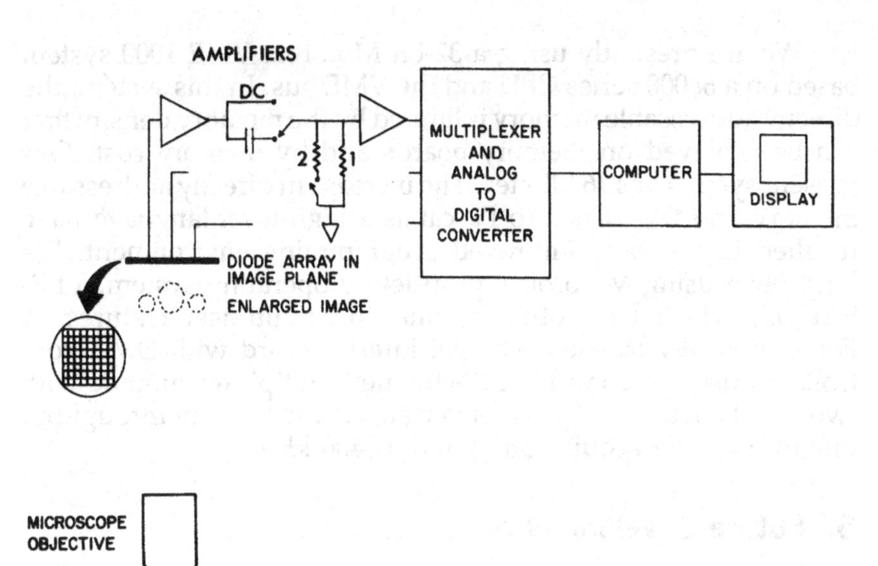

Fig. 9. Schematic drawing of the amplifiers and computer hardware used to make a parallel recording from a 12 × 12 photodiode array. This is the system currently used in our laboratory. The output of each diode, used in a photovoltaic mode, is converted into a voltage signal by the first LF356 amplifier. There are two fixed high-frequency cutoff RC filters at this stage (not shown). The output of the first amplifier is fed to the second via one of three switch-selectable pathways. The DC coupling mode allows the resting light intensity to be recorded. A short time-constant AC coupling (both resistors in) is used to allow quick settling of the input to the second amplifier just after the light is turned on. The long time-constant AC coupling (only one resistor) is used during the signal measurement. The second amplifier is used to provide the gain (250–5000x). Amplifiers of this type can be purchased from Vic Pantani, Department of Physiology, Yale University School of Medicine, New Haven, CT.

gram development difficult and meant that considerable effort had to be expended to achieve recording times of tens of seconds. To obtain recording times of up to 20 min, Hirota et al. (1985) had to interpose a set of tape recorders between the analog-to-digital converters and the computer.

We are presently using a 32-bit Motorola VME 1000 system based on a 68000 series CPU and the VME bus. In this system, the directly addressable memory is limited by the memory density that can be achieved on memory boards and by memory cost. Our present system has 16 Mbytes. The increase in directly addressable memory and the switch to Pascal as a high-level language have resulted in a greatly improved programming environment. We have been using Motorola's proprietary operating system, VERSADOS, which has both real-time and multi-user features. A Force, optically isolated, parallel interface card with DMA controller is used to control a 512-channel multiplexer module with two 16-bit analog-to-digital converters. The maximum throughput rate of this data acquisition system is 400 kHz.

6. Future Developments

It is clear that improvements in the signal-to-noise ratios of optical measurements of membrane potential would be useful. A number of avenues remain partially or completely unexplored.

Only three optical properties of stained membranes have been examined for signals in response to changes in membrane potential. The possibilities of finding large changes in other optical properties—for example, energy transfer, circular dichroism (optical rotation), or absorption-enhanced Raman scattering—have been largely neglected. Ehrenberg and Berezin (1984) have used resonance Raman to study surface potential, and Ehrenberg and Loew (personal communication) are planning to investigate its use in measuring transmembrane potential. There was a report of holographic signals in leech neurons (Sharnoff et al., 1978a), but signals were not found in subsequent experiments on squid axons (Sharnoff et al., 1978b). There were also reports of changes in intrinsic infrared absorption (Sherebrin, 1972; Sherebrin et al., 1972), but these have not been pursued further. Thus, one approach to looking for larger signals would be to investigate new types of optical phenomena.

A second approach involves improvement of the apparatus. One useful improvement would be further quieting of arc and laser light sources and investigation of new kinds of light sources. The successful implementation of a confocal microscope for preparations with substantial scattering or thickness would greatly improve spatial resolution.

The third appraoch is finding or designing better dyes. All of the dyes in Fig. 1 and the vast majority of those synthesized in recent years are of the general class named cyanines (Hamer, 1964), a class that is used to extend the wavelength response of photographic film. It is certainly possible that improvements in signal size can be obtained with new cyanine dyes (*see* Waggoner and Grinvald [1977] for a discussion of maximum possible fractional changes in absorption and fluorescence). On the other hand, the fractional change on squid axons has not increased much in recent years (Gupta et al., 1981; L. B. Cohen, A. Grinvald, K. Kamino, and B. M. Salzberg, unpublished results), and most improvements (Grinvald et al., 1982a) have involved synthesizing analogs that work well on new preparations. Radically different synthetic approaches or a modicum of cooperation from corporations like Eastman Kodak might prove to be very useful.

The measurements that have been made with optical methods provide new and previously unobtainable information about cell and organ function. Clearly, there has been dramatic progress since the first recording of an action potential in a leech neuron (Salzberg et al., 1973). We hope that additional improvements will further increase the utility of these methods.

Acknowledgments

The authors are indebted to their collaborators Vicencio Davila, Amiram Grinvald, Kohtaro Kamino, Bill Ross, Brian Salzberg, and Alan Waggoner for numerous discussions about optical methods. We also thank Pancho Bezanilla, David Kleinfeld, Ana Lia Obaid, Jerome Pine, Guy Salama, Peter Saggau, and David Senseman, who have allowed us to cite unpublished results. The experiments carried out in our laboratory were supported by NIH grant NS-08437.

References

Blasdel G. G. and Salama G. (1986) Voltage-sensitive dyes reveal a modular organization in monkey striate cortex. *Nature* **321,** 579–585.

Boyde A., Petran, M. and Hadravsky, M. (1983) Tandem scanning reflected light microscopy of internal features in whole bone and tooth samples. *J. Microsc.* **132,** 1–7.

Boyle M. B. and Cohen L. B. (1980) Birefringence signals that monitor membrane potential in cell bodies of molluscan neurons. *Fed. Proc.* **39,** 2130.

Braddick H. J. J. (1960) Photoelectric photometry. *Rep. Prog. Physics* **23,** 154–175.

Cohen L. B. and Lesher S. (1986) Optical monitoring of membrane potential: methods of multisite optical measurement. *Soc. Gen. Physiol. Ser.* **40,** 71–99.

Cohen L. B. and Salzberg B. M. (1978) Optical measurement of membrane potential. *Rev. Physiol. Biochem. Pharmacol.* **83,** 35–88.

Cohen L. B., Davila H. V. and Waggoner A. S. (1971) Changes in axon fluorescence. *Biol. Bull.* **141,** 382.

Cohen L. B., Keynes R. D. and Hille B. (1968) Light scattering and birefringence changes during nerve activity. *Nature* **218,** 438–441.

Cohen L. B., Landowne D., Shrivastav B. B. and Ritchie J. M. (1970) Changes in fluorescence of squid axons during activity. *Biol. Bull.* **139,** 418–419.

Cohen L. B., Salzberg B. M., Davila H. V., Ross W. N., Landowne D., Waggoner A. S. and Wang C. H. (1974) Changes in axon fluorescence during activity: molecular probes of membrane potential. *J. Membr. Biol.* **19,** 1–36.

Conti F. (1975) Fluorescent probes in nerve membranes. *Ann. Rev. Biophys. Bioeng.* **4,** 287–310.

Conti F. and Tasaki I. (1970) Changes in extrinsic fluorescence in squid axons during voltage clamp. *Science* **169,** 1322–1324.

Conti F., Tasaki I. and Wanke E. (1971) Fluorescence signals in ANS-stained squid axons during voltage clamp. *Biophys. J.* **8,** 58–70.

Dainty J. C. (1984) *Laser Speckle and Related Phenomena* (Springer-Verlag, New York).

Dasheiff R. M. (1988) Fluorescent voltage sensitive dyes: Applications for neurophysiology. *J. Clin. Neurophysiol.* **5,** 211–235.

Davidovitz, P. and Egger M. D. (1969) Scanning laser microscope. *Nature* **223,** 831.

Davila H. V., Cohen L. B., Salzberg, B. M. and Shrivastav B. B. (1974) Changes in ANS and TNS fluorescence in giant axons from *Loligo. J. Membr. Biol.* **15,** 29–46.

Dillon S. and Morad M. (1981) Scanning of the electrical activity of the heart using a laser beam with acousto-optics modulators. *Science* **214,** 453–456.

Egger M. D. and Petran M. (1967) New reflected light microscope for viewing unstained brain and ganglion cells. *Science* **157,** 305–307.

Ehrenberg B. and Berezin Y. (1984) Surface potential on purple membranes and its sidedness studied by resonance Raman dye probe. *Biophys. J.* **45,** 663–670.

Grinvald A. and Farber I. C. (1981) Optical recording of calcium action potentials from growth cones of cultured neurons with a laser microbeam. *Science* **212**, 1164–1167.

Grinvald A., Ross W. N. and Farber I. (1981a) Simultaneous optical measurements of electrical activity from multiple sites on processes of cultured neurons. *Proc. Natl. Acad. Sci. USA* **78,** 3245–3249.

Grinvald A., Cohen L. B., Lesher S. and Boyle M. B. (1981b) Simultaneous optical monitoring of activity of many neurons in invertebrate ganglia using a 124-element photodiode array. *J. Neurophysiol.* **45,** 829–840.

Grinvald A., Hildesheim R., Farber I. C. and Anglister L. (1982a) Improved fluorescent probes for the measurement of rapid changes in membrane potential. *Biophys. J.* **39,** 301–308.

Grinvald A., Manker A. and Segal M. (1982b). Visualization of the spread of electrical activity in rat hippocampal slices by voltage-sensitive optical probes. *J. Physiol. (Lond.)* **333,** 269–291.

Grinvald A., Frostig R. D., Lieke E. and Hildesheim R. (1988) Optical imaging of neuronal activity. *Physiol. Rev.* **68,** 1285–1366.

Grinvald A., Lieke E., Frostig R. D., Gilbert C. D. and Wiesel T. N. (1986) Functional architecture of cortex revealed by optical imaging of intrinsic signals. *Nature* **324,** 361–364.

Gupta R. K., Salzberg B. M., Grinvald A., Cohen L. B., Kamino K., Lesher S., Boyle M. B., Waggoner A. S. and Wang C. H. (1981) Improvements in optical methods for measuring rapid changes in membrane potential. *J. Membr. Biol.* **58,** 123–137.

Hamer F. M. (1964) *The Cyanine Dyes and Related Compounds* (Wiley, New York).

Hill B. C. and Courtney K. R. (1985) Optical monitoring of myocardial conduction: Observations of reentry and examples of lidocaine action. *Biophys. J.* **47,** 496a.

Hirota A., Kamino K., Komuro H., Sakai T. and Yada T. (1985) Optical studies of excitation-contraction coupling in the early embryonic chick heart. *J. Physiol. (Lond.)* **366,** 89–106.

Kauer J. S. (1988) Real-time imaging of evoked activity in local circuits of the salamander olfactory bulb. *Nature* **331,** 166–168.

Kauer J. S., Senseman D. M. and Cohen L. B. (1987) Odor elicited activity monitored simultaneously from 124 regions of the salamander olfactory bulb using a voltage sensitive dye. *Brain Res.* **418,** 255–261.

Kazan B. and Knoll M. (1968). *Electronic Image Storage* (Academic, New York).

Lev-Ram V. and Grinvald A. (1986) Ca^{2+} and K^{+}-dependent communication between central nervous system myelinated axons and oligodendrocytes revealed by voltage-sensitive dyes. *Proc. Natl. Acad. Sci. USA* **83**, 6651–6655.

Loew L. M., Cohen L. B., Salzberg B. M., Obaid A. L. and Bezanilla F. (1985) Charge-shift probes of membrane potential. Characterization of aminostyrylpyridinium dyes on the squid giant axon. *Biophys. J.* **47**, 71–77.

London J. A., Zecevic D. and Cohen L. B. (1987) Simultaneous optical recording of activity from many neurons during feeding in *Navanax*. *J. Neurosci.* **7**, 649–661.

Malmstadt H. V., Enke C. G., Crouch S. R. and Harlick G. (1974) *Electronic Measurements for Scientists* (Benjamin, Menlo Park, CA).

Nakajima S. and Gilai A. (1980) Action potentials of isolated single muscle fibers recorded by potential sensitive dyes. *J. Gen. Physiol.* **76**, 729–750.

Orbach H. S. and Cohen L. B. (1983) Optical monitoring of activity from many areas of the in vitro and in vivo salamander olfactory bulb: a new method for studying functional organization in the vertebrate central nervous system. *J. Neurosci.* **3**, 2251–2262.

Orbach H. S., Cohen L. B. and Grinvald A. (1985) Optical mapping of electrical activity in rat somatosensory and visual cortex. *J. Neurosci.* **5**, 1886–1895.

Parsons T. D., Kleinfeld D., Raccuia-Behling G. F. and Salzberg B. M. (1989a). Recording of the electrical activity of synaptically interacting *Aplysia* neurons in culture using potentiometric probes. *Biophys. J.*, in press.

Parsons T. D., Obaid A. L., Kleinfeld D. and Salzberg B. M. (1989b). Continuous long-term optical recording from identified *Aplysia* neurons in culture. *Biophys. J.* **55**, 174a.

Patrick J., Valeur B., Monnerie L. and Changeux J.-.P. (1971) Changes in intrinsic fluorescence intensity of the electroplax membrane during electrical excitation. *J. Membr. Biol.* **5**, 102–120.

Petran M. and Hadravsky M. (1966) Czechoslovakian patent 7720.

Ross W. N. and Krauthamer V. (1984) Optical measurements of potential changes in axons and processes of neurons of a barnacle ganglion. *J. Neurosci.* **4**, 659–672.

Ross W. N. and Reichardt L. F. (1979) Species-specific effects on the optical signals of voltage-sensitive dyes. *J. Membr. Biol.* **48**, 343–356.

Ross W. N., Salzberg B. M., Cohen L. B. and Davila H. V. (1974) A large change in dye absorption during the action potential. *Biophys. J.* **14,** 983–986.

Ross W. N., Salzberg B. M., Cohen L. B., Grinvald A., Davila H. V., Waggoner A. S. and Wang C. H. (1977) Changes in absorption, fluorescence, dichroism, and birefringence in stained giant axons: optical measurement of membrane potential. *J. Membr. Biol.* **33,** 141–183.

Russell J. T., Beeler T. and Martonosi A. (1979) Optical probe responses on sarcoplasmic reticulum. Merocyanine and oxonol dyes. *J. Biol. Chem.* **254,** 2047–2052.

Sakai T., Hirota A., Komuro H., Fujii S. and Kamino K. (1985) Optical recording of membrane potential responses from early embryonic chick ganglia using voltage-sensitive dyes. *Brain Res.* **349,** 39–51.

Salzberg B. M. (1983) Optical recording of electrical activity in neurons using molecular probes, in *Current Methods in Cellular Neurobiology* (Barker J. L. and McKelvy J. F., eds.), Wiley, New York, pp. 139–187.

Salzberg B. M. and Bezanilla F. (1983) An optical determination of the series resistance in *Loligo*. *J. Gen. Physiol.* **82,** 807–817.

Salzberg B. M., Davila H. V. and Cohen L. B. (1973). Optical recording of impulses in individual neurones of an invertebrate central nervous system. *Nature* **246,** 508–509.

Salzberg B. M., Obaid A. L., Senseman D. M. and Gainer H. (1983) Optical recording of action potentials from vertebrate nerve terminals using potentiometric probes provides evidence for sodium and calcium components. *Nature* **306,** 36–40.

Salzberg B. M., Grinvald A., Cohen L. B., Davila H. V. and Ross W. N. (1977) Optical recording of neuronal activity in an invertebrate central nervous system: simultaneous monitoring of several neurons. *J. Neurophysiol.* **40,** 1281–1291.

Senseman D. M. and Salzberg B. M. (1980) Electrical activity in an exocrine gland: optical recording with a potentiometric dye. *Science* **208,** 1269–1271.

Sharnoff M., Henry R. W. and Belleza D. M. J. (1978a) Holographic visualization of the nerve impulse. *Biophys. J.* **21,** 109A.

Sharnoff M., Romer, N. J., Cohen L. B., Salzberg B. M., Boyle M. B. and Lesher S. (1978b) Differential holography of squid giant axons during excitation and rest (abstract). *Biol. Bull.* **155,** 465–466.

Shaw R (1979) Photographic detectors. *Appl. Optics Optical Eng.* **7,** 121–154.

Sherebrin M. H. (1972) Changes in infrared spectrum of nerve during excitation. *Nature* **235,** 122–124.

Sherebrin M. H., MacClement B. A. E. and Franko A. J. (1972) Electric-field induced shifts in the infrared spectrum of conducting nerve axons. *Biophys. J.* **12,** 977–989.

Shrager P., Chiu S. Y., Ritchie J. M., Zecevic D. and Cohen L. B. (1987) Optical measurement of propagation in normal and demyelinated frog nerve. *Biophys. J.* **51,** 351–355.

Tank D. and Ahmed Z. (1985) Multiple-site monitoring of activity in cultured neurons. *Biophys. J.* **47,** 476A.

Tasaki I., Watanabe A. and Hallett A. (1972) Fluorescence of squid axon membrane labeled with hydrophobic probes. *J. Membr. Biol.* **8,** 109–132.

Tasaki I., Watanabe A., Sandlin R. and Carnay L. (1968) Changes in fluorescence, turbidity, and birefringence associated with nerve excitation. *Proc. Natl. Acad. Sci. USA* **61,** 883–888.

Waggoner A. S. (1979) Dye indicators of membrane potential. *Ann. Rev. Biophys. Bioeng.* **8,** 47–68.

Waggoner A. S. and Grinvald A. (1977) Mechanisms of rapid optical changes of potential sensitive dyes. *Ann. NY Acad. Sci.* **303,** 217–241.

Zecevic D., Wu J.-.Y., Cohen L. B., London J. A., Höpp H.-.P. and Xiao C (1989) Hundreds of cells in the *Aplysia* abdominal ganglion are active during the gill-withdrawal reflex, *J. Neurosc.* **9,** 3681–3689.

From: ***Neuromethods, Vol. 14: Neurophysiological Techniques: Basic Methods and Concepts*** **Edited by: A. A. Boulton, G. B. Baker, and C. H. Vanderwolf**

Fabrication and Implementation of Ion-Selective Microelectrodes

Walter G. Carlini and Bruce R. Ransom

1. Introduction

Techniques for measuring ion concentrations in tissue have been vitally important for critically evaluating a broad range of physiological phenomena. Refinements that allow ion activities to be monitored with considerable spatial and temporal precision have been particularly valuable for this purpose. These methods are widely applied to the study of the nervous system, whose characteristic feature of excitability results in a complex set of dependencies on highly regulated ionic gradients.

Many techniques have been developed for making ion measurements; this chapter focuses mainly on the use of ion-selective microelectrodes (ISMs), because this method has been the one most often employed to study the nervous system. Other methods that are utilized to make biological ion measurements include electron probe microanalysis, nuclear magnetic resonance spectroscopy, microspectrophotometry, and optical methods using ion-sensitive dyes; each of these methods has its respective advantages and disadvantages (Ammann, 1986). ISM and ion-sensitive dye techniques have the advantage of being able to measure ion activity rather than total ion content; it is usually ion activity rather than ion concentration or quantity of ion that is the biological parameter of importance (Ammann, 1986). ISMs also have the advantage of being able to measure ion activity with quite high spatial resolution. One relative disadvantage of ISMs is their moderate temporal resolution; unless special precautions are taken (*see* section 4.3.3.), the ISMs have response times of hundreds of milliseconds. An important practical consideration should be mentioned in discussing the many methods available for measuring ions; the ISM method, compared to many of the others, is relatively easy to employ. ISMs can be manufactured using commonly available laboratory equipment, and standard electrophysiological equipment will often suffice for implementation.

Several excellent monographs on the topic of ISM methods are extant (Thomas, 1978; Zeuthen, 1981; Koryta and Stulík, 1983; Ammann, 1986). Ammann's book (1986) is especially recommended; it addresses many aspects of ISM theory and practice that will not be given full coverage here. Morf (1981) gives a detailed exposition of the physical chemistry of ion-selective electrodes. The monograph by Thomas (1978), although necessarily dated, is still a useful and succinct account of ISM manufacture and implementation. The present treatment of this subject gives a short introduction to the theory of ion-selective electrodes, and then focuses on the practical aspects of ISM manufacture and use. Suggestions are offered about which type of ISM construction should be chosen for which type of application, and modified methods for special circumstances are discussed.

2. Theory of Ion-Selective Electrodes

A basic understanding of the electrochemical theory connected with ion-selective electrodes is a necessary prerequisite to properly constructing, calibrating, and using ISMs. To that end, the following section of this chapter presents a summary discussion of the relevant electrochemistry; given the limitations imposed by the scope of this chapter, the discussion is perforce telegraphic. The theory is presented with an eye to its application to ISMs, but because much of the theory deals with the properties of ion-selective liquid membranes, it is also applicable to the biological membranes with which neuroscientists are more familiar. Presented first is an outline of ionic solution theory, drawn primarily from Hamer (1986), Koryta et al. (1970), Bates and Robinson (1978), Koryta (1982), Starzak (1984), Horvath (1985), and Goodisman (1987). Thereafter follows a discussion of the physical chemistry of ion-selective membranes that draws from the writings of Schwartz (1971), Bauer (1972), Buck (1978), Bailey (1980), Morf (1981), Koryta (1982), Koryta and Stulík (1983), Starzak (1984), Silver (1985), and Ammann (1986). Readers interested in the chemistry of ion carriers are referred to Morf (1981), Pretsch et al. (1985), Ammann (1986), Ammann et al. (1987), and Bartsch et al. (1987).

2.1. Ions in Solution

2.1.1. Dissociation of Strong Electrolytes

It is worthwhile to begin the discussion of ionic solution theory with a review of the process of electrolyte dissociation in water, because this process is central to aqueous electrochemistry. The importance of aqueous electrolyte solutions in biological systems need hardly be pointed out; moreover, aqueous electrolyte solutions are exactly the object of interest to users of ISMs. Horvath (1985) offers a far more detailed, yet readable treatment of this topic.

When a strong electrolyte salt, such as KCl, is added to water, all of the salt dissolves. The energy change occurring during the dissolution of the salt is given by the total enthalpy, ΔH_{tot}:

$$\Delta H_{tot} = \Delta H_{subl} + \Delta H_{diss} + \Delta H_{e,K} + \Delta H_{e,Cl} + \Delta H_{solv} \quad (1)$$

ΔH_{subl} is the enthalpy of sublimation, the energy required to sublimate KCl from the salt into molecular gas: $KCl_{(solid)} \rightarrow KCl_{(gas)}$. ΔH_{diss} is the energy of dissociation of the KCl gas molecules into neutral gaseous K and Cl atoms: $KCl_{(gas)} \rightarrow K_{(gas)} + Cl_{(gas)}$. $\Delta H_{e,K}$ is the energy of ionization for $K_{(gas)} \rightarrow K^{+}{}_{(gas)}$; similarly, $\Delta H_{e,Cl}$ is the energy of ionization for $Cl_{(gas)} \rightarrow Cl^{-}{}_{(gas)}$. ΔH_{solv} is the energy of solvation for both $K^{+}{}_{(gas)} \rightarrow K^{+}{}_{(aq)}$ and $Cl^{-}{}_{(gas)} \rightarrow Cl^{-}{}_{(aq)}$. All of the enthalpy terms are endothermic, with the exception of the enthalpy of solvation, which is usually exothermic. In most cases, the total enthalpy is endothermic, which accounts for the cooling of water observed upon the addition of strong electrolyte salts. The processes of sublimation, dissociation, and ionization will not be discussed here.

Solvation is a complex process, especially with water as the solvent (Koryta et al., 1970; Horvath, 1985). Because of the complexity of water as a solvent, it is necessary to very briefly discuss its structure prior to describing the process of hydration of ions; this discussion is based on Koryta et al. (1970), Kirk-Othmer (1984), and Horvath (1985). The structure of liquid water is still a topic of active research, but it is apparent that the view of water as a simple liquid is inadequate. Like molecules in all liquids, water molecules interact via van der Waals interactions, which are Coulombic dipole–dipole interactions between induced dipoles; van der Waals interactions are weak, and the force from these interactions falls off as the inverse seventh power of the distance separating molecules in the liquid. In addition to the van der Waals interactions present in all fluids, because H_2O is an asymmetric dipole, water molecules experience the much stronger direct dipole–dipole interactions, the force of which falls off only as the inverse cube of the separation between molecules. The presence of strong dipoles in water accounts for its high dielectric constant. Furthermore, even in the liquid state, transient hydrogen bonds are constantly forming, breaking, and reforming among water molecules. At short distances, these interactions effectively order neighboring water molecules into ice-like aggregates. The Pauling model of liquid water posits that these crystal-like clusters of highly ordered water molecules are surrounded by less ordered, thermally agitated "liquid" water molecules, and that there is a continual exchange of water molecules between the crystalline clusters and the relatively disordered background of "liquid" water.

Upon introduction of an ionic solute into water, either of two consequences follow. If the ion is small and has a high surface charge density, then water dipoles orient themselves relative to the ion and form an inner hydration shell about the ion. This shell consists of water molecules "bound" to the ion by ion–dipole interaction; the inner shell of water molecules, in turn, aligns a series of outer shells of water molecules. The entire aggregate, centered about the ion, acts as a larger and more tightly bound version of the quasi-crystalline clusters in pure water. The ion with its hydration shells is a more stable aggregate than the ordinary local water clusters, because the additional energy of dipole–ion binding must be overcome in order to exchange hydration-shell water molecules with exterior "liquid" water molecules. Thus, small ions are structure stabilizing. If, on the other hand, the ion is large and has a low surface charge density, as would happen if it had a large electron cloud shielding excess nuclear electric charge, then the ion is structure disrupting. The ion prevents water molecules in its vicinity from forming local quasi-crystalline aggregates; structure-disrupting ions increase the local entropy of nearby water molecules, favoring the disordered "liquid" structure. For example, both the cation and anion in lithium chloride are structure stabilizing; both the cation and anion in cesium perchlorate are structure disrupting.

When the concentration of strong electrolyte in water is low, the average separation of hydrated ions is great, and there is virtually no ion–ion interaction; the solution behaves in a nearly ideal fashion. Thermodynamic calculations are simplified for ideal solutions; it is of particular interest to ion-selective electrode (ISE) theory that the activities of ions in ideal solutions are equal to their concentrations. Recall that the thermodynamic definition of chemical potential of a species, n_i, in any solution is the partial derivative of the Gibbs free energy with respect to n_i, with all other variables held constant:

$$\mu_i = (\partial G/\partial n_i)_{T,P,\psi,n_j} \tag{2}$$

G is the Gibbs free energy, T is temperature, Ψ is the electrical potential, P is pressure, and the n_j are the chemical potentials of all other species ($j \neq i$). In an ideal solution, the differential of the chemical potential, $d\mu_i$, of ion n_i in the solution can be related simply to its molar concentration, c_i, via:

$$-d\mu_i = RT\, d(\ln c_i) \tag{3}$$

However, biological solutions are seldom well-approximated as being ideal, because the sum concentration of all species can be quite high. In nondilute solutions, hydrated ions approach each other closely enough for ion–ion interactions to take effect.

2.1.2. Ion Activities

In nonideal solutions, ion–ion interactions destroy the assumption of ion independence required to derive Eq. (3); hence, the simple relationship between chemical potential and concentration can no longer be used to relate ion concentration to the numerous thermodynamic parameters dependent upon chemical potential. Ion–ion interactions cause the chemically effective concentration of an ion to be less than its molar concentration. To correct for ion–ion interactions, it is necessary to introduce the idea of ion activity. The activity of an ion is equal to its chemically effective concentration; that is, the activity of an ion in a solution (in mM) is equivalent to the concentration (in mM) the ion would have in an ideal solution to yield the same chemical potential. In an ideal solution, obviously, this definition states that the activity of an ion is precisely equal to its concentration. The molar activity, a_c, of an ion is simply the molar concentration of that ion, c, scaled by an appropriate molar activity coefficient, γ_c: $a_c = \gamma_c c$. Many formulas for γ_c are extant (Hamer, 1968). In general, γ_c is a function of ionic strength (defined below) and temperature of the solution; in all formulations, $\gamma_c \rightarrow 1$ as $c \rightarrow 0$. Hereafter, all activities and concentrations will refer to the molar scale, the subscript c will be implied and will no longer appear explicitly.

Suppose that there is a solution consisting of n_i moles of N different ions, i, each with concentration, c_i, and charge, z_i. Electroneutrality requires that:

$$\sum_{i=1}^{N} z_i n_i = 0 \tag{4}$$

The ionic strength of this solution, I, is given by:

$$I = \frac{1}{2} \sum_{i=1}^{N} z_i^2 c_i \tag{5}$$

It is impossible, using the methods of classical thermodynamics, to derive γ theoretically for individual charged species; thermodynamics can only evaluate the free energy of neutral ion combinations (Bates and Robinson, 1978). Thus, purely thermodynamic activities are only average electrolyte activities, $\hat{a}_\pm$, and no comparable value for single ions can be obtained via pure thermodynamics. Various approximations for the individual molar activity coefficients of cations and anions, γ_+ and γ_-, exist (Hamer, 1968; Koryta et al., 1970). The Debye-Hückel, Güntelberg, Davies, Modified Güntelberg, Scatchard, Extended Scatchard, and Bjerrum equations for activity coefficients are all applicable to solutions of ionic strength less than about 0.1M (Hamer, 1968); however, most biological solutions have an ionic strength greater than this. The Stokes-Robinson method must be used in these instances. Koryta and Stulík (1983) recommends the use of the Bates and Robinson form of the Stokes-Robinson activity coefficient equations (Bates and Robinson, 1978):

$$\ln \gamma_\pm = |z_+ z_-| \ln f_{DH} - \frac{h}{v} \ln a_{wt} - \ln [1 + 0.018(v - h)] \tag{6}$$

In Eq. (6), $\gamma_\pm$ is the mean molar electrolyte activity coefficient, z_+ and z_- are the charges of the cation and anion, respectively, f_{DH} is the Extended Debye-Hückel activity coefficient (Eq. [7]), h is the electrolyte hydration number (Bates and Robinson, 1978), a_{wt} is the activity of water (nearly unity), and v is the number of ions formed by the dissociation of one mulecule of electrolyte. The Extended Debye-Hückel activity coefficient is given by (Hamer, 1968):

$$\ln f_{DH} = -\frac{|z_+ z_-| A_c [I]^{1/2}}{1 + B_c a_i [I]^{1/2}} \tag{7}$$

where a_1 is the ionic size, A_c is given by:

$$A_c = \left(\frac{2\pi N}{1000}\right)^{1/2} \frac{e^3}{2.302585 k^{3/2}} \left(\frac{1}{T^{3/2} \epsilon^{3/2}}\right) \tag{8}$$

and B_c is given by:

$$B_c = \left(\frac{8\pi N}{1000}\right) \frac{e}{k^{1/2}} \left(\frac{1}{T^{1/2} \epsilon^{1/2}}\right) \tag{9}$$

With N being Avogadro's constant ($\cong 6.02 \times 10^{23}$), e being the elementary charge ($\cong$ 4.81 esu), k being the Boltzmann constant ($\cong 1.38 \times 10^{-16}$ erg K^{-1}), T being the absolute temperature, and ϵ being the dielectric constant of water (78.3 at 25°C.)

Now, using $\gamma_\pm$, the mean molar electrolyte activity coefficient obtained from Eq. (6), Bates and Robinson (1978) derived an expression for the individual cation, γ_+, and anion, γ_-, activity coefficients. Equations (10) and (11) are applicable to uniunivalent electrolytes; Eqs. (12) and (13) are applicable to unidivalent electrolytes. Let h_M and h_X be the hydration numbers of the cation and anion, respectively, c be the solution molarity, and ϕ be the electrolyte osmotic coefficient (Bates and Robinson, 1978):

$$\ln \gamma_+ = \ln \gamma_\pm + 0.00782(h_M - h_X)c\phi \tag{10}$$

$$\ln \gamma_- = \ln \gamma_\pm + 0.00782(h_X - h_M)c\phi \tag{11}$$

$$\ln \gamma_+ = 2 \ln \gamma_\pm + 0.00782 h_M c\phi + \ln[1 + 0.018(3 - h_M)c] \tag{12}$$

$$\ln \gamma_- = \frac{1}{2} (\ln \gamma_\pm - 0.00782 h_M c\phi - \ln[1 + 0.018(3 - h_M)c] \tag{13}$$

These equations are the most useful for calculating the activities of ions in biological solutions. Koryta and Stulík (1983) reproduced selected values of activity coefficients for alkali and alkaline earth chlorides, calculated following Bates and Robinson's procedures.

2.2. Physical Chemistry of Ion-Selective Membranes

2.2.1. Introduction

Ion-selective electrodes are electrochemical assemblies that permit potentiometric determination of the activity of a given ion in a solution also containing other ions (Buck, 1978; Morf, 1981; Koryta and Stulík, 1983). Modern ISEs are based on ion-selective membranes of one type or another; therefore, the theory of modern ion-selective electrodes is the theory of ion-selective membranes. For the physiologist, it is noteworthy that much of the theoretical analysis of ion-selective membranes in ISEs can be applied equally well to the lipid bilayer ion-selective membranes of living cells.

A membrane is a phase, usually thin, that separates two other phases in such a manner that material movement between the contacting phases is altered compared to the movement that would occur between the bulk phases were they directly adjacent to each other (Morf, 1981; Koryta and Stulík, 1983). Material movement can include neutral as well as charged species. A semipermeable membrane is one that has different permeabilities to different components of the two contacting phases. A semipermeable membrane allows certain components to pass from one bulk phase through the membrane phase into the other bulk phase more readily than it allows other components to pass; the semipermeable membrane is said to be selective for those certain components. A semipermeable membrane that separates two electrolyte solutions and is selective for certain of the ion species in solution is termed an ion-selective membrane.

Often, theoretical discussions of membranes treat them simply as phase partitions—as hypothetical two-dimensional surfaces; in these treatments, the internal electrochemistry of membranes and the electrochemistry of the membrane–solution interfaces are neglected. It is obvious that real ion-selective membranes, such as pH-sensitive glass or the K^+-selective membranes of glial cells, differ substantially in their physicochemical characteristics from the bulk electrolyte phases bordering them. All real membranes have an inside region and two boundary regions that are adjacent to the two bulk phases on either side. Physical and chemical properties do not change abruptly from the outer phases to the membrane; rather, many physical and chemical properties, such as charge density, ion activity, and potential distribution, vary continuously from outer phase to membrane boundary to membrane interior.

2.2.2. Thermodynamics

The physicochemical structure of ion-selective membranes and the kinetic electrochemistry of the ion-transfer processes that such membranes mediate in ISEs and cells will be taken up further on; for now, the ion-selective membrane will be considered more simply, in order to present basic concepts necessary to the later discussion. A thermodynamic—as opposed to kinetic—analysis of electrolyte solutions partitioned by a membrane obviates the need to consider the details of membrane structure.

Inasmuch as ion-selective membranes will always be considered as part of galvanic cells, some discussion of galvanic cells is appropriate. Following Koryta and Stulík (1983), define the Galvani potential of a phase α, $\phi(\alpha)$, as being the electrical work needed to transfer 1 Faraday of positive charge into the phase from infinity. The Volta potential of phase α, $\Psi(\alpha)$, is the electrical work needed to transfer 1 Faraday of positive charge from infinity into the neighborhood of phase α. The Galvani potential difference between two phases, α and β, is given by:

$$\Delta_{\beta}^{\alpha}\phi = \phi(\alpha) - \phi(\beta) \tag{14}$$

Similarly, the Volta potential difference between α and β is given by:

$$\Delta_{\beta}^{\alpha}\Psi = \Psi(\alpha) - \Psi(\beta) \tag{15}$$

Starzak (1984) provides a more detailed explanation of the distinction between $\phi(\alpha)$ and $\Psi(\alpha)$. Because of the complications of induced image charges, surface dipole potentials, and the requirement for chemical identity between phases in measuring Galvani potentials, for all practical purposes, the Volta potential is the only experimentally accessible potential. However, theory usually is framed in terms of Galvani potentials to discount surface potentials.

Consider a reversible Galvanic equilibrium cell; for definiteness, consider the example schematized below in IUPAC convention (phase boundaries are denoted by vertical lines):

$$\left| \begin{array}{c|c|c|c|c|c|c} Cu & Pt & H_2 & HCl,H_2O & AgCl & Ag & Cu \\ 1 & 2 & 3 & 4 & 5 & 6 & 7 \end{array} \right| \tag{16}$$

The total electromotive force (EMF) that would be measured by a potentiometer between copper electrodes 1 and 7 is given by:

$$E = \phi(7) - \phi(1) \tag{17}$$

The EMF of a Galvani cell in equilibrium is related to the Gibbs energy of the cell reaction, ΔG, by:

$$E = -zF\Delta G \tag{18}$$

where z is the charge number of the cell reaction and F is Faraday's constant (96,500 Coulombs/mol). In the case of the Galvani cell in Eq. (16), the cell reaction is the dissociation by hydrogen of silver chloride into elemental silver and hydrogen chloride, and it has $z = 2$; the EMF at standard state can easily be found by reference to standard tables of inorganic half cell reactions (Tinoco et al., 1978). To compute the energy of Galvani cell reactions at conditions other than standard state, the partial molar Gibbs energy of a given component in a phase is a useful quantity:

$$\tilde{\mu}_i(\alpha) = \mu_i^0 + RT \ln a_i + \pi v_i + z_i F\phi(\alpha) \tag{19}$$

where α is the phase, μ_i^0 is the standard chemical potential of the ith component, a_i is the activity of the ith component in phase α, π is the solvent osmotic pressure, V_i is the partial molar volume of the ith component (*see* Eq. [24]), z_i is the charge number of the ith component, F is the Faraday constant, and $\phi(\alpha)$ is the Galvani potential of phase α. The partial molar Gibbs energy of a component in a charged phase is also known as the electrochemical potential of that component. When Ψ in Eq. (2) is nonzero, Eq. (2) is an alternative representation for the electrochemical potential of a component in a phase. Neglecting osmotic pressure, Starzak (1984) presents a derivation of Eq. (19) in differential form (Eq. [20]) from first principles.

$$d\tilde{\mu}_i = RT\, d(\ln a_i) + z_i F\, d\phi \tag{20}$$

Suppose that two electrolyte solutions, *cis* and *trans*, are instantaneously separated by a membrane as shown in Eq. (21):

$$|cis|trans| \tag{21}$$

Assume further that the system is electroneutral, it is at constant temperature, and initially the membrane is permeable only to water; then, at equilibrium, the chemical potentials of the two electrolyte phases must be equal:

$$d\mu_{cis} = d\mu_{trans} \quad (22)$$

Prior to equilibration (the kinetics, obviously, are of no concern here), the chemical potentials of the two phases were not equal. According to assumption, only water can pass through the membrane; therefore, equilibration occurs by water permeation until the activity of water on the two sides of the membrane is equal. Water permeation creates an osmotic pressure, the equation for which (Eq. [23]) can be derived from the free energy expression for the system.

$$\bar{V}\Pi = RT(a_{\text{water, } trans} - a_{\text{water, } cis}) \quad (23)$$

Here Π is the osmotic pressure, $a_{\text{water, } side}$ is the activity of water on the given side of the membrane, and

$$\bar{V} = d\mu/dP \quad (24)$$

Now assume that the membrane is equally permeable to both cations and anions; again, at equilibrium, the chemical potentials of all species must be equal in the *cis* and *trans* phases. Equation (22) can be applied to each component; this, together with the requirement for electroneutrality of each phase, allows for the computation of the Donnan equilibrium distribution of the species in the system. Finally, assume that the membrane is ion-selective; in this case, differential permeability to cations and anions causes an electrical potential to be formed across the membrane at equilibrium.

At equilibrium, the electrochemical potentials of all components, i, in both phases must be equal across the ion-selective membrane. Thus, using Eq. (20):

$$RT\, d(\ln a_{i,cis}) + z_i F\, d\phi_{cis} = RT\, d(\ln a_{i,trans}) + z_i F\, d\phi_{trans} \quad (25)$$

Here $a_{i,cis}$, $a_{i,trans}$, ϕ_{cis}, and $d\phi_{trans}$ are the final equilibrium values. Integrating Eq. (25):

$$\int_a^{a_{i,cis}} d(\ln a_{i,cis}) + \int_0^{\phi_{cis}} z_i F\, d\phi_{cis} = \int_a^{a_{i,trans}} d(\ln a_{i,trans}) + \int_0^{\phi_{trans}} z_i F\, d\phi_{trans} \quad (26)$$

Elimination of common terms and rearrangement yield the Nernst equation:

$$\Delta\phi = \frac{RT}{z_i F} \ln\left(\frac{a_{i,cis}}{a_{i,trans}}\right) \tag{27}$$

The more general form of the above equation is (Koryta and Stulík, 1983):

$$\Delta_\beta^\alpha \phi = \frac{\mu_i^0(\beta) - \mu_i^0(\alpha)}{z_i F} + \frac{RT}{z_i F} \ln\left(\frac{a_i(\beta)}{a_i(\alpha)}\right) \tag{28}$$

α and β have been substituted for *cis* and *trans*, respectively.

2.2.3. Phases with Different Solvents

So far, it has been assumed that the two phases have a common solvent, namely water. Ion-selective membranes need not have the same solvent as the adjoining electrolyte phases; indeed, they often do not. It is of interest to examine the case of two phases with unlike, primarily immiscible solvents. If α and β have different solvents, then, in general, Eq. (28) is useless, because the difference in standard chemical potentials (the numerator of the first term on the right side of Eq. [28]) cannot be derived thermodynamically. This difference is the result of the difference in the standard Gibbs energies of solvation of component i in phases α and β, also known as the standard Gibbs energy of transfer for component i from α to β, and is written:

$$\Delta G_{tr,i}^{0,\beta\rightarrow\alpha} = \mu_i^0(\alpha) - \mu_i^0(\beta) \tag{29}$$

The standard Gibbs energy of transfer can only be computed thermodynamically for nonionic species by using distribution equilibria between phases; the distribution coefficient for a substance, S, between phases α and β is given by:

$$k_S^{\beta,\alpha} = \exp(\Delta G_{tr,S}^{0,\beta\rightarrow\alpha}/RT) \tag{30}$$

Unfortunately, additional, nonthermodynamic, assumptions must be introduced to divide the standard Gibbs energy of transfer of an electrolyte into the separate contributions from the cation and the anion.

There exists, however, a salt whose component anion and cation can be assumed to have approximately equal energies of solvation in most solvents. Tetraphenylarsonium tetraphenylborate (TPATPB) is made up of the cation tetraphenylarsonium (TPA^+) and the anion tetraphenylborate (TPB^-); because both of these ions are large, and hence, both have low surface charge density, their solvation by most solvent molecules is equally diffuse, their interaction with solvent dipoles being almost entirely electrostatic. Furthermore, because both ions face the solvent with the same benzene ring structures, they both share the same benzene ring–solvent interactions. Assuming, then, that the energies of solvation of TPA^+ and TPB^- are the same in any pair of solvents, the standard Gibbs energy of transfer for TPA^+ and TPB^- can be determined from the distribution of TPATPB salt in any two immiscible solvents. Then, knowing the standard Gibbs energy of transfer for TPA^+ and TPB^-, the standard Gibbs energy of transfer for any cation, C^+, and any anion, A^-, can be calculated from the distribution coefficients of the salts CTPB and TPAA.

If a salt formed of a cation, C, with positive charge z_+, and an anion, A, with negative charge, z_-, is in equilibrium between contacting phases α and β with immiscible solvents, then, in general, a distribution potential exists and is given by:

$$\Delta_\beta^\alpha \phi_{distrb} = \frac{1}{2F}\left(\frac{z_-}{z_+}\Delta G_{tr,C}^{0,\beta\rightarrow\alpha} - \frac{z_+}{z_-}\Delta G_{tr,A}^{0,\beta\rightarrow\alpha}\right) \tag{31}$$

At equilibrium, the concentrations of the anion and cation in the two phases are equal because of the requirement for bulk electroneutrality. At the phase boundary, however, bulk electroneutrality need not hold and, in fact, an electrical double layer exists if C and A have differential solubilities in the two solvents.

2.2.4. Electrical Double Layers

Electrical double layers form at interfaces between solution phases containing charged species that partition differently into the two-phase solvents; they also form at interfaces between solution phases if one of the phases contains amphiphilic species. In addition, electrical double layers may form between a solid electrolyte phase and a solution electrolyte phase by differential dissolution of a component of the solid phase into the solution phase. Chapter 8 of Starzak (1984) discusses electrical double layers in detail. Here the electrical double layer will be discussed only brief-

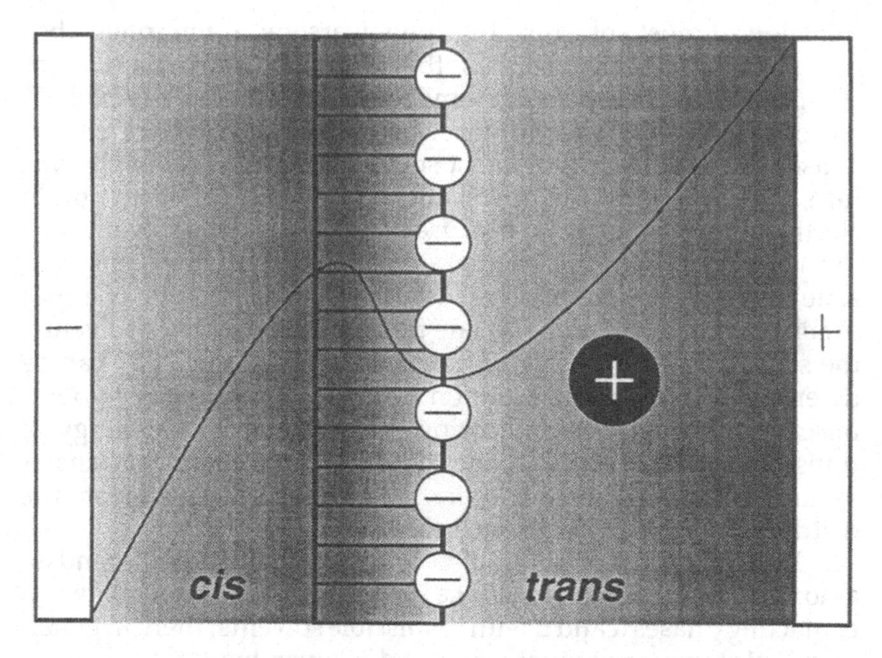

Fig. 1. Electrical double layer near a planar membrane. A planar membrane (diagrammed as a ladder-like object) separates two solutions: *cis* and *trans*. The *trans* face of the membrane has fixed negative charges (white circles). Current is passed from the positive electrode in the *trans* solution to the negative electrode in the *cis* solution, inducing a potential field between the electrodes. The energy of the field with respect to a cationic test particle (black circle) is represented in the figure by the density of the shading, denser shading representing higher energy. The curve in the figure is an alternative representation of the energy of the field relative to a cation. Near the fixed negative sites on the *trans* face of the membrane, the field energy is at a local minimum because of the interaction of the fixed negative sites with the cationic probe. On the *cis* side of the *trans* solution–membrane interface, a local maximum in the field energy exists because of the energy of charge separation between the cationic probe and the fixed negative sites.

ly. Imagine a planar membrane (such as in Fig. 1) that has fixed negative surface charges on its right-hand surface (facing the *trans* solution). If electrodes are introduced into the *cis* and *trans* phases, and a positive potential is extraneously applied through the electrodes (so that the *cis* phase is made more negative than the *trans*), then a cation in the *trans* phase will tend to approach the right-hand membrane surface. If the membrane has no surface charges, the cation will simply pass through the membrane, and its potential energy will dissipate linearly. However, because of the fixed negative surface charges, as the cation nears the surface, it will experience a drop in potential before entering the membrane. Just inside the right side of the membrane, the cation will be in a local potential well, because of favorable charge interaction with the negative surface charges. To proceed further down the imposed potential gradient, the cation will need to pass an activation energy barrier caused by uncoupling from the interactions with the anionic surface. Once sufficient activation energy from, say, thermal collisions is imparted to the cation, it will then continue to drop down the potential gradient linearly through the membrane. Note that, thermodynamically, the cation will eventually wind up on the *cis* side, because of the imposed potential gradient. However, the kinetics of the ion-transfer process are highly dependent on the structural properties of the membrane. If the anionic surface charge on the membrane is great enough, cation passage will be greatly slowed.

In the absence of an imposed potential, the overall membrane phase must be electroneutral. The membrane–solution interface can still be charged, however; adsorbed counterions on the membrane surface can dissolve into the solution, leaving behind a surface charge. The membrane surface charge can be neutralized by oppositely charged ions in the solution, which are not necessarily the same counterion species as originally present in the membrane. The solution counterions may not transfer into the membrane phase, either because they are surrounded by solvation shells or because they are intrinsically insoluble in the membrane-phase solvent (which may be solid or liquid). In that case, the solution counterions and membrane charges pair with each other to form a double layer of charge, hence the term electrical double layer. Helmholtz theory maintains that the double layer of charge can be treated as two parallel surfaces of charge separated by a fixed distance (the sum of the ionic radii of the membrane ions and solution ions, or the sum of the ionic radii of the membrane ions and the solvation shell radii of the solution counterions). This arrangement is a capacitor. Moreover, because the separation between the charged surfaces is quite small, the capacitance of the electrical double layer is substantial:

$$C = A\epsilon_r\epsilon_0/d \tag{32}$$

where A is the area of the surfaces, d is their separation, ϵ_r is the relative dielectric constant of the solution solvent, and ϵ_0 is the permittivity constant = 8.85×10^{-12} F m^{-1}. Recalling that the potential difference, V, is related to the charge, q, and capacitance by:

$$V = q/C \tag{33}$$

The expression for the Helmholtz double-layer potential is obtained from Eqs. (32), (33), and the relation for surface charge density, $\sigma = q/A$:

$$V = d\sigma/\epsilon_r\epsilon_0 \tag{34}$$

Helmholtz theory is inadequate in reality, because its assumptions simplify matters too drastically, especially in the case of two liquid phases in contact. Surface charge in a membrane rarely exists as a homogenous sheet of charge, counterions in a solution do not pair irreversibly with the membrane charges, and electrical interactions can extend beyond the single layer of charges assumed in the Helmholtz model. In the case of a solid phase in contact with a liquid phase, or two solid phases in contact, the idea of a compact electrical double layer is a reasonable approximation; for the case of two liquid phases, the Gouy-Chapman model describing diffuse electrical double layers is better. Bypassing the derivation (*see* Chapter 8 of Starzak, 1984), the equation for the Gouy-Chapman potential, $V(x)$, at a distance x from the membrane surface, is:

$$V(x) = \frac{\sigma}{\epsilon\kappa} \exp -\kappa(x - a) \tag{35}$$

with σ being the unit charge density, $\epsilon = \epsilon_r\epsilon_0$, a being the closest distance to which a counterion can approach the membrane surface, and κ being the inverse of the Debye length, κ^{-1}. Note that, at $x = a$, the Helmholtz potential and Gouy-Chapman potential are identical when $d = \kappa^{-1}$.

Suppose that a membrane phase is liquid and has a solvent immiscible with that of an electrolyte solution adjoining it. An electrical double layer forms at the membrane–solution interface, and the electrical potential between the membrane phase, α, and the electrolyte phase, β, is given by:

$$\Delta^{\alpha}_{\beta}\phi = \phi_2(\alpha) + \phi_2(\beta) \tag{36}$$

where $\phi_2(\alpha)$ and $\phi_2(\beta)$ are the potential differences between the inside of phase α and the interface and between the inside of phase β and the interface, respectively. Gouy-Chapman theory permits the calculation of $\phi_2(\alpha)$ and $\phi_2(\beta)$ from the following equations:

$$q(\alpha) = [8\epsilon_\alpha RT\ c(\alpha)]^{1/2} \sinh(F\phi_2(\alpha)/2RT) \tag{37}$$

$$q(\beta) = -\ [8\epsilon_\beta RT\ c(\beta)]^{1/2} \sinh(F\phi_2(\beta)/2RT) \tag{38}$$

$$\frac{\sinh(F\phi_2(\alpha)/2RT)}{\sinh(F\phi_2(\beta)/2RT)} = \left[\frac{\epsilon_\alpha c(\alpha c(\alpha)}{\epsilon_\beta c(\alpha c(\beta)}\right]^{1/2} \tag{39}$$

where ϵ_α and ϵ_β are the permittivities of the solvents of phases α and β, respectively; $q(\alpha)$ and $q(\beta)$ are the total electric charges in the part of the electrical double layer in phases α and β, respectively; $c(\alpha)$ and $c(\beta)$ are the total electrolyte concentrations in phases α and β, respectively.

The equations presented above are linearized approximate solutions to the one-dimensional Poisson-Boltzmann differential equation describing the potential over the diffuse electrical double layer. The Grahame equation is the exact solution of this Poisson-Boltzmann equation:

$$\sigma = \left(2RT\epsilon \sum_i\ a_i \exp -\ (1\ +\ z_i F\phi(x)/RT)\right)^{1/2} \tag{40}$$

where all the symbols are as defined previously, with the exception that here $\phi(x)$ is the Grahame diffuse double-layer potential at a distance x from the interface. *See* Bauer (1972) or Starzak (1984) for the derivation of the Grahame equation as the solution to the one-dimensional Poisson-Boltzmann differential equation.

2.2.5. Kinetics

The importance of the electrical double layer to ISE theory is that the nonlinear potential of the electrical double layer influences the dynamics of charge transfer through the ion-selective membrane. Further dynamical analysis of the ion-selective membrane requires consideration of component transfer through the membrane; the case of interest for an ion-selective membrane is that of a mobile component with charge. Components diffusing down a concentration gradient obey Fick's laws (Tinoco et al., 1978). Assuming isotropy of concentration in the planes parallel to the membrane, the steady-state flux, J, (across a unit cross section) of a diffusing uncharged species with diffusion coefficient D through a membrane of thickness L with activities a_{cis} and a_{trans} on its two faces is given by:

$$J = (D/L)\,(a_{trans} - a_{cis}) \tag{41}$$

The dynamical diffusion of the species in the direction normal to the membrane is described by the one-dimensional form of Fick's second law, where x is the distance into the membrane and $C(x, t)$ is the activity of the species at time t at position x:

$$\frac{\partial C(x, t)}{\partial t} = \frac{D\partial^2 C(x, t)}{\partial x^2} \tag{42}$$

The solutions of this equation are Fourier sums (Starzak, 1984). The diffusion of charged species is more complex, especially in the presence of a potential field, such as, for instance, the interfacial potential of the diffuse double layers on either face of the membrane. Moreover, charged species diffusing from one phase to another give rise to a potential, the diffusion potential, also known as the junction potential.

Again, assuming isotropy of concentration in the planes parallel to the membrane, the steady-state flux (across a unit cross section) for a diffusing species with charge z, diffusion coefficient D, and mobility u is given by the Nernst-Plank electrodiffusion equation:

$$J = -\,uRT\,\frac{da}{dx} - zFuc\,\frac{d\phi}{dx} \tag{43}$$

To find the activity, a, and potential, ϕ, as a function of position is difficult in the general case; complications introduced by nonlinearities in the potential gradient can by eliminated by fiat. The Goldman assumption is that the electric field is constant and is unaffected by charged species within the membrane. If the total potential across the membrane is $\Delta\phi$, then the field is simply:

$$E = -\,d\phi/dx = -\,\Delta\phi/L \tag{44}$$

Under this assumption, for a membrane of thickness L with activities of the charged species a_{cis} and a_{trans} on either side, Eq. (43) is solved by the Goldman equation:

$$J = -\left(\frac{zF\Delta\phi}{LRT}\right)\frac{a_{trans} - a_{cis}e^{zF\Delta\phi/RT}}{1 - e^{zF\Delta\phi/RT}} \tag{45}$$

Morf (1981) presents a version of the Goldman-Hodgkin-Katz equation that generalizes to N noninteracting charged species and converts species flux into current. Schwartz (1971) and Silver (1985) discuss the limitations of the Goldman equation; Schwartz (1971) showed that a Goldman-Hodgkin-Katz type of expression remains valid, even if the requirement for a constant field in the membrane is relaxed to a requirement that the electric field within the membrane merely have an odd symmetry about the midplane of the membrane.

The potential created by the diffusion of charged species from one liquid phase to another is termed the liquid-junction potential. Morf (1981), Koryta (1982), Koryta and Stulík (1983), and Goodisman (1987) present derivations of various expressions for the liquid-junction potential. The Planck liquid-junction model is based on the Nernst-Planck equation (Eq. [43]) and, in general form, is quite complicated. If simplifying assumptions are made, then a more tractable expression is obtained. Suppose that two solutions of a single electrolyte are adjoined, and the activities of electrolyte in the solutions are a_{cis} and a_{trans}. Furthermore, let the transport number of the cation, t_+, and anion, t_-, in the electrolyte be given by:

$$t_+ = z_+u_+/z_+u_+ - z_-u_- \tag{46}$$

$$t_- = z_-u_-/z_+u_+ - z_-u_- \tag{47}$$

The z_+, z_- and u_+, u_- denote the charges and mobilities of the cation (+) and anion (–), respectively. Then the liquid-junction potential between the two solutions is:

$$\Delta\phi_L = \frac{RT}{F}\left(\frac{t_-}{z_+} - \frac{t_+}{z_+}\right)\ln\frac{a_{trans}}{a_{cis}} \tag{48}$$

The liquid-junction potential is minimized by electrolytes with cations and anions of similar mobilities. For more complex (multicomponent) solutions, the Morf version of the Planck liquid-junction potential expressions (Morf, 1981) or the Henderson approximation (Koryta, 1982; Koryta and Stulík, 1983; Silver, 1985) must be used:

$$\Delta\phi_L = \frac{\sum_i z_i u_i (a_{i,trans} - a_{i,cis})}{\sum_i z_i^2 u_i (a_{i,trans} - a_{i,cis})} \frac{RT}{F} \ln \frac{\sum_i z_i^2 u_i a_{i,trans}}{\sum_i z_i^2 u_i a_{i,cis}} \tag{49}$$

where the z_i are the signed charges, the u_i are the mobilities, and the a_i are the activities of species i.

2.2.6. ISE Cells

A typical ion-selective electrode consists of an ion-selective membrane with a reference solution and reference electrode connected to an amplifier on one side. The other side of the membrane faces the solution to be tested. The following diagram describes this situation:

Reference electrode	Test solution	Membrane	Reference solution	Reference electrode	(50)

In practice, the reference electrodes are usually solid, in order to connect the system to measuring circuitry, which uses electrons for charge transfer. Thus, in most instances, reversible electrodes of the first or second kind are used to convert ionic charge flux in solution into electron charge flux in metal. Two common choices for reference electrodes are the calomel ($Hg|Hg_2Cl_2|KCl$) and $Ag|AgCl$ electrodes. This system contains a number of different potentials. The interface between the reference electrode and test solution has associated with it a solid-to-liquid-junction potential and a compact electrical double layer with its attendent boundary potential. To reduce the junction potential, practical ISE systems often use a KCl bridge from the reference electrode to the test solution. The test solution|membrane interface has either a compact (if the membrane is solid phase) or diffuse (if the membrane is liquid phase) electrical double layer with a space charge region extending into both the membrane and the test solution. If the system is not static, then a number of interfacial and intramembrane potentials can exist. For example, diffusing charged species within the membrane create diffusion potentials (liquid-junction potentials if the membrane is in liquid phase and shares solvent with the test solution), or differential partitioning of charged species between the membrane and the test solution creates distribution potentials. The reference solution, so-called because it contains a "known" activity of the species to be analyzed (called the analysate or determinand), gives rise to yet another set of electrical double-layer boundary potentials, diffusion (junction) potentials, and distribution potentials at the reference solution|membrane interface. The reference solution|reference electrode interface possesses interfacial potentials similar to those discussed for the test solution|reference electrode interface. A substantial body of theory addresses the resolving of the total system EMF into its component potentials (Morf, 1981; Buck, 1978; Koryta and Stulík, 1983).

In practice, the interesting potentials in this system are those that change as the test solution changes; because the reference electrode and reference solution remain constant, the potentials contributed by that part of the system can be assumed to remain constant and can be neglected in the discussion that follows. The junction potential between the external reference electrode and the test solution will also be neglected, under the assumption that a suitable stable bridge (such as 3*M* KCl in agar) is incorporated between the reference electrode and the test solution. The remaining potentials all involve the membrane in one way or another, and can be summarized as a transmembrane potential measured between the two reference electrodes:

$$\Delta\phi_M = \Delta\phi_{left} + \Delta\phi_{inside} + \Delta\phi_{right} \quad (51)$$

$\Delta\phi_M$ is the transmembrane potential, $\Delta\phi_{left}$ is the sum of the potentials of the test solution–membrane boundary, $\Delta\phi_{inside}$ is the sum of the intramembrane potentials, and $\Delta\phi_{right}$ is the sum of the potentials of the reference solution–membrane boundary. The classical Nicolsky equation is a standard approximation for $\Delta\phi_M$ at steady-state; this expression takes into account the Nernstein potential of the analysate, the differing distribution coefficients of components of the outer (solution) phases, and the differing mobilities of these components in the membrane phase. Let $k_h^{sol \to mem}$ be the distribution coefficient from the solutions into the membrane (assumed equal for the two solutions) for the hth species, and let u_i be the mobility of the hth species in the membrane; then, for any two different components *i, j* in the solution phases, the selectivity coefficient for *j* with respect to *i* is defined as:

$$K_{i,j}^{pot} = u_j k_j^{sol \to mem} / u_i k_i^{sol \to mem} \quad (52)$$

If the charged component to which the membrane is primarily selective is i, and if ϕ_i^0 is the standard potential for this component in the system, then the Nicolsky equation for this system is:

$$\Delta\phi_M = \phi_i^0 + \frac{RT}{z_i F} \ln \left(a_i + \sum_{j=i} a_j K_{i,j}{}^{pot} \right) \quad (53)$$

A major shortcoming of this form of the Nicolsky equation is that it does not accurately provide for phases containing mixtures of components with differing valencies; Buck (1978) presents some Nicolsky-like equations that correct for such mixtures. Although selectivity coefficients for given ion-selective membranes are often reported pairwise (that is, between the primary ion to which the membrane is sensitive and interferent ions), these coefficients, in reality, also depend on the other components present in the contacting phases, and, as is easy to see, the definition of the selectivity coefficient ignores the difference between the distribution coefficients for a component on either side of the membrane.

2.2.7. Types of Ion-Selective Membranes

The Nicolsky equation is an extremely practical approximate expression for the ion-selective membrane potential (neglecting liquid-junction potentials); more realistic expressions for the ion-selective membrane potential are difficult to obtain and demand that the details of the membrane itself be taken into account. Ion-selective membranes are a distinct and separate phase in ISE systems; as such, the ion-selective membrane must be at least partially immiscible with the adjoining phases. Commonly used ion-selective membranes either are solid or are based on a hydrophobic liquid solvent. These membranes are permeable to one or more different species; usually, but not always, practical ion-selective membranes are nonporous and do not allow solvent transport to occur from the contacting phases. Within the ion-selective membrane, whether it be solid or liquid, there exists a variety of neutral, charged, and ionizable species, the states of which depend on the milieu enveloping the membrane. Some ion-selective membranes have spatially fixed, charged species that function as ion-exchange sites for the primary determinand; other ion-selective membranes have mobile ion-exchange sites that are free to roam within the membrane, but that cannot partition into the adjoining phases. Yet other ion-selective membranes lack ion-exchange sites, and function by dint of neutral carriers or via bulk transport. In addition to charged ion-exchange sites or neutral carriers, ion-selective membranes contain two other important components: coions and counterions. Counterions are ions of opposite charge sign to any charged site species or charged neutral carrier–ion complex in the membrane that balance the charge of the fixed sites, mobile sites, or neutral carrier–ion complex to maintain bulk electroneutrality of the membrane phase. Counterions need not be restricted to the membrane phase, and can diffuse into and out of the adjoining phases. Coions are ions of the same charge sign as any charged site species or charged neutral carrier–ion complex in the membrane; coions, like counterions, are free to pass into and out of the membrane phase; indeed, typically, the majority of coions originate in the adjoining phases.

According to Buck (1978), ion-selective membranes can be grouped into three broad classes: Class I—site-free membranes, which utilize neutral carriers or solvent flow to transfer analysate ions; Class II—fixed-site membranes, which use fixed ion-exchanges sites to transfer analysate ions; and Class III—mobile-site membranes, which utilize mobile ion-exchange sites to transfer analysate ions. Each of these membrane classes includes ion-selective membranes that are applied in biological research. Ion neutral carriers dissolved in organic solvents (such as valinomycin in dibutyl sebacate) are examples of site-free membranes; glass membranes (such as La_2O_3-doped lithia glass) are examples of fixed-site membranes; liquid ion-exchangers (such as lauryl [trialkylmethyl] ammoniate in benzene) are examples of mobile-site membranes.

The general theory of ion-selective membranes presented above is applicable to liquid neutral-carrier-based membranes, liquid ion-exchanger-based membranes, and glass-based membranes alike; extensions of this general theory to the particulars of each of these categories of membrane will now be presented in brief form; for futher details, *see* Morf (1981), Koryta and Stulík (1983), and Buck (1978).

Liquid neutral-carrier-based membranes consist of a hydrophobic (usually organic) solvent, a neutral, lipophilic, ion-complexing agent (the ionophore), and hydrophobic counterions. To date, fully functional ionophores for anions have not been devised (Ammann et al., 1987); the neutral complexing agents in

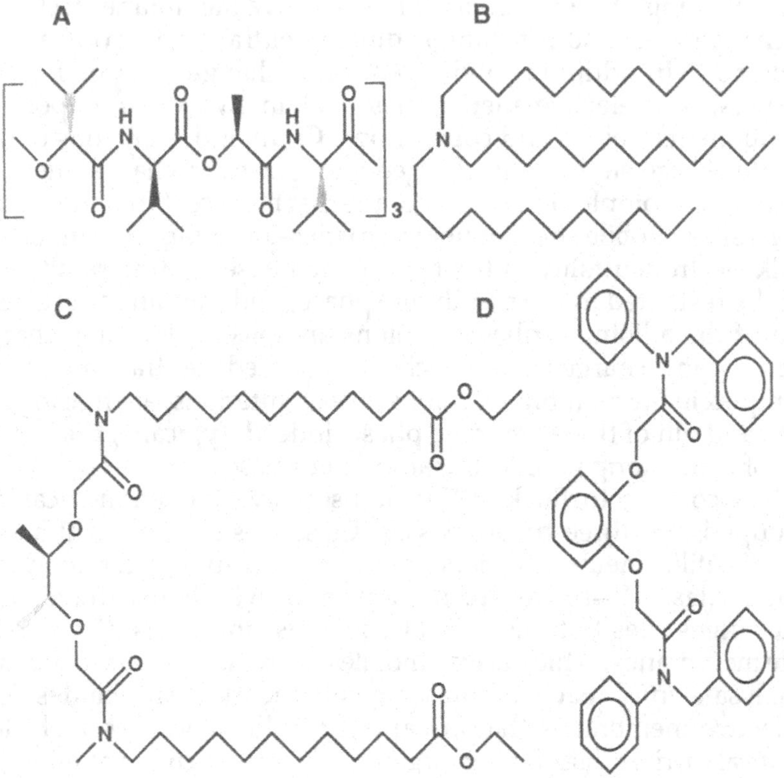

Fig. 2. Structures of some cation-selective neutral carriers. **A:** Valinomycin—selective for K^+. **B:** Tri-*n*-dodecylamine—selective for H^+. **C:** (–)-(R, R)-*N*, *N*'-bis-[11-(ethoxycarbonyl)-undecyl] *N*,*N*',4,5-tetramethyl-diocaoctane diamide–selective for Ca^{2+}. **D:** *N*,*N*'-dibenzyl-*N*,*N*'-diphenyl-1,2-phenylene-dioxy acetamide—selective for Na^+.

use are all selective for group IA or IIA cations. Morf (1981) discusses the chemistry of the A-cation-selective ionophores. Apart from the antibiotic valinomycin, the majority of practical neutral carriers have been synthesized by Simon and coworkers (Güggi et al., 1976; Oehme et al., 1976; Ammann et al., 1981). Typically, neutral carriers are multifunctional molecules of moderate size (fewer than about 100 atoms) (*see* Fig. 2; Ammann, 1986). In these neutral carrier molecules, polar functional groups line a constrained, but not entirely rigid, cavity and provide multiple oxygen coordinating sites for complexing cation; nonpolar functional groups line the exterior of the carrier and form a lipophilic shell for the entire carrier–ion complex.

The selectivity of these ionophores for one cation vs another is based upon size discrimination and the energetics of complexing. A lithium-selective neutral carrier, for instance, has a relatively small complexing cavity into which Li^+ can fit, but from which larger cations are excluded. On the other hand, a potassium-selective neutral carrier discriminates against other cations on the basis of the energetics of association with the complexing functional groups. The K^+-selective neutral-carrier cavity, being semi-rigid, holds its complexing functional groups in a certain spatially fixed configuration; this configuration is such that it allows for the formation of the maximum number of noncovalent bonds between the carrier and K^+. The formation of these bonds compensates for the free-energy increase occurring as the potassium ion sheds its solvation shell. If a sodium ion should attempt to occupy the K^+-selective neutral-carrier cavity it would not be hindered on the basis of ionic radius, because it is smaller than K^+; rather, Na^+ would be discriminated against on the basis of unfavorable energetics. Na^+ has a shorter bonding radius than does K^+; hence, it would not be able to form the same number of free-energy-reducing noncovalent bonds simultaneously with the spatially fixed complexing functions in the carrier cavity as could potassium. Hille (1984) discusses these concepts of ion selectivity in the context of ion channels.

The derivation of expressions for membranes based on ion neutral carriers is attributable to Morf (1981). Assume that one or two neutral carrier molecules can bind to an analysate (determinand) ion at once. If J^+ represents the determinand ion, K^+ represents the interferent ion, and X represents the neutral carrier ionophore, then the potential at the test solution–membrane interface, from which $\Delta\phi_M$ can be obtained via Eq. (51), is given by:

$$\Delta\phi_{left} = -\Delta_{mem}^{sol}\phi_{J^+}^{0} + \frac{RT}{F}\left[\ln\left(a_{J^+,sol} + \frac{(\beta_{K^+X}a_X\gamma_{K^+X}^{-1}) + (\beta_{K^+X_2}a_X^2\gamma_{K^+X_2}^{-1})}{(\beta_{J^+X}a_X\gamma_{J^+X}^{-1})(\beta_{J^+X_2}a^2_X\gamma_{J^+X_2}^{-1})}\right)K_{exch}a_{K^+,sol} + \ln\left(\frac{(\beta_{J^+X}a_X\gamma_{J^+X}^{-1}) + (\beta_{J^+X_2}a^2_X\gamma_{J^+X_2}^{-1})}{c_{total\ J}}\right)\right] \quad (54)$$

where the stability constants, β, are:

$$\beta_{J^+X} = \frac{a_{J^+X}}{a_{J^+}a_X},\ \beta_{J^+X_2} = \frac{a_{J^+X_2}}{a_{J^+}a_X^2}\ \beta_{K^+X} = \frac{a_{K^+X}}{a_{K^+}a_X},\ \text{and}\ \beta_{K^+X_2} = \frac{a_{K^+X_2}}{a_{K^+}a_X^2} \quad (55)$$

K_{exch} is the equilibrium constant of the phase exchange reaction in which the interferent and determinand change places, J^+ going from membrane to solution, and K^+ going from solution to membrane:

$$K_{exch} = \exp(\Delta G^{0,sol\to mem}_{tr,J^+} - \Delta G^{0,sol\to mem}_{tr,K^+})/RT \quad (56)$$

and $c_{total\ J}$ is the total concentration of the determinand in the membrane phase. Expressions for selectivity coefficients for z_{J^+}–valent determinands v z_{K^+}–valent interferents can be derived from Eq. (54); these show that reasonable neutral carrier ionophores yield liquid ion-selective membranes with far better selectivities than most liquid ion-exchanger-based ion-selective membranes. However, neutral-carrier-based ion-selective membranes are frequently of much higher resistance than ion-exchanger-based ion-selective membranes, because they tolerate a lower concentration of counterions; thus, the total number of charge carriers is fewer in neutral carrier membranes.

Ion-exchanger-based liquid ion-selective membranes incorporate strongly lipophilic mobile ion-exchange sites, which themselves are ions of opposite charge to that of their determinand. The solvent of these membranes is hydrophobic and usually organic. In addition, most ion-exchanger-based liquid membranes contain various species and quantities of coions and counterions. The selectivity of these membranes is based on the differential partitioning of the determinand v interferent ions from the solution-phase solvent into the membrane-phase solvent, on the formation of ion-exchanger–determinand ion pairs within the membrane, and on the partitioning of the ion-exchanger–determinand ion salt between the membrane and solution phases. Other factors contributing to the transmembrane potential achieved by these membranes include intramembrane diffusion in the presence of counterions, differences in charge number between interferent ions and the determinand ion, and the extent of coion penetration from the solution phase into the membrane phase.

The equation for the transmembrane potential for an ion-exchanger-based liquid ion-selective membrane with intramembrane diffusion is (Koryta and Stulík, 1983):

$$\Delta\phi_M = \frac{RT}{F} \ln \frac{a_{J^+,\text{test sol}} + (\gamma_{J^+} u_{K^+}/\gamma_{K^+} u_{J^+})\, K_{exch}\, a_{K^+,\text{test sol}}}{a_{J^+,\text{ref sol}}} \quad (57)$$

which implies that the Nicolsky equation (Eq. [53]) is valid when a diffusion potential exists within the liquid ion-exchanger-based membrane if the selectivity coefficient is given by:

$$K^{pot}_{J^+,K^+} = (\gamma_{J^+} u_{K^+}/\gamma_{K^+} u_{J^+})\, K_{exch} \quad (58)$$

The most familiar example of a fixed site ion-exchanger membrane is the common hydrogen-selective glass membrane found on bench-top pH electrodes. A panoply of theories (based on different assumptions) exists for solid-state ion-exchanger membranes. If it is assumed that the thickness of the membrane is less than the Debye length, that I cations (all of valence z) are in the test solution, that J cations (also all of valence z) are in the ion-standard reference solution, that interfacial ion-exchange is perfectly reversible, that inside any space charge region of the membrane:

$$z_{\text{sites,mem}} + a_{\text{sites,mem}} + z \sum_{i=1}^{I} a_{i,\text{mem}} = 0 \quad (59)$$

(where $z_{\text{sites,mem}}$ is the valence of the ion-exchange sites in the membrane and $a_{\text{sites,mem}}$ is the activity of the ion-exchange sites in the membrane) and that the diffusion of cation i in the membrane is given by:

$$D_i = RTu_{i,\text{mem}} \quad (60)$$

then the transmembrane potential can be expressed as:

$$\Delta\phi_M = \frac{RT}{zF} \ln \left(\sum_{i=1}^{I} D_i k_i^{\text{sol,mem}} a_i \gamma^{-1}_{i,\text{mem}} / \sum_{j=1}^{J} D_j k_j^{\text{sol,mem}} a_j \gamma^{-1}_{j,\text{mem}} \right) \quad (61)$$

The detailed electrochemistry of the three different types of ion-selective electrodes is thus seen to be different, although all three types are theoretically similar in many respects. ISE theory, as presented above, is founded primarily on experiments done with ion-selective macroelectrodes. The theory has been used for ion-selective microelectrodes (ISMs), and for the most part, the extension to smaller scales holds good, *mutatis mutandis*. However, the smaller tip size of ion-selective microelectrodes, and the specialized uses to which they are put, gives rise to a number of theoretical and practical consequences that are discussed in section 4.3.2. below.

3. Construction of ISMs

3.1. Introduction

Any researcher embarking upon the manufacture of ISMs must read the books by Thomas (1978) and Ammann (1986), and would be very well advised to read Purves' (1981) book as well. Ammann (1986), in his excellent chapter on the construction of ISMs, restricts his treatment to liquid-membrane ion-selective microelectrodes. Thomas' (1978) and Koryta and Stulík's (1983) monographs cover both solid-state ISMs (i.e., ion-selective glass-membrane microelectrodes) and liquid-membrane ISMs. Thomas' (1978) account of liquid-membrane ISMs is necessarily somewhat dated, but his technical description of ion-selective glass-membrane microelectrodes remains unsurpassed. Koryta and Stulík's account of the fabrication of ISMs is terse—especially in the case of glass-membrane ISMs—but this abbreviated description of construction protocols is well supplemented by a valuable discussion of the conditions for the correct application of ISMs. Purves (1981) does not discuss ISMs, but superbly and concisely summarizes reference microelectrode recording methods; the insights offered by a careful reading of this monograph will repay the electrophysiologist manyfold.

The experimental use to which an ISM will be put dictates its construction. For example, if the experimental calls for a long-lived, rugged pH microelectrode for extracellular use, a glass-membrane-design ISM might be best; by contrast, a simultaneous intracellular measurement of, say (Ca^{2+}) and membrane potential would call for a side-by-side, double-barreled, neutral carrier, liquid-membrane ISM. Throughout the following outline of ISM fabrication techniques, reference will often be made to the different requirements made upon the ISM depending on its intended use; appropriate designs will be suggested to meet those requirements. It is hoped that, by emphasizing this aspect of ISM manufacturing, that is, the matching of the construction of ISMs to the performance demanded of them, this exposition of ISM fabrication adds a gainful slant to the wealth of practical instruction presented by Thomas (1978), Koryta and Stulík (1983), and Ammann (1986).

Before discussing the fabrication of ISMs, a few practical comments applicable to the fabricating of both glass-membrane and liquid-membrane ISMs are worthwhile at this point. It is extremely helpful to examine the ISM microscopically throughout the manufacturing process (Ammann, 1986). Often, a fault in the fabrication can be seen under the microscope and appropriate corrective action taken to salvage the ISM; many fabrication errors, if undetected until the ISM is completed and tested, cannot be corrected. All solutions used in the fabrication and calibration of ISMs should be made up with deionized water, and solutions to be placed directly into microelectrodes should be filtered—the most convenient way to do this is to use syringe filters.

3.2. Liquid-Membrane ISMs

Liquid-membrane ISMs were first made by Orme (1969); since then, a wide variety of new fabrication techniques and liquid-membrane sensors has been developed (Ammann, 1986). The ready availability of a wide and ever-expanding collection of liquid-membrane sensors, together with the relative ease with which liquid-membrane ISMs can be fabricated, has made them the overwhelming favorites of electrophysiologists (Ammann, 1986). There remain applications for which glass-membrane ISMs are preferable (*see* section 3.3.), but for the majority of ISM applications, liquid-membrane ISMs are the microelectrodes of choice.

Almost as many different fabrication techniques for liquid-membrane ISMs exist as there are research groups using ISMs (Ammann, 1986). Although numerous different manufacturing protocols are in use, all of them share the same basic steps. First, the glass tubing is cut to size, and cleaned if necessary. Next, the glass pieces are pulled into micropipets. Then the micropipets are silanized; that is, their interiors are chemically rendered more hydrophobic and thus better able to retain the liquid membrane solution. Double- or multi-barreled micropipets usually have one barrel left unsilanized; this barrel later becomes the reference microelectrode. Finally, liquid membrane solution is introduced into the silanized barrels, followed by the appropriate ion-standard solution, and, in the case of double- or multi-barreled ISMs, the reference barrel is filled with reference electrolyte. A typical completed double-barreled ISM is schematically shown in Fig. 3. To clarify the following discussion, Fig. 3 also indicates the usual names of the parts of a liquid-membrane ISM.

3.2.1. Preparation of the Glass Micropipet

Micropipets for liquid-membrane ISMs are usually fashioned from pieces of commercially available glass tubing. The properties and structure of silicate glasses used in the manufacture of micropipets are reviewed in Hebert (1969), Elmer (1980), Greaves et al. (1981), Corey and Stevens (1983), and Ammann (1986). A wide variety of glass stock is available, differing in composition, size, and configuration. The type of glass stock to use depends on the style of ISM desired by the investigator. The effects of glass composition, tubing size, and tubing configuration on ISM manufacture will be considered in turn.

The glass used for ISM construction must be of as high a resistance as possible; otherwise shunt conductance through the glass at the micropipet tip can interfere with ISM performance (Lavallée and Szabo, 1969; Armstrong and Garcia-Diaz, 1980; Lewis and Wills, 1980; Ammann, 1986; Carlini and Ransom, 1987; *see also* section 4.3.5. below). Because of suggestions that hydration of the thin glass (Agin, 1969) near the micropipet tip could cause appreciable shunt conductance (Thomas, 1978; Lewis and Wills, 1980; Tsien and Rink, 1981), it would seem wise to select glass with high water resistivity (Ammann, 1986); however, the importance of glass hydration in microelectrode function has been questioned (Coles et al., 1985). The glass must also be resistant to microcrack formation, since such cracks at the micropipet tip have been implicated in the degradation of ISM function (*see* section 4.3.5.; Coles et al., 1985; Ammann, 1986). Glasses used in the construction of liquid-membrane ISMs obviously must not have intrinsic ion selectivity, but some degree of ion selectivity may be unavoidable (section 4.3.5.; Agin, 1969; Eisenman, 1969; Ammann, 1986; Carlini and Ransom, 1987). Finally, it is desirable that the glass be workable—that it have a relatively low softening point (Corey and Stevens, 1983; Ammann, 1986) and that it be stiff, but not excessively brittle (Coles et al., 1985). Aluminosilicate and borosilicate (Pyrex™) glasses generally fulfill the above requirements (Corey and Stevens, 1983; Ammann, 1986); in particular, both of

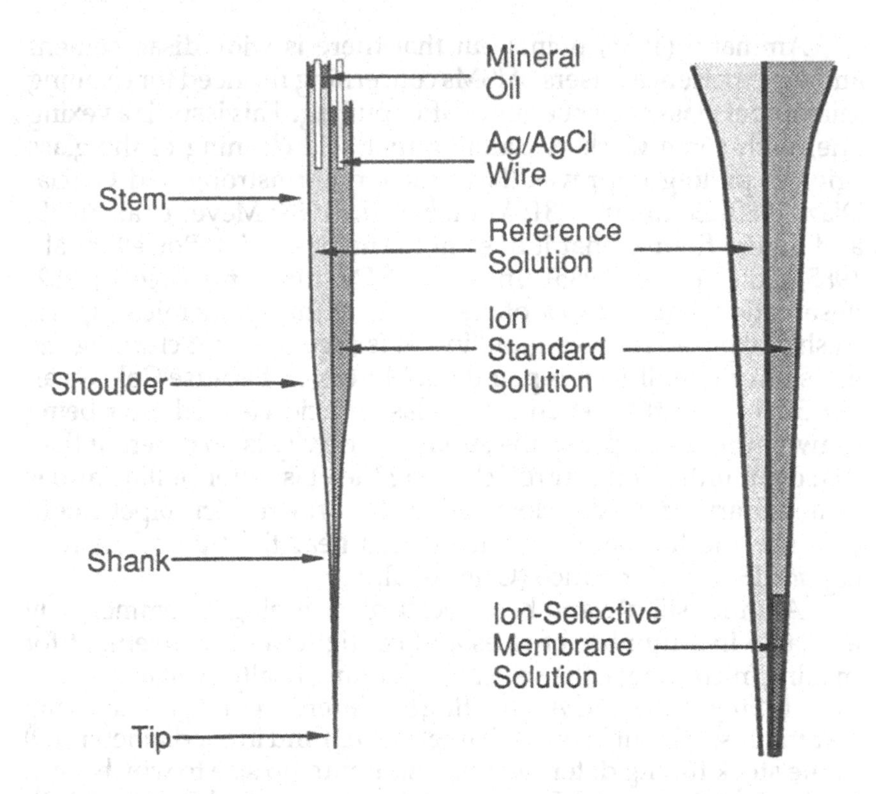

Fig. 3. Diagrammatic cross-sections of a typical double-barreled, liquid-membrane ISM. On the left is a stylized diagram of the entire ISM, and on the right is a magnified view of the tip. The diagram is not drawn to scale.

these types of glass have high specific resistances—at least 10^{16} Ω cm (Hebert, 1969; Lavallée and Szabo, 1969)—low softening points—less than 1000°C (Hebert, 1969; Ammann, 1986)—and low intrinsic ion-selectivity (but *see* Agin, 1969). In some instances, the greater stiffness and microcrack resistance of quartz glass warrants its use, despite the difficulties imposed by its high softening point (1480° C); quartz glass micropipets have the practical advantages of greater longevity and mechanical strength (Munoz and Coles, 1987), and the theoretical advantage of less susceptibility to microcrack-related shunt conductance (Coles et al., 1985). In addition, quartz glass has the highest specific resistivity, and lowest coefficient of thermal expansion, dielectric constant, and surface charge of any silicate glass (Corey and Stevens, 1983).

Ammann (1986) points out that there is wide disagreement among experienced users of ISMs concerning the need for cleaning micropipet glass either before or after pulling. This issue is a vexing one, with some workers claiming that acid cleaning of the glass prior to pulling improves ISM function (Armstrong and Garcia-Diaz, 1980; Zeuthen, 1981; Munoz et al., 1983; Meyer et al., 1985) and others finding that it does not (Thomas, 1978; Borrelli et al., 1985; Coles et al., 1985). In view of Deyhimi and Coles' (1982) observation that over 99% of the glass near the tip of a micropipet is freshly exposed following pulling, it is arguable that cleaning the glass prior to pulling it is pointless. Moreover, because Coles et al. (1985) showed that, even if the glass is acid cleaned after being drawn into a micropipet, the surface resistivity is no different than that of an untreated control, cleaning the glass after pulling also is unnecessary. Indeed, acid treating the drawn micropipet might enhance the formation of microcracks near the tip and thereby *degrade* ISM performance (Coles et al., 1985).

Aluminosilicate and borosilicate glass tubing is commercially available in a number of sizes and configurations convenient for making micropipets; the selection of commercially available quartz glass tubing is sparser. Abstracting for differences in glass-drawing techniques, the initial outer diameter (OD) and inner diameter (ID) of the stock tubing determine the minimum tip size to which it can be drawn (Brown and Flaming, 1977; Flaming and Brown, 1982). To obtain micropipets with the smallest tip sizes—and yet retain practical tip opening diameters—narrow diameter (OD $\leq$ 1 mm), thin-walled (thickness $\leq$ 300 μm) glass tubing should be used (Flaming and Brown, 1982; Borrelli et al., 1985; Ammann, 1986). On the other hand, to reduce interbarrel coupling in double- or multi-barreled ISMs, micropipets should be made with a thick septum and/or thick-walled glass (Sakmann and Neher, 1983; Ammann, 1986; Carlini and Ransom, 1987).

Obviously, the configuration of the tubing also influences the size and shape of the micropipet tip (Brown and Flaming, 1977). The original liquid-membrane ISMs devised by Orme (1969) and Walker (1971) were made from plain, single-barreled, glass tubing. Since then, ISMs have been made from single-barreled tubing with and without internal filling fibers, from single-barreled tubing of triangular and star-shaped cross-section, double-barreled tubing of side-by-side design or theta cross-section, coaxial tubing, and from various triple- and quadruple-barreled tubings (Fig. 4) (Lux

Fig. 4. Cross-sections of glass stock with a variety of barrel configurations. The top row shows single-barreled configurations: from the left, a plain capillary tube, a capillary tube with an internal filling fiber, a thick-walled capillary tube with an internal filling fiber, triangle glass, and star glass. The second row shows side-by-side double-barreled configurations. The third row shows theta-style double-barreled configurations. The bottom row shows multi-barreled configurations.

and Neher, 1973; Kessler et al., 1976; Fujimoto and Honda, 1980; Ujec et al., 1981; Harris and Symon, 1984; Borrelli et al., 1985; Ammann, 1986). Some workers prepare their own double-barreled glass (Zeuthen, 1981; Schlue and Thomas, 1985). The sharp angles provided by internal glass filaments or by the geometry of theta, triangular, and star-shaped cross-section tubing significantly simplify and accelerate the filling of micropipets via injection from the back (Purves, 1981); indeed, the high viscosity of most liquid membrane solutions makes successful back-injection virtually impossible in micropipets lacking fibers or sharp internal angles. Freshly pulled quartz glass has such high hydrophobicity that theta glass micropipets constructed therefrom take an inordinately long time to back-fill with reference electrolyte solution, thus, Munoz and Coles (1987) use quartz theta glass tubing with internal glass fibers in both barrels.

Most ISMs used in neurophysiological research are double-barreled, so as to permit the simultaneous measurement of reference voltage and ion-specific potential, and thereby permit the determination of ion activity by subtraction (*see* section 4.1.). Thus, attention must be drawn to the special problems posed by double- and multi-barreled micropipets. Interbarrel coupling—resistive as well as capacitative—can become significant if the barrels are insufficiently isolated (*see* section 4.3.5., Ammann, 1986; Carlini and Ransom, 1987). To prevent unwanted interbarrel coupling, side-by-side double- and multi-barreled configurations can be used instead of theta or multi-partitioned tubings (Borrelli et al., 1985), or else tubing with thick internal septum(s) can be used (Meyer et al., 1985; Ammann, 1986; Carlini and Ransom, 1987). However, it must also be kept in mind when selecting tubing for intracellular ISMs that side-by-side designs may cause more impalement damage to cells than internally partitioned designs (*see* section 4.3.6.).

Having selected a barrel configuration, tubing diameter, and glass composition fitting the experimental mission of the ISM, the stock tubing must be cut into pieces of the necessary length and pulled into an appropriate micropipet. If multi-barreled tubing is used, it is often advantageous to cut the barrels to different lengths; by so doing, the chance of confusing the barrels later in the fabrication process is lessened, the danger of inadvertent interbarrel coupling at the back end of the ISM is decreased, and connecting the ISM to amplifier headstages is facilitated (Borrelli et al., 1985). Zeuthen's (1981) silanization protocol requires that barrels be cut to different lengths, so as to allow the selective silanization more easily. Munoz and Coles' (1987) protocol also requires that barrels be cut to different lengths and that the end of the barrel destined to become the reference barrel be molten shut prior to pulling.

Next, the prepared pieces of tubing stock must be drawn into the desired micropipets. Modern micropipet pullers offer considerable control over pulling parameters, allowing a wide spectrum of micropipets to be pulled from any given stock tubing (Brown and Flaming, 1977; Flaming and Brown, 1982; Ammann, 1986). To pull quartz glass tubing, it is necessary to modify commercial pullers, because the melting point of commercial heating elements is approximately the softening point of quartz glass (Munoz and Coles, 1987). Usually, it is advisable that the investigator determine what micropipet shape and tip characteristics best fit the experimental application by using "dry run" micropipets, that is, micropipets filled only with reference electrolyte solutions (Borrelli et al., 1985). For example, it is senseless to go to the bother of using completed ISMs to test whether a given tip shape can be used to impale a cell, when the same information is more easily obtained with "dry run" microelectrodes. There is no guarantee that a micropipet shape satisfactory for reference microelectrode use will perform as well when used for ISMs, of course, but in most cases, if the tip opening is not too small (Ammann, 1986), this is not a problem. In general, micropipet with the largest tip opening compatible with the experimental application should be used, because of the relatively high resistance of most currently available liquid membrane solutions (Ammann, 1986). *Ceteris paribus,* micropipets with shorter rather than longer shanks are preferred for ISM manufacturing, since they are less prone to interbarrel coupling, they back-fill more easily, and they are sturdier (Flaming and Brown, 1982).

Some ISM fabrication protocols require manipulations during micropipet pulling. The protocol of Schlue and Thomas (1985) necessitates twisting two barrels about each other during the pulling of the micropipet. Pulling can also be done in multiple stages; double-pulling produces micropipets with short shanks yet fine tips (Snell, 1969; Sakmann and Neher, 1983; Tripathi et al., 1985), which is a favorable geometric combination (Flaming and Brown, 1982). Meyer et al. (1985) silanize their micropipets as they are being pulled.

Many protocols advise that micropipet tips be beveled (Kailia and Voipio, 1985; Munoz and Coles, 1987) or broken back (Tsien and Rink, 1980; Walden et al., 1984; Tripathi et al., 1985) immediately after pulling (so-called "dry beveling"), although this is certainly not always necessary (Borrelli et al., 1985; Meyer et al., 1985; Schlue and Thomas, 1985; Chesler and Kraig, 1987; Siebens and Boron, 1987). Beveling can also be done after silanization (Tsien and Rink, 1980). A number of different techniques for beveling (Brown and Flaming, 1979; Kaila and Voipio, 1985; Munoz and Coles, 1987) or breaking back (Briano, 1983; Tripathi et al., 1985) micropipets exist. Beveled micropipets filled with reference electrolyte have lower resistances than unbeveled micropipets of the same external tip diameter (Brown and Flaming, 1979; Purves, 1981), are less likely to clog (Purves, 1981), and often cause less tissue trauma (Purves, 1981; Ammann, 1986). In addition, ISMs made from beveled or broken back micropipets can have improved selectivity and faster response times (Tripathi et al., 1985). It is important that micropipet tips be cleaned of any debris created by beveling or breaking (Brown and Flaming, 1979) prior to being silanized, to prevent incomplete silanization or blockage. In view of the influence that tip size can have on ISM performance (*see* section 4.3.5.; Carlini and Ransom, 1987; Ransom et al., 1987), it is important that investigators achieve consistent tip sizes during ISM fabrication—beveling and breaking back of tips should be done in a controlled and reproducible fashion. Light microscopic observation of the micropipet throughout the manufacturing process is recommended, to assure that the micropipet tip remains at its intended size, that beveling or breaking, if done, is satisfactory, and—later in the fabrication process—that silanization and filling are being properly carried out. Although it is often difficult, using air objectives, to visualize the very tip of fine-tipped micropipets satisfactorily, it is usually possible to determine if any important change in tip diameter has occurred as a result of beveling or breaking.

3.2.2. Silanization

Silicate glasses have a high density of free hydroxyl groups at their surface (Elmer, 1980; Deyhimi and Coles, 1982; Corey and Stevens, 1983; Munoz et al., 1983; Ammann, 1986). Under normal laboratory conditions, these free hydroxyl groups adsorb water and therefore cause the surface to become hydrophilic (Munoz et al., 1983.) Liquid ion-selective membranes, being hydrophobic, are easily displaced from the tips of untreated glass micropipets by aqueous solutions (Corey and Stevens, 1983); hence, both the standard electrolyte solution behind the liquid ion-selective membrane and the test or biological aqueous solution into which the tip is immersed tend to insinuate between the liquid membrane and the glass (Munoz et al., 1983). The displacement of the liquid membrane by aqueous solutions can cause a number of problems, including formation of short-circuiting leak conductances along the glass walls (Coles and Tsacopoulos, 1977; Munoz et al., 1983; Ammann, 1986), formation of a dead volume at the tip, and formation of unstable electrolyte–membrane phase interfaces (*see* section 4.3.1.).

To prevent aqueous displacement of the hydrophobic membrane phase, facilitate the introduction of the membrane phase into the micropipet tip, and prolong the lifetime of the ISM by stabilizing membrane–glass interactions, the inner surface of the glass near the micropipet tip is usually rendered hydrophobic by reacting the surface hydroxyl groups with a silane derivative. Details of the reactions of silicate glasses with silanes can be found in Plueddeman (1980), Deyhimi and Coles (1982), Corey and Stevens (1983), Munoz et al. (1983), and Ammann (1986). A number of different silanes and silanization techniques have been used to silanize micropipets (Walker, 1971; Lux and Neher, 1973; Coles and Tsacopoulos, 1977; Tsien and Rink, 1980; Munoz et al., 1983; Borrelli et al., 1985; Meyer et al., 1985; Tripathi et al., 1985; Ammann, 1986; Munoz and Coles, 1987). Silanes react with water to form reaction products that can polymerize (Munoz et al., 1983; Ammann, 1986); these unwanted polymerization products can plug small-tipped micropipets (Munoz et al., 1983; Ammann, 1986). Side reaction of silanes with adsorbed surface water can be prevented by drying the glass micropipets prior to silanization (Coles and Tsacopoulos, 1977; Ammann, 1986). If micropipets are silanized immediately after being pulled, the freshly exposed glass surface (Deyhimi and Coles, 1982) near the tip is usually free of adsorbed water, and the drying step can be omitted (Borrelli et al., 1985).

To summarize, there are two main methods to silanizing micropipets: liquid-phase silanization and vapor-phase silanization. The first silanization technique (Walker, 1971) used to prepare liquid-membrane ISMs was a liquid-phase method. The original liquid-phase silanization method consists simply of dipping the micropipet into a silane solution (e.g., trimethylchlorosilane dissolved in carbon tetrachloride) and sucking the solution up and down the shank of the micropipet a few times. A simple modification of this method allows the selective silanization of a single barrel of double- or multi-barreled micropipet (Lux and Neher, 1973). After the final expellation of silane solution, either the silanized barrel can be filled immediately (Lux and Neher, 1973), or else the micropipet can be baked so as to ensure the reaction of the remaining thin film of silane with the glass and to drive off any remaining solvent (Walker, 1971). Liquid-phase methods, in which the tip of the micropipet is dipped into a silane solution, also silanize the exterior of the micropipet tip. External silanization may be useful in improving the durability of intracellular recording (Aicken and Brading, 1982) and in reducing surface conduction along the exterior of the micropipet (Orme, 1969; Munoz et al., 1983). The liquid-phase suction method of silanization is adequate for preparing the relatively large-tipped (≥ 3-μm tip diameter) extracellular ISMs, but does not work well with the fine-tipped micropipets necessary for intracellular investigations, because of the difficulty of generating sufficient suction through a small tip opening.

Vapor-phase silanization was pioneered by Coles and Tsacopoulos (1977) to permit the construction of fine-tipped ISMs. Essentially, the method consists of blowing a stream of silane-impregnated nitrogen gas into a micropipet barrel and then baking the micropipet. In the case of double- or multi-barreled micropipets, clean N_2 gas is simultaneously blown through the barrel(s) not to be silanized, to protect them from being silanized by vapor entering through the tip (Fig. 5). The technique has undergone steady improvement and modification since then (Tsien and Rink, 1980; Schlue and Thomas, 1985; Tripathi et al., 1985; Munoz and Coles, 1987). Tsien and Rink's (1980) method is particularly suited to preparing batches of single-barreled ISMs; it has the additional advantage of silanizing the exterior of the micropipet while the interior is silanized. Deyhimi and Coles (1982) and Munoz et al. (1983) studied the effect of varying the type of glass, the silane reagent, the temperature of baking, and the time of baking on the degree of hydrophobicity achieved. In particular, they determined that aminosilanes are more reactive than chlorosilanes and impart the greatest degree of hydrophobicity to the glass. Dimethylamino-trimethylsilane is the aminosilane of choice, because it does not form a polymerization product with water, it has a high vapor pressure, and it has a noncorrosive hydrolysis byproduct (Munoz et al., 1983).

A hybrid liquid-phase–vapor-phase silanization technique is advocated by Borrelli et al. (1985), Meyer et al. (1985), and Chesler and Kraig (1987). The hybrid methods generally are simpler and more efficient than vapor-phase methods, yet permit the construction of fine-tipped ISMs. Briefly, silane solution is introduced into a barrel of glass tubing from the stem end just before (Meyer et al., 1985) or just after (Borrelli et al., 1985; Chesler and Kraig, 1987) it is

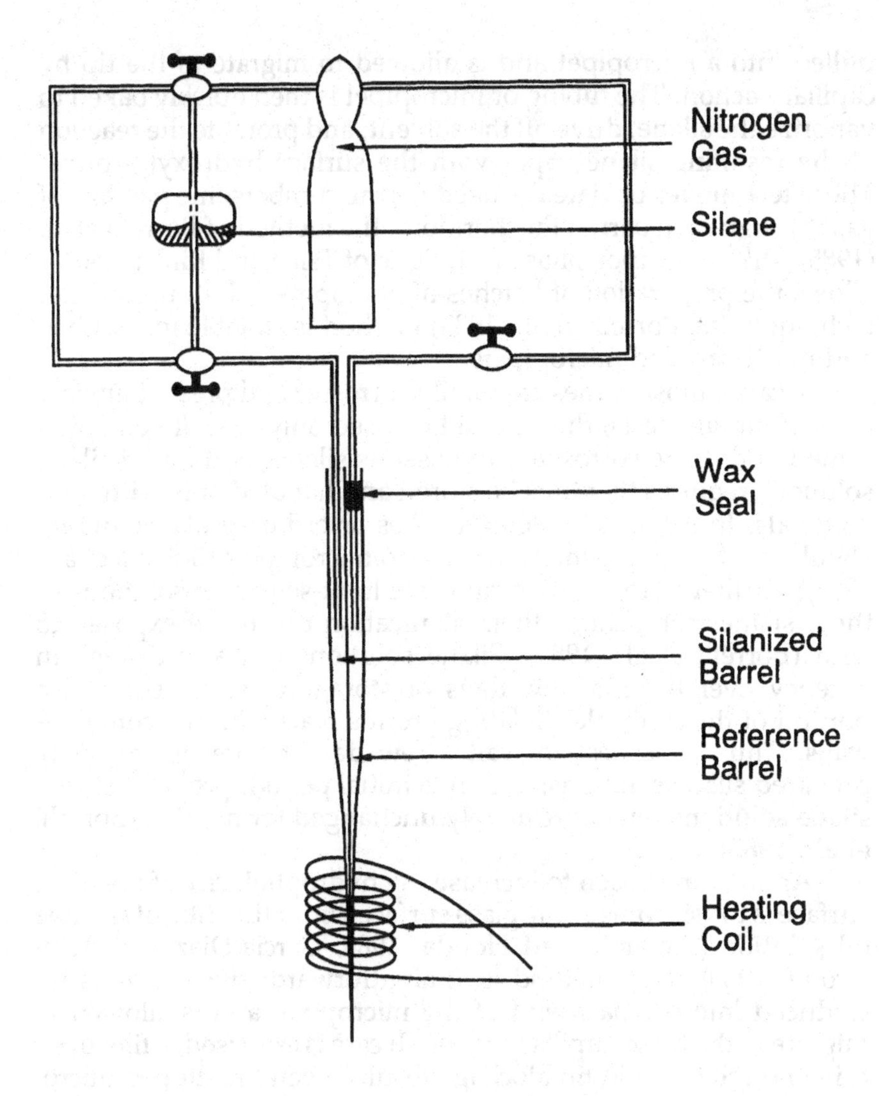

Fig. 5. Diagram of a Coles and Tsacopoulos (1977) style vapor-phase silanization setup. Valves control the flow of gas. Clean N_2 gas is blown through the reference barrel on the right while silane-impregnated N_2 is blown through the other barrel. The heating coil bakes the tip of the micropipet.

pulled into a micropipet and is allowed to migrate to the tip by capillary action. The tubing or micropipet is then quickly baked to vaporize the silane, drive off the solvent, and promote the reaction of the resulting silane vapor with the surface hydroxyl groups. These techniques obviate the need for the cumbersome passage of gas through each barrel. Furthermore, the method of Borrelli et al. (1985)—like the vapor-phase technique of Tsien and Rink (1980)—allows the preparation of batches of micropipets, but, unlike that technique, the Borrelli et al. (1985) method is suitable for double- and multi-barreled micropipets.

Because most silanes are highly corrosive and present unclear risks of mutagenesis, they should be used only in well-ventilated fume hoods. The corrosivity of gaseous silanes and liquid silane solutions requires that they be stored and handled only with inert materials. In particular, metal needles or ordinary plastic tubing should not be used to inject silane into micropipets (Borrelli et al., 1985). Furthermore, many silanes are light-sensitive; solutions of these silanes can change their silanization potency if exposed to light (Borrelli et al., 1985). Silane solutions can also change in potency over the first few days of storage in glass, because a portion of the available silylating groups react with the container walls. Thus, it is best to wait a few days before using newly prepared silane solutions; after this initial period, properly stored silane solutions remain relatively unchanged for months (Borrelli et al., 1985).

Another approach to increase the hydrophobicity of the glass surface is to "siliconize" the glass surface with a thin film of silicone oil solution (Fujimoto and Honda, 1980; Garcia-Diaz and Armstrong, 1980). This method is straightforward: silicone oil is introduced into the back end of the micropipet and is allowed to migrate to the tip by capillary action. If care is exercised, a film of oil thin enough to avoid tip blockage results; even fine-tipped micropipets can be siliconized (Garcia-Diaz and Armstrong, 1980; Ammann, 1986), but Munoz et al. (1983) warn that it may sometimes be difficult to control the thickness of the siliconizing film, and Zeuthen (1981) finds that tips with ID ≤ 3 μm clog. Silicone oil does not form silicone polymers with water; thus, the problem of polymerization products blocking the tips of micropipets—which can occur with the more reactive silane monomers (Ammann, 1986)—is sidestepped.

3.2.3. Filling

Following silanization, the next steps in the fabrication process are the introduction of liquid ion-selective solution into the silanized barrel(s) of the micropipet, the filling of the reference barrel, if any, and the layering of ion-standard solution(s) atop the liquid-membrane column(s). The first of these steps to be described will be the formation of an ion-selective liquid membrane from a short column of ion-selective solution placed into the silanized tip of the micropipet (Fig. 3). Although this step will be described first, it need not necessarily be the first filling step in practice; often, the reference barrel is filled just before the ion-selective barrel is filled *(see below)*.

The organic ion-selective solution, like any other micropipet filling solution, can be introduced through either the back end of the stem (back-filling) or through the tip (front-filling) of the micropipet (Purves, 1981). Front-filling a silanized single-barreled micropipet with ion-selective solution is straightforward. The tip is dipped into a reservoir containing ion-selective solution; with sufficiently wide tips, capillary forces and hydrophobic interactions between the silanized glass and the organic liquid ion-selective solution then suffice to draw up a 200–500-μm-long column of ion-selective solution (Orme, 1969; Walker, 1971). Suction can be applied to the back end to hasten filling (Orme, 1969). Similarly, individual barrels of double- and multi-barreled micropipets can be selectively front-filled if the other barrels are protected from suction or capillary filling (Lux and Neher, 1973). Although front-filling works well for extracellular-sized tip openings, front-filling micropipets with small tip openings (≤2 μm) is much more difficult; even with an aqueous electrolyte solution (Purves, 1981), front-filling is tedious and inefficient—let alone with a highly viscous, organic liquid-membrane solution. Front-filling also presents another problem—that of bringing the internal ion-standard solution into contact with the liquid membrane (Ammann, 1986). After forming the liquid-membrane column, Walker's (1971) method requires the injection of internal filling solution through the back end of the micropipet using fine tubing; unfortunately, it is frequently difficult to achieve a satisfactory filling-solution–liquid-membrane interface using this technique, because the top of the liquid membrane is far down into the narrow part of the shank. Lux and Neher's (1973) method avoids this interface problem, because the ion-selective solution is drawn up only after the active barrel is first filled with internal ion-standard solution (either by front- or back-filling); however, inasmuch as this technique requires that all barrels be filled with internal electrolyte solutions prior to drawing up ion-selective solution, there may be cross-contamination of internal filling solutions at the tip ("crawl around" of the electrolyte solutions). Another potential problem of this method is the formation of a very small column of ion-selective solution at the reference tip when the micropipet is dipped into the ion-selective solution reservoir.

The limitations of front-filling proscribe its use for fine-tipped micropipets; instead, back-filling, in which filling solutions are introduced into the micropipet from the stem end, is usually the technique of choice for fine-tipped micropipets (Purves, 1981). When introduced from the stem end of the micropipet, highly viscous ion-selective solutions migrate to the tip very slowly, unless the barrel has an internal fiber or sharp dihedral angles (Fig. 4; *see also* Purves, 1981). The usual procedure is to inject a droplet of ion-selective solution into the shank of the micropipet through a fine tube or needle (Thomas, 1978; Borrelli et al., 1985; Ammann, 1986); then, if the tip is properly silanized and not blocked, the solution migrates to the tip by capillary action, especially if an internal filling fiber is present. It is preferable to use glass or plastic tubing (Borrelli et al., 1985) rather than metal needles for injecting solutions, to avoid potential problems resulting from the reaction of the filling solutions with the metal (Corey and Stevens, 1983). The resulting liquid-membrane column is usually several millimeters in length (Borrelli et al., 1985; Ammann, 1986). In order to significantly reduce the resistance of the ion-selective column, the liquid-membrane column must be made shorter than 100 μm (Tsien and Rink, 1980; Ujec et al., 1981; Ammann, 1986). To achieve such short columns, the coaxial inner micropipet technique can be used (*see below;* Ujec et al., 1979).

The descent of ion-selective solution into the tip can be speeded by several expedients. Positive pressure at the stem end can be used to force the ion-selective solution into the tip, or vacuum can be applied at the tip (Ammann, 1986). Meyer et al. (1985) inject diethyl ether into the micropipet tip ahead of the column of ion-selective solution; the micropipet is then put into a 100° C oven for 2 min. This treatment usually causes the ion-selective solution to fill to the tip quickly; they hypothesize that the fast evaporation of ether creates a suction effect at the tip, thereby drawing the ion-selective solution into it. Often, small air bubbles are present in the column of ion-selective solution; with time, these usually dissipate spontaneously (Borrelli et al., 1985). If air bubbles do not dissipate spontaneously, or do so too slowly, a fine glass, tungsten, or cleaned animal whisker can be inserted through the stem and into the membrane solution (Thomas, 1978); by wiggling the whisker to and fro, bubbles usually can be dislodged. Schlue (personal communication) dislodges bubbles by briefly warming the micropipet at a point just tip-side of the bubble(s) using a fine-tipped soldering iron. One must not overheat the micropipet, lest the solvent of the ion-selective solution vaporize. If quartz glass or thick-walled micropipets are being employed, vigorous tapping of the micropipet often suffices to dislodge bubbles trapped in the mid-shank (Corey and Stevens, 1983).

With the exception of micropipets prepared by some of the liquid-phase silanization methods—which require that the reference barrel be filled before the introduction of ion-selective solution into the silanized barrel (Lux and Neher, 1973)—it is immaterial whether the reference barrel of double-and multi-barreled micropipets is filled before or after the silanized barrel(s) are filled. For practicality, however, it is suggested that reference barrels be filled first, generally just before introducing ion-selective solution(s) into the silanized barrel(s); by so doing, the expense and effort of filling with ion-selective solution(s) are spared should it happen that the reference barrel is blocked—as occasionally occurs when the reference barrel tip is inadvertantly silanized. In such instances, the aqueous reference electrolyte solution will usually not fill the tip, and instead, a portion of the hydrophobic ion-selective solution frequently will creep from its barrel(s) into the tip of the reference barrel (Borrelli et al., 1985); such "crawl around" of the liquid membrane effectively ruins the ISM. On the other hand, if the silanization of the ion-selective barrel(s) is inadequate, the opposite "crawl around" can occur, wherein the ion-selective solution is prevented from filling or displaced from the tip of its barrel by the reference electrolyte.

The reference barrel may be either front-filled or back-filled; because of the ease and speed of back-filling, and because most double- and multi-barreled glass tubing used for manufacturing

ISMs is available with internal fibers and/or sharp internal dihedral angles, back-filling has become the method of choice. For the reasons outlined above for the case of ion-standard solutions, the tubes or needles with which the reference solution is introduced should preferably be made of glass or plastic, and should be flushed prior to each use (Corey and Stevens, 1983). The same strategems useful for removing bubbles from columns of ion-selective solution can be employed for removing bubbles from the reference solution column.

The choice of reference solution depends on a number of factors, including the design of the ISM, its intended application, and the characteristics of the experimental preparation. In double- or multi-barreled ISMs, the proximity of the reference barrel's tip to the ion-selective barrel's tip increases the likelihood of exposing the ion-selective membrane to leaking reference solution; thus, the reference solution ought not to contain more of the determinand ion than is expected in the experimental fluid, especially if the ISM has large tip opening(s) or if it is to be used in a restricted volume (i.e., intracellularly). For similar reasons, the ionic strength and pH of the reference solution should be close to that of the experimental fluid. The prospect of reference solution leakage (Thomas, 1978; Fromm and Schultz, 1981) further demands that the reference solution be biologically compatible (*see* section 4.3.6.). The inner half-cell in virtually all ISM reference barrels is Ag/AgCl, hence the reference solution must contain some chloride ion (Bates, 1969; Purves, 1981; Ammann, 1986; Raynauld and Laviolette, 1987). Many ISM applications require accurate absolute measurements of reference potentials; reference solutions for these applications need to be composed to minimize electrolyte leakage, interference with the ion-selective barrel(s), and liquid-junction and tip potentials (*see* section 4.3.; Agin, 1969; Purves, 1981; Thomas and Cohen, 1981; Ammann, 1986). It is usually impossible to optimize the reference solution with respect to all of these variables at once (Ammann, 1986). Thomas and Cohen (1981) used an organic reference solution to avoid these difficulties (*see* section 4.3.2.).

Once the reference barrel, if any, has been filled and the liquid-membrane column has stabilized, the exterior of the ISM can be cleaned if desired, and the appropriate ion-standard solution can be layered upon the liquid membrane. A jet of distilled water squirted down the shank towards the ISM tip usually washes away any debris or fluids adhering to the micropipet's exterior without harming its tip (Orme, 1969); so long as each barrel of the micropipet is filled to its tip, there is no danger of the water jet displacing any of the filling solutions. It is good practice to rinse the ISM tip in this manner whenever it is removed into air from immersion in a solution—otherwise the rapid evaporation of thin films of adherent solution may leave salt crystals that can block the ISM tip. It is especially important to do this with side-by-side barrel designs, because of their propensity to hold fluids in the grooves between their barrels. Thick-walled and quartz glass ISMs are strong enough to withstand cleaning by being wiped between one's wetted fingers (Munoz and Coles, 1987).

Ion-standard solutions are layered upon the ion-selective liquid membrane by being injected through a fine glass or plastic tube into the stem end of the ion-selective barrel. The tip of the injecting tube should reach at least as far as the top of the column of ion-selective solution; the ion-standard solution is then forcefully ejected into the barrel, resulting in the flushing away of a quantity of the ion-selective solution from the top of its column and the formation of a distinct, sharp ion-standard solution–liquid-membrane interface. Good interfaces are free of bubbles and present a smooth meniscus that is flat or convex, with the apex of the convexity directed towards the ISM tip; concave menisci usually indicate inadequate silanization (Tripathi et al., 1985).

Ideally, the activity of the determinand ion in the ion-standard solution should fall well within the linear portion of the EMF vs log ion-activity relationship. In some cases, however, low activity ion-standard solutions may be preferable; for example, Ca^{2+} ISMs maintain their calibration for longer when low (Ca^{2+}) (100 n*M*) ion-standard solutions are used instead of 100 m*M* Ca^{2+} solutions (Tsien and Rink, 1980). The ionic strength and pH of the ion-standard solution should approximate that of the experimental fluid, to assure that both sides of the ion-selective liquid membrane are exposed to solutions that are as similar as possible to each other in every respect except one—their determinand ion activity (Koryta and Stulík, 1983). In critical cases, if the activity of interferent ions in the experimental fluid is known, the ion-standard solution can be prepared to have the same activity of interferent ions, thereby making the transmembrane gradient of these ions as small as possible.

After the ion-standard solution(s) are layered atop their respective columns of ion-selective solution(s) and the reference barrel, if present, is filled with reference electrolyte, most workers put a 3–10-mm-thick layer of mineral oil into the top of each barrel, to improve the electrical isolation of each barrel (Walker, 1971; Koryta and Stulík, 1983; Borrelli et al., 1985; Ammann, 1986). The layer of mineral oil also prevents the evaporation of filling solutions and protects against cross-contamination of the filling solutions when the internal Ag/AgCl wire electrodes are inserted.

3.2.4. Assembling and Storing

To complete the construction of the liquid-membrane ISM, inner reference electrodes are put into each barrel. The almost universal choice of inner reference electrode is Ag/AgCl wire (Ammann, 1986; Raynauld and Laviolette, 1987). The Ag/AgCl wire can be either sealed into place with wax (Schlue and Thomas, 1985) or left loose. Fixing the wires into place with wax has the additional benefit of sealing off the barrels, ensuring that no interbarrel electrolyte connection forms at the stem end, and preventing the evaporation or contamination of the filling solutions. If the wires are to be left unsecured, they are usually not inserted until the ISM is actually to be used. A typical double-barreled ISM is diagrammed in Fig. 3.

Various additional modifications can be made to some specific ISM designs. For example, side-by-side, double- or multi-barreled ISMs have deep exterior grooves running between their barrels. During the preparation or use of such ISMs, droplets of solution may fall into these grooves, or solution may be wicked up from the immersed tip by capillary action; solution trapped in the grooves can then form a continuous column of solution from the stem end to the tip end of the ISM. Columns of solution in the grooves can short-circuit any or all of the ISM barrels or cause interbarrel coupling (Borrelli et al., 1985). To prevent this, a dollop of silicone grease or molten wax is applied to the exterior midstem of side-by-side double- or multi-barreled ISMs, so as to occlude the grooves (Borrelli et al., 1985).

To reduce the longitudinal resistance of ion-selective barrel(s) in an ISM, thin micropipet(s) filled with ion-standard solution can be inserted coaxially into the barrel(s) and far up into the liquid-membrane column(s) (Ujec et al., 1978, 1981). These coaxial ISMs have faster response times, lower noise levels, and greater resolving power than ordinary liquid-membrane ISMs, and are therefore useful for measuring fast ion transients (Ujec et al., 1979, 1981; Ammann, 1986). Unfortunately, it is technically difficult to insert the inner filling micropipet so that its tip is within 30 μm of the ISM tip, which is necessary in order to optimize the response time; moreover, thus far, coaxial construction has been restricted to ISMs with tip sizes of 1 μm or larger (Ammann, 1986).

Liquid-membrane ISMs can be stored dry or wet, in partially assembled form or as completed microelectrodes (Orme, 1969; Tsien and Rink, 1980; Tripathi et al., 1985; Ammann, 1986). When stored dry, ISMs are usually placed in a dessicator; it is imperative that filled ISMs stored in this manner have mineral oil or wax seals at their back ends. When stored wet, liquid-membrane ISMs are stored with their tips immersed in a conditioning solution (*see* section 4.2.1.) protected from evaporation and dust. Giving theoretical or experimental bounds on the lifetimes of completed ISMs is difficult (Oesch et al., 1985; Ammann, 1986). Variables influencing the lifetime of the ISM include its type, its method of construction, and the means of storage. If diffusion of carrier from the membrane phase into the adjoining aqueous phases is the primary determinant of ISM longevity, calculations show that liquid-membrane ISMs stored with their tips immersed should remain sensitive for over 2 mo (Oesch et al., 1985; Ammann, 1986). A similar estimate is obtained for liquid membrane ISMs stored with tips dry, but liquid membranes in contact with back-filled ion-standard solutions. However, it is not the diffusive loss of liquid-membrane components that is the primary limitation on ISM lifetimes. ISMs in storage can suffer a number of insults, including development of leak conductances by prolonged hydration (Lanford, 1977); breakage by handling; tip-plugging by particulates in the air or storage solution; tip-plugging by crystals evaporated from the tip of the reference barrel; surface contamination and membrane poisoning by impurities in the storage or back-filling solutions; and displacement of liquid membrane from the tip by reference solution, storage solution, or back-filling solution (Ammann, 1986).

To avoid these problems, liquid-membrane ISMs can be stored in partially constructed form. Silanized but unfilled micropipets will keep for months in a dessicator (Tripathi et al., 1985; Munoz and Coles, 1987). It sometimes happens that double- or multi-barreled micropipets stored in this fashion will be found to have hydrophobic reference barrel tips—perhaps because residual

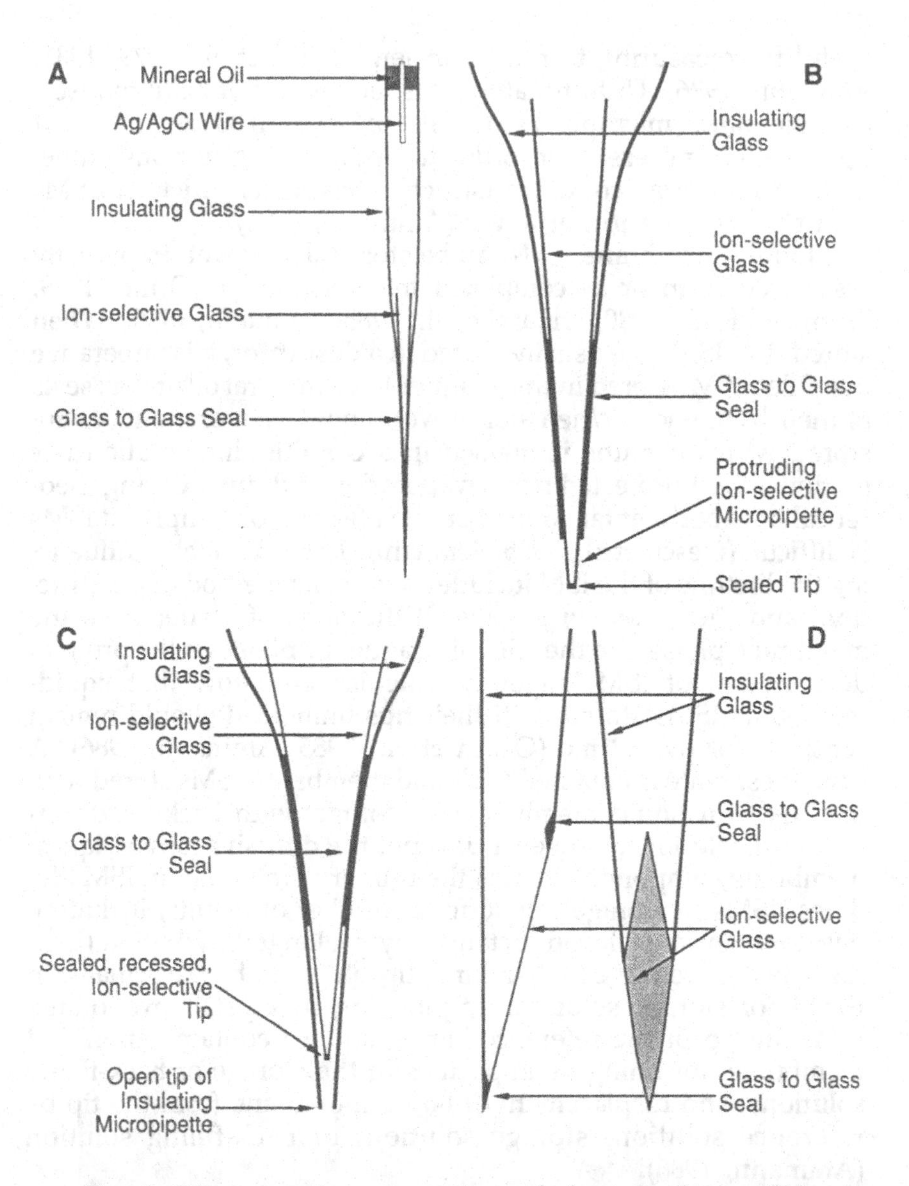

Fig. 6. Diagrammatic cross-sections of glass membrane ISMs. **A:** Stylized cross-section of an entire single-barreled, spear, or recessed-tip-design glass-membrane ISM. **B:** Magnified view of the tip of a spear-design ISM. The sealed inner micropipet tip protrudes from the outer, insulating micropipet for about 5–20 μm. The glass-to-glass seal extends for 50–200 μm. **C:** Magnified view of the tip of a recessed-tip ISM. The

silane in the silanized barrels diffuses around the tip to silanize the reference barrel partially (Borrelli et al., 1985). To combat this difficulty, double-barreled ISMs can be stored dry, with just the ion-selective solution column in place; a potential problem with this storage technique is that the ion-selective solution might "crawl around" to the reference side; however, this minute column of ion-selective solution in the reference tip is usually readily displaced, once reference solution is put into the reference barrel and the ISM is immersed into calibrating or conditioning solutions.

3.3. Glass-Membrane ISMs

The first ISMs were fabricated with glass H^+-selective membranes (Caldwell, 1954). Numerous improvements in the design of glass-membrane ISMs have been implemented over the years since (Thomas, 1978; Zeuthen, 1981; Koryta and Stulik, 1983; Pucacco et al., 1986; Hinke, 1967; 1987). Although glass-membrane ISMs are more difficult to make and are less easily miniaturized than are liquid-membrane ISMs, they have certain favorable characteristics that recommend them for particular applications (Thomas, 1978; Hinke, 1987). In general, compared to liquid-membrane ISMs, glass-membrane ISMs are more rugged and longer lived; less susceptible to interference, membrane poisoning, and electrical shunt problems; and do not leak components into the preparation (*see* section 4.3.6.). Glass-membrane ISMs are recommended for applications in which fine-tip size is unnecessary (i.e., for extracellular use or use in large cells), for use in mechanically adverse circumstances (i.e., active muscle tissue), and for applications in which leakage of membrane components cannot be tolerated (e.g., in vivo human tissues).

Glass-membrane ISMs suitable for neurophysiological applications fall into three main design categories: spear, recessed-tip, and stretched-membrane types (Fig. 6). The recessed-tip and

←

sealed inner micropipet tip is recessed by 10–30 μm from the orifice of the outer micropipet tip. The glass to glass seal extends for 40–100 μm. **D:** Side (left) and front (right) views of the tip of a stretched membrane ISM. The thin, ion-selective glass membrane (shaded) seals over the beveled opening in the insulating micropipet.

stretched-membrane types of glass-membrane ISM can also be paired with reference barrels to construct double-barreled ISMs (Zeuthen, 1981; Pucacco et al., 1986). A variety of H^+-, K^+-, and Na^+-selective glasses are available for use in the construction of any of these designs (Hebert, 1969; Pucacco et al., 1986; Hinke, 1987). Suitable insulating glasses include the borosilicate, aluminosilicate, and quartz glasses already mentioned in section 3.2.1.

The central problem in constructing glass-membrane ISMs is that of miniaturizing their tips while retaining their sensitivity to the determinand ion and keeping their resistivity within manageable bounds; technically, this problem reduces to achieving simultaneously adequate exposure of the ion-selective glass membrane at the active tip and good insulation of the ion-selective membrane throughout the inactive length of the microelectrode (Hinke, 1987). The spear design (Fig. 6A,B) aims to achieve this by having a small extent of the ion-selective glass membrane protrude out from its insulating glass sheath. The recessed-tip design allows a relatively large area of active membrane to contact the sample solution, while keeping a small tip size by having a length of ion-selective glass membrane recessed within a small-tipped insulating micropipet (Fig. 6C). The stretched-membrane design solves the problem of maintaining sufficient exposure of the ion-selective glass membrane in the face of tip size limitations by having the membrane stretched over the beveled tip of an insulating micropipet (Fig. 6D). The construction of each of these designs will be described in turn. The details of cleaning the stock glass tubing, cutting it to size, pulling it into micropipet, and beveling the micropipets do not differ materially from those described for liquid membrane ISMs and will not be repeated here.

3.3.1. Spear-Design ISMs

To construct single-barreled spear-design ISMs, two micropipets must be pulled: an outer, insulating micropipet made from an inactive glass capillary tubing *without* internal filling fiber and an inner, active micropipet made from ion-selective glass tubing. Obviously, the ID of the inactive glass capillary tubing must be greater than the OD of the active glass tubing. The tip of the active micropipet is sealed by melting it shut with a microforge; it is important that only the extreme tip be molten, because a big molten glob at the tip will prevent the active micropipet from being properly positioned in the external, insulating micropipet and too thick an active membrane will increase the resistance of the ISM to unacceptable levels. The active micropipet is then coaxially inserted into the outer micropipet, with care being taken not to damage its tip, until the tip of the active micropipet produces from the tip opening of the insulating micropipet. It is easier to insert and position the inner micropipet if the outer micropipet is first filled with distilled water; after insertion, this water must be evaporated (Zeuthen, 1981). The tip opening of the outer, inert micropipet must be sufficiently wide that a 5–20 μm length of the inner micropipet protrudes.

Next, the insulating micropipet must be fused to the active micropipet to prevent electrical shunting. There are a number of techniques for doing this (Hinke, 1987), of which the best is probably a direct glass–to–glass seal (Zeuthen, 1981; Hinke, 1987). The coaxial micropipet assembly is positioned in the heating element of a microforge so that the end of the outer micropipet, but not the tip of the inner microelectrode, is within the heating element. If the ion-selective glass melts at a lower temperature than the insulating glass, a gas-tight tube is sealed into stem of the inner micropipet, and about 10 atm of pressure are applied and maintained with a hand syringe at the distal end of the tubing; if the reverse is the case, this need not be done. Enough heat is then applied to melt a 50–200-μm-wide zone (measured back from the end of the outer micropipet) of the glass with the lower melting point, without melting the glass with the higher melting point. If the inner micropipet is made of the lower-melting-point glass, then the combination of the heat and internal pressure will cause it to balloon outwards and fuse with the external micropipet; if the reverse is the case, then the outer micropipet will collapse around and fuse with the inner one (Zeuthen, 1981). By tugging on the inner micropipet, it can be made to break off just behind the glass-to-glass seal, so that only a short segment of ion-selective micropipet remains in the ISM assembly.

Filling solutions for glass-membrane ISMs are the same as those for liquid-membrane ISMs. The methods of filling and storing spear-design glass-membrane ISMs are similar to those described for liquid-membrane ISMs in section 3.2.4. However, because the tip of the spear-design ISM is sealed, front-filling is impossible. Furthermore, because the tip is sealed and the outer micropipet lacks a filling fiber (otherwise a tight glass-to-glass seal cannot be obtained), back-filling is much more difficult. It may be necessary to fill these micropipets using the boiling and solution-replacement methods described in Purves (1981). Another expedient is to insert after the glass-to-glass seal has been formed, a thin glass, tungsten, or animal whisker, to function as a filling fiber (Purves, 1981); bubbles in the filling solution can be dislodged by heating in a microforge (Thomas, 1972) or by any of the methods described for dislodging bubbles in liquid-membrane ISMs (*see* section 3.2.3). Spear-design ISMs are completed by assembling and storing them following the protocols outlined for liquid-membrane ISMs (section 3.2.4.). Conditioning is not optional for glass-membrane ISMs; it is required. Ion-selective glasses must be well hydrated before they can function properly; therefore, they must be soaked in conditioning solutions (section 4.2.1.). This is especially true for ISMs whose glass membranes may have become dehydrated during the heating steps of the fabrication process (Thomas, 1972; Zeuthen, 1981).

3.3.2. Recessed-Tip ISMs

Recessed-tip ISMs admit more miniaturization than do spear-type ISMs, and do so without compromising the area of exposed active glass membrane (Thomas, 1978; Zeuthen, 1981; Hinke, 1987). The construction of single-barreled, recessed-tip ISMs is similar to that of spear-design ISMs (Thomas, 1972; Thomas, 1978). An inner micropipet made of ion-selective glass is threaded into an outer micropipet drawn from filamentless inert glass tubing, but, instead of protruding from the outer micropipet, the sealed tip of the inner micropipet remains ensheathed within the outer, insulating micropipet, positioned about 10–30 μm back from the open, 0.5–2 μm OD tip of the outer micropipet (Fig. 6C). To obtain the correct fit of inner and outer micropipets requires adjustment of their respective tapers; correct tapers will be arrived at only after some trial and error at the micropipet puller. A glass-to-glass seal is made as described for the spear-type ISM, so that approximately 40–100 μm of ion-selective glass remains exposed on the tip side of the seal (Fig. 6C). The techniques for filling, assembling, and storing recessed-tip ISMs are the same as those for spear-type ISMs.

3.3.3. Stretched-Membrane ISMs

If fine-tipped glass-membrane ISMs are sought, if ruggedness is not required, and if the investigator is not daunted by technical difficulties, then the stretched-membrane design is worth considering. The design is straightforward, even if the construction method is not: a beveled micropipet made of inert glass has a thin sheet of ion-selective glass sealed over its opening (as shown in Fig. 6D). Briefly, a micropipet is pulled from inert glass stock; then it is given a very flat bevel. In a custom-made microforge, the beveled tip of the inert micropipet is pushed through a thin sheet of molten ion-selective glass (created by blowing a glass microbubble at the end of a sealed micropipet made of ion-selective glass), causing the ion-selective glass to seal across the beveled opening; to obtain ion-selective glass with the optimum visco-elastic, selectivity, and electrical properties, the glass must be custom mixed and fired (Pucacco and Carter, 1976, 1978; Pucacco et al., 1986). The fragility of the thin ion-selective glass membrane prevents the use of vacuum and steam-filling methods (Pucacco and Carter, 1978), so that these ISMs must be filled by inserting a very fine tube to near the tip of the micropipet and then pressure-ejecting filling solution from this tube—a difficult and time-consuming task. Details of the stretched-membrane-design fabrication procedure cannot be presented here; the reader is referred to Pucacco and Carter (1976, 1978) and Pucacco et al. (1986) for further instruction.

3.3.4. Double-Barreled Glass ISMs

Double-barreled glass ISMs consist of a reference barrel and an ion-selective barrel. The ion-selective barrel can be of recessed-tip design (Zeuthen, 1981) or modified recessed-tip design (Pucacco et al., 1986). Double-barreled micropipets made from inert glass, side-by-side, double-barreled glass capillaries without filling fibers are pulled in the usual manner (section 3.2.1.); alternatively, double-barreled micropipets can be fabricated by twisting two single-barreled capillaries about each other and then pulling them (Zeuthen, 1981; Schlue and Wuttke, 1983). In Zeuthen's method (1981), a well-tapered micropipet drawn from inert glass is inserted into one of the barrels, and a glass-to-glass seal is made as described for single-barreled recessed-tip ISMs; before heating the glass to form the seal, a supporting micropipet (made from high-melting-point glass) must be inserted into the reference barrel, so that the reference barrel does not collapse upon itself from the heat of sealing. In Pucacco et al.'s (1986) method, a stretched-glass ISM is given a coating of plastic near its tip. After the coating cures, the ISM is carefully guided into one barrel of a double-barreled, inert-glass micropipet until it can advance no farther; it is then epoxied into place at the back end. This latter method is so extravagently expensive in term of both time and effort—both to make the stretched glass ISM and to insert it into its double-barreled shell—and its advantages over the simple recessed-tip, double-barreled ISM are so minor that it cannot be strongly recommended to anyone.

Double-barreled glass ISMs are completed by filling their reference barrels in the same manner by which liquid-membrane ISM reference barrels are filled (section 3.2.3). The ion-selective barrel, being sealed, is filled as described in section 3.3.1. With glass membrane ISMs, "crawl around" problems are nonexistent, so the order of filling the barrels is unimportant. Assembly and storage procedures for glass-membrane double-barreled ISMs are the same as those for their liquid-membrane counterparts (section 3.2.4.).

4. Implementation of ISMs

The danger of misapplying ISMs or misinterpreting experimental data obtained with ISMs is ever present. To avoid these dangers, ISMs must be used with the correct electronic and mechanical equipment, they must be thoroughly tested and calibrated, their limitations must be known and kept in mind, and the data obtained with them must be analyzed and interpreted carefully. In the following sections, each of these topics will be taken up in turn.

4.1. Equipment

4.1.1. Electronics

The ion-selective barrel(s) of ISMs have a high resistance, ranging from several hundred megaohms for large-tipped, liquid-ion-exchanger-based ISMs to a few hundred gigaohms for fine-tipped, neutral-carrier-based ISMs (Armstrong and Garcia-Diaz, 1980; Lewis and Wills, 1980; Ujec et al., 1981; Tripathi et al., 1985; Ammann, 1986); reference barrel resistances are typically many orders of magnitude less (Purves, 1981; Borrelli et al., 1985). The input impedance of signal amplifiers used with ISMs should be at least 100 × the ion-selective barrel impedance, to decrease high-frequency noise (Purves, 1981; Finkel, 1987) and prevent undue attenuation of the measured potential (Lewis and Wills, 1980; Purves, 1981; Koryta and Stulík, 1983). Ion-selective amplifier bandwidth is not as critical as input impedance because, apart from coaxial-design ISMs (Ujec et al., 1981), the response time of an ISM is slow enough to obviate instrumentation bandwidth as a factor in recording fidelity (*see* section 4.2.4.; Armstrong and Garcia-Diaz, 1980; Ujec et al., 1981; Ammann, 1986).

It is desirable to have the headstage of the ion-selective amplifier close to the ISM, to minimize the distance over which the low-current, high-input impedance signal must travel; reducing this distance reduces leakage currents, input capacitance, and radiated noise reception (Purves, 1981; Koryta and Stulík, 1983; Ammann, 1986; Finkel, 1987). The wire(s) connecting the inner reference electrode(s) of the ion-selective barrel(s) to their respective headstage(s) should therefore be as short as possible, and unless low-impedance (i.e., coaxial) design ISMs are used, it is inadvisable to shield the connecting wires. Shielded cables can add input capacitance that will increase high-frequency electrode noise despite the use of shield-driving circuitry (Purves, 1981; Finkel, 1987). They can also cause signal attenuation, by degrading amplifier input resistance through the addition of leakage conductances between the shield and the inner conductor (Finkel, 1987).

The high impedances of ISMs render them highly susceptible to interference from radiated electronic noise (Purves, 1981; Koryta and Stulík, 1983; Ammann, 1986). It is virtually impossible to obtain satisfactory experimental data without first shielding the ISMs and the preparation from electromagnetic radiation. A Faraday cage is highly recommended (Tsien and Rink, 1980; Koryta and Stulík, 1983; Ammann, 1986).

In order to measure the true potential caused by ion activity in the preparation, it is necessary that any voltages recorded by the ion-selective barrel(s) not resulting from ion activity be subtracted out; examples of such non-ion-specific potentials commonly contributing to the total potential sensed by ISMs include the membrane potentials of cells and the field potentials of active tissues. In most neurophysiological applications, common mode potentials are eliminated from the ion-specific signal by measuring reference potential at the same time that the ion-selective potentials are being measured, and then subtracting this reference signal from the ion-specific signal (Ammann, 1986). The reference microelectrode can be incorporated as one of the barrels in a double- or multi-barreled ISM, or else it can be an entirely separate microelectrode. It is critical that the reference microelectrode measure the same common mode potential as is measured by the ion-selective barrel(s) of the ISM (*see* section 4.3.5.); to assure this, it is usual to place the reference and ion-selective microelectrodes near each other (Ammann, 1986). The advantages of physical proximity of the reference and ion-selective barrel(s) account for the popularity of double- and multi-barreled ISMs.

The electronics required for reference microelectrode recording are quite familiar to most neurophysiologists and are well summarized in Purves' book (1981). Reference amplifier input-impedance should be 100× the reference barrel impedance (Purves, 1981; Finkel, 1987). Note that, if asymmetric reference cells are used, the amplifier should have voltage offset compensation correction at the input, rather than later in the circuitry, to avoid the problem of DC leakage current being injected into the preparation from the reference barrel (Finkel, 1987; Raynauld and Laviolette, 1987). To record fast transients with the reference barrel, it is helpful to have capacitance and resistance compensation built into the amplifier (Purves, 1981). For certain applications, it is desirable to have the option of injecting controlled pulses of current into the preparation. Current can be passed either through a dedicated current-passing microelectrode or else through the reference microelectrode itself if the amplifier is equipped with the appropriate bridge circuitry (Purves, 1981).

The ionic changes encountered in neurophysiological experiments are usually small in relation to the baseline ionic activities (*see* Ammann 1986 for numerous references); the potential changes measured by ISMs are correspondingly small. Thus, the amplifiers used for both ion-selective and reference signals should have high resolution (0.1 mV), and be free from temperature- and time-dependent circuit drift (Koryta and Stulík, 1983). For the same reasons, the amplifier circuitry should be inherently low-noise and protected from line current hum.

Potentials measured by reference microelectrodes or barrels may be subtracted from the net potential sensed by ion-selective barrel(s) by either analog (Borrelli et al., 1985; Ammann, 1986;

Finkel, 1987) or digital means (Ellerman et al., 1985), to yield a differential signal proportional to the logarithm of ion activity (*see* section 2.2.6.). Because the response speeds of ion-selective and reference barrels are usually different (*see* section 4.2.4.; Koryta and Stulík, 1983; Ammann, 1986), the common mode signals will be out of phase between the two barrels (Ellerman et al., 1985; Finkel, 1987); therefore, subtraction of the raw reference and ion-selective barrel signals will not faithfully cancel common mode signals, and a false differential signal will result. There are a number of ways to address this problem (section 4.3.3.), including speeding up the ion-selective barrel using coaxial techniques (Ujec et al., 1981), slowing down the reference barrel by using a liquid ion-exchanger reference solution (Thomas and Cohen, 1981), or electronically matching the response speeds of the two barrels (Finkel, 1987). Independent capacitance neutralization control over the ion-selective and reference barrels permits the response speed of the ion-selective barrel to be increased and that of the reference barrel to be decreased (Finkel, 1987). Another electronic means to matching the response speeds is to enhance the coupling capacitance electronically from the faster barrel to the slower (Finkel, 1987).

4.1.2. Mechanical and Perfusion Systems

Most neurophysiological applications of ordinary, reference microelectrodes require a fair degree of mechanical stability and isolation from vibration for the preparation and the electrodes (Purves, 1981). ISM work requires, if anything, even more careful vibration isolation and mechanical stability; it is also often necessary to isolate the ISM and the experimental preparation from air currents that can change the fluid level in which the ISM is immersed, and thereby affect its function (Vaughan-Jones and Kaila, 1986). It is recommended that the ISMs, the preparation, and all the experimental apparatus mechanically connected to the ISMs and preparation be mounted on vibration isolation tables. The experimental setup should be protected from drafts, and investigators should refrain from sudden motions within a meter or so of ISMs engaged in measurement.

Perfusion systems used in conjunction with ISMs must not generate macroscopic streaming of the perfusate around the ISM tip, since this could reduce the stability of measurements by at least three mechanisms:

1. Erratic bulk movement of the analysate solution at the ISM tip may generate streaming potentials that could affect the reference or ion-selective barrels of the ISM (Goodisman, 1987)
2. Fluctuations in the level of the perfusate will change the depth of ISM immersion; under some common experimental conditions, such fluctuations can unpredictably alter ISM responses (*see* section 4.2; Vaughan-Jones and Kaila, 1986)
3. Changes in the depth of ISM immersion will change the transmural capacitance of the ISM; these capacitance changes can significantly increase the time constant of the reference barrel and thereby degrade recordings of transients (Finkel and Redman, 1983).

If the temperature of the perfusate is changed during the course of an experiment in which ISMs are utilized, appropriate corrections for the predicted temperature-dependent change in ISM performance need to be applied to the measurements (*see* section 4.3.4.; Ammann, 1986; Vaughan-Jones and Kaila, 1986).

4.2. Calibration and Testing

4.2.1. Conditioning

ISMs can be tested, calibrated, and used immediately, or else they may be conditioned before being tested and used (Koryta and Stulík, 1983; Borrelli et al., 1985; Ammann, 1986). ISMs are conditioned by soaking their tips in a calibrating solution (*see* section 4.2.2.) in which the determinand ion activity is near the middle of its measurable range; if necessary, ISMs can be further conditioned by repeatedly dipping their tips in calibrating solutions containing various activities of the determinand ion (Koryta and Stulík, 1983; Borrelli et al., 1985). If, after repeated soaking, a recalcitrant ISM continues to function erratically, it can sometimes be salvaged by controlled breaking or beveling of its tip (Tripathi et al., 1985). The conditioning process can straightforwardly be adapted as part of the calibration procedure. Conditioned ISMs tend to have stabler, faster, and more reproducible responses (Koryta and Stulık, 1983); indeed, some ultrafine-tipped ISMs will not function at all unless they have been conditioned (Borrelli et al., 1985).

4.2.2. Calibration

Individual ISMs, even if they are of the same design and the same batch, may differ quantitatively in their ion activity vs EMF relationship, thus, each ISM must be individually tested and calibrated prior to being used (Thomas, 1978; Tsien and Rink, 1980; Lee, 1981; Koryta and Stulík, 1983; Ammann, 1986). Moreover, because the characteristics of ISMs can change during use, ISMs should also be calibrated after each use (Thomas, 1978; Borrelli et al., 1985; Hinke, 1987). In routine calibrations of ISMs used in physiological experiments, it is usually sufficient to determine the slope of the linear portion of the ion activity vs EMF relationship over the range of ion activities expected in the experimental solution(s). It is impractical to test each ISM for its selectivity, response speed, temperature sensitivity, and so on. Of course, investigators should check these ISM performance parameters whenever they attempt an unfamiliar ISM design or a novel manufacturing protocol.

ISMs are best calibrated in calibrating solutions that resemble the experimental solution(s) as closely as possible (Koryta and Stulík, 1983; Ammann, 1986). At the minimum, calibrating solutions should have an ionic strength and pH equal to or near that of experimental solution(s) (Kriz and Syková, 1981), they should contain activities of major interferent ions similar to those of experimental solution(s), and, when in use, their temperature should be within ±0.5°C of that of the experimental solution(s) (Koryta and Stulík, 1983; Ammann, 1986). In the following discussion, reference will be made to ion activities rather than to ion concentrations; the proper conversion formulas are presented in section 2.1.2. To calibrate an ISM, a series of such calibrating solutions are prepared with different activities of the determinand ion, such that the span of activities in the series brackets the expected range of determinand activities in the experimental solution(s); the ISM is then tested in each calibrating solution, starting with the one having the lowest determinand ion activity and proceeding up to the one having the highest (Koryta and Stulík, 1983; Levy et al., 1985; Ammann, 1986). For some determinand ions (i.e., Ca^{2+}), it is necessary to prepare calibrating solutions of very low activity (i.e., $10^{-8}M$); for greater accuracy, it is recommended that in these cases the calibrating solutions be prepared with specific buffers of the

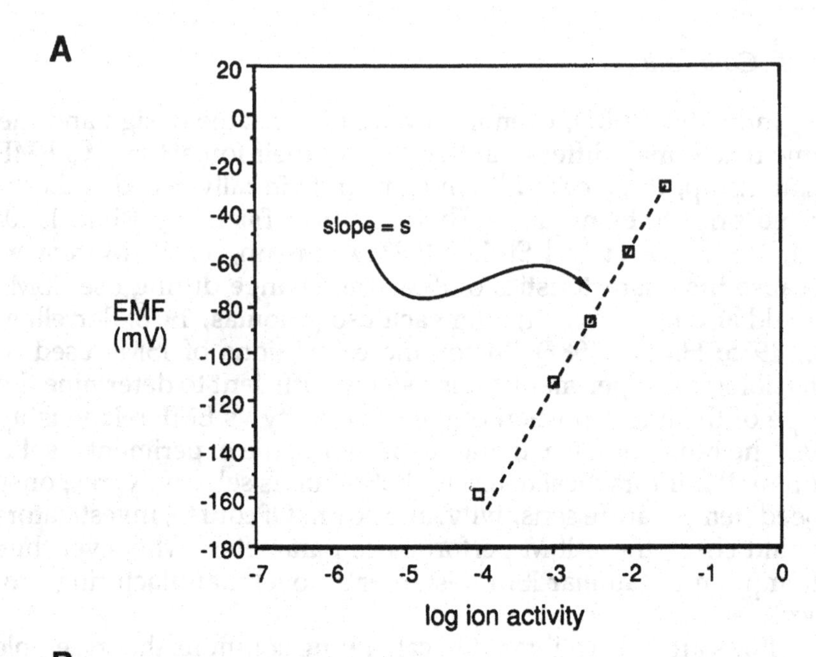
A
slope = s
EMF
(mV)
20
0
-20
-40
-60
-80
-100
-120
-140
-160
-180
-7
-6
-5
-4
-3
-2
-1
0
log ion activity

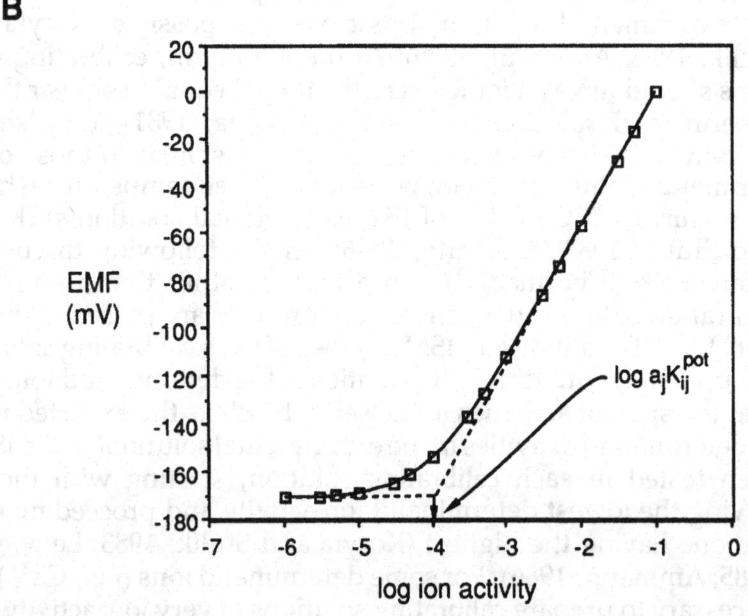
B
EMF
(mV)
log $a_jK_{ij}^{pot}$
20
0
-20
-40
-60
-80
-100
-120
-140
-160
-180
-7
-6
-5
-4
-3
-2
-1
0
log ion activity

determinand ion (Tsien and Rink, 1980; Levy et al., 1985; Ammann, 1986).

To determine the slope-response of the ISM, a curve is plotted that relates the differential EMF measurements taken by the ISM to the common logarithm of the determinand ion activities in the corresponding calibrating solution (the ISM response curve). A representative curve for a hypothetical ISM is shown in Fig. 7A. Let a_{low} denote the lowest ion activity on the linear part of the curve, and let a_{high} denote the highest ion activity on the linear part of the curve. Let EMF_{low} be the EMF recorded by the ISM in the a_{low} calibrating solution; similarly, let EMF_{high} be the EMF recorded in the a_{high} calibrating solution. Then the slope-response, s, of the ISM is given by:

$$s = (EMF_{high} - EMF_{low})/(a_{high} - a_{low}) \qquad (62)$$

If the experimental solutions and calibrating solutions are alike in the variables listed above, if EMF_{cal} is the voltage recorded in a calibrating solution with determinand ion activity a_{cal} anywhere on the linear part of the ISM response curve, and if EMF_{exp} is the measured EMF in the experimental solution, then the corresponding experimental determinand ion activity, a_{exp}, is given by:

$$a_{exp} = a_{cal} \exp(EMF_{exp} - EMF_{cal})/s \qquad (63)$$

Levy et al. (1985) developed a computer-controlled system to automate the calibrating of ISMs and to translate experimental EMF recordings into ion activity or concentration recordings.

Fig. 7. Calibration and selectivity curves for a hypothetical ISM. **A:** Calibration curve. EMFs are measured over a range of activity values spanning the expected determinand ion activity range in the sample solution. A best fit regression line is fitted to the EMF vs $\log_{10}$(ion activity) data; the slope of the line gives the slope-response of the ISM. **B:** Selectivity curve. EMF vs $\log_{10}$(ion activity) is plotted over a wide range of determinand ion activities; each calibrating solution has the same fixed activity of the interferent, a_j, in it. Asymptotes to the two branches of the curve are constructed; the intersection of the asymptotes corresponds to $\log_{10} a_j K^{pot}_{ij}$.

If the experimental and calibrating solutions differ significantly in their composition, then the aforementioned method of calibration is inadequate. Corrections must be made for any differences in ionic strength, liquid-junction potentials, temperature, pH, and interferent ion activities (Ammann, 1986). To correct for differences in ionic strength between calibrating and experimental solutions, determinand and interferent ion concentrations in the calibrating and experimental solutions must be expressed as activities; the activity coefficients necessary for this purpose can be derived using Eqs. (6)–(13) of section 2.1.2. (*see also* Meier et al., 1982). Liquid-junction potential differences between calibrating and experimental solutions predominantly affect the reference microelectrode or barrel (Meier et al., 1982). To correct for liquid-junction potential differences, use the Henderson equation (Eq. [49]) of section 2.2.5.

If the temperatures of the calibrating and experimental solutions are different, it is necessary to introduce a temperature correction factor (Koryta and Stulík, 1983). The correction factor can be determined empirically or estimated theoretically. Theoretically, temperature can affect ISM responses by changing the activity coefficients of the determinand or interferent ions (Eqs. [8] and [9], the liquid-junction potential (Eq. [49]), the selectivity coefficients of the ion-selective membrane (Morf, 1981; Ammann, 1986), and the standard potential or the *RT/zF* term in the Nicolsky equation (Eq. [53]). Deriving an expression to take into account all these temperature dependencies is difficult; instead, by measuring the change in differential EMF between a calibrating solution at one temperature and that at another, an empirical temperature correction factor can be determined.

If the activity of significant interferent ion(s) (including H^+) in the experimental solution is unknown, it is necessary to measure the determinand and interferent ion(s) simultaneously in the experimental solution (Ammann, 1986). Using these measurements, the appropriate selectivity coefficients (determined experimentally as described below), and the Nicolsky equation (Eq. [53]), the determinand ion activity, $a_{i,\mathrm{exp}}$ is given by the solving the following system of equations for $a_{i,\mathrm{exp}}$:

$$\mathrm{EMF}_{m,\mathrm{exp}} - \mathrm{EMF}_{m,\mathrm{cal}} = s_m \left[\log \left(a_{m,\mathrm{exp}} + \sum_{n=m} a_{n,\mathrm{exp}} K^{\mathrm{pot}}_{m,n} \right) - \log \left(a_{m,\mathrm{cal}} + \sum_{n=m} a_{n,\mathrm{cal}} K^{\mathrm{pot}}_{m,n} \right) \right] \quad (64)$$

It is to be understood that this is a system of M equations, where there are $M - 1$ interferent ion(s) and one determinand ion, m takes all values from 1–M, $i = m$ for some m, the $EMF_{m,exp}$ and $EMF_{m,cal}$ are the measured differential EMFs of the ion_m-selective ISM in the experimental and calibrating solutions, respectively, the s_m are the slope-responses of the ion_m-selective ISMs, $a_{m,cal}$ are the activities of ion_m in the calibrating solutions, and the $K_{m,n}{}^{pot}$ are the experimental selectivity coefficients determined as outlined below. From this system of M equations, the M unknown activities, $a_{m,exp}$, including $a_{i,exp}$, can be extracted (the case for a determinand and a single interferent ion is discussed in Lev, 1964). Other calibrating formulas useful in special experimental situations are presented in Koryta and Stulík (1983).

4.2.3. Selectivity Testing

There are three methods for determining selectivity coefficients: the separate-solution method, the fixed-interference method, and the fixed-primary-ion method (Durst, 1978; Ammann, 1986). The method recommended by IUPAC is the fixed-interference method (Koryta and Stulík, 1983; Ammann, 1986); only this method will be explained here. The reader is referred to Koryta and Stulík (1983) and Ammann (1986) for an exposition of the other methods. In the fixed-interference method, the ISM is tested in a series of calibrating solutions containing a constant activity (a_j) of the interferent ion and varying activities (a_i) of the determinand ion. The range of determinand activities in the selectivity-testing solutions should extend low enough to assure that the detection limit for the ISM is included; to guarantee accuracy at such low determinand activities, it may be necessary to prepare solutions buffered with respect to the determinand (Ammann, 1986). The differential EMF values are then plotted vs the common logarithm of the determinand ion activity. Fig. 7B shows a hypothetical selectivity curve. Asymptotes to the two branches of the curve are constructed as in Figure 7B; the intersection of the asymptotes defines a point with abscissa corresponding to $\log_{10} a_j K_{i,j}{}^{pot}$ (Koryta and Stulík, 1983; Ammann, 1986). Selectivity factors derived for a given ISM are not independent of the method used for determining them; investigators reporting selectivity factors ought to specify the technique used and conditions under which an ISM were tested (Ammann, 1986).

4.2.4. Response-Speed Testing

Suppose that the determinand ion activity bathing the ISM tip is suddenly switched from activity a_1 to a_2. The ISM and associated instrumentation will not instantaneously register a corresponding change in the EMF; rather, the ISM system will respond to the change in determinand ion activity with a certain time course (Fig. 8; *see also* Koryta and Stulík, 1983; Ammann, 1986). If the EMF measured in the a_1 solution is designated EMF_1 and the EMF measured in the a_2 solution (at a very long time after the solution switch) is designated EMF_2, then the IUPAC definition of ion-selective electrode response time, τ_{90}, is that amount of time taken for the EMF to go from EMF_1 to 90% of EMF_2 (Fig. 8). Koryta and Stulík (1983) and Ammann (1986) emphasize that the overall response time of an ISM system depends not only on the intrinsic response speed of the ISM itself, but also on the characteristics of the ISM instrumentation and the measuring technique. Because of their high internal resistances and resultant large RC time constants, most ISMs have response times far exceeding instrumentation response times (Ammann, 1986). Thus, if the solution bathing the ISM tip is switched quickly and thoroughly (with rapid stirring), then the overall response time measured will faithfully reflect the response time of the ISM itself (Ammann, 1986). High-resistance ISMs have response time courses that are described by the following equation:

$$EMF(t) = EMF_2 - (EMF_2 - EMF_1)e^{-t/RC} \tag{65}$$

where $EMF(t)$ is the EMF at time t, EMF_1 and EMF_2 are the steady-state EMFs in the test solutions, R is the ISM resistance, and C is the ISM system capacitance. Because of the widely different conventions in measuring and reporting ISM response times, investigators should be careful to specify the method used in determining the response times of their ISMs and the definition of their response time parameter (Ammann, 1986).

4.3. Pitfalls, Limitations, and Caveats

4.3.1. Silanization Problems

Aside from designing and synthesizing the ion-selective liquid membranes themselves—a task beyond the province of most

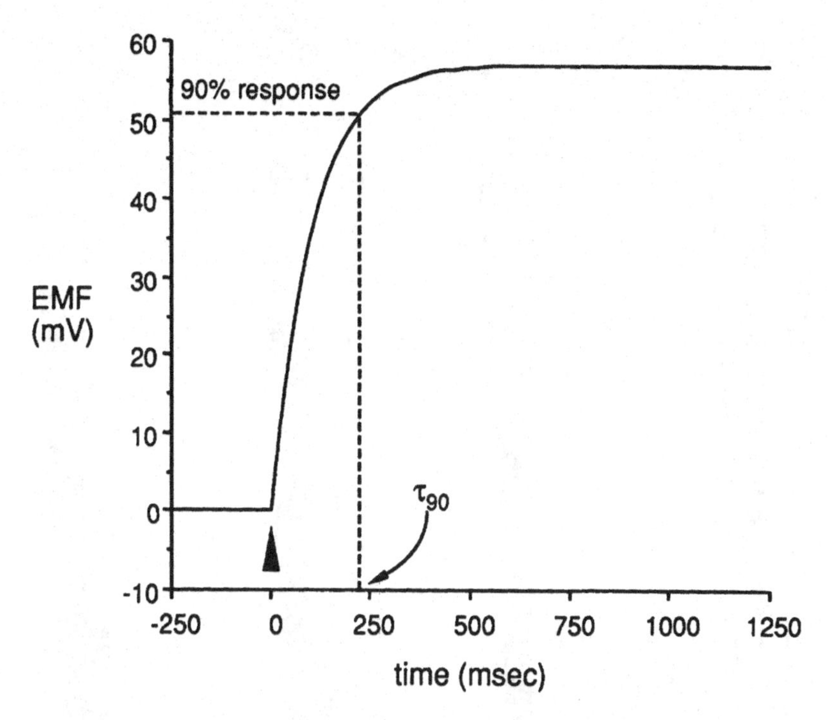

Fig. 8. Response time curve for a hypothetical ISM. At time zero (big arrowhead), the solution bathing the ISM tip is switched from one containing determinand ion activity a_1 to one containing determinand ion activity a_2. The time taken for the ISM's differential EMF to go from its initial value, EMF_1, to 90% of its final value, EMF_2, is termed the 90% response time, τ_{90}.

neurophysiologists in any case—the bulk of the difficulties encountered in constructing liquid-membrane ISMs is the result of the silanization step. Whenever an ISM responds erratically or has a sub-Nernstein calibration after conditioning, improper silanization should be the first suspicion; microscopic observation of the ISM tip will frequently confirm that silanization is at fault by showing the typical signs of improper silanization (Borrelli et al., 1985; Fig. 9). Silanization problems can be divided into two categories: undersilanization and oversilanization.

Under the microscope, the liquid-membrane–ion standard solution interface of an adequately silanized ion-selective barrel should appear concave or straight (Fig. 9C; *see* Borrelli et al., 1985;

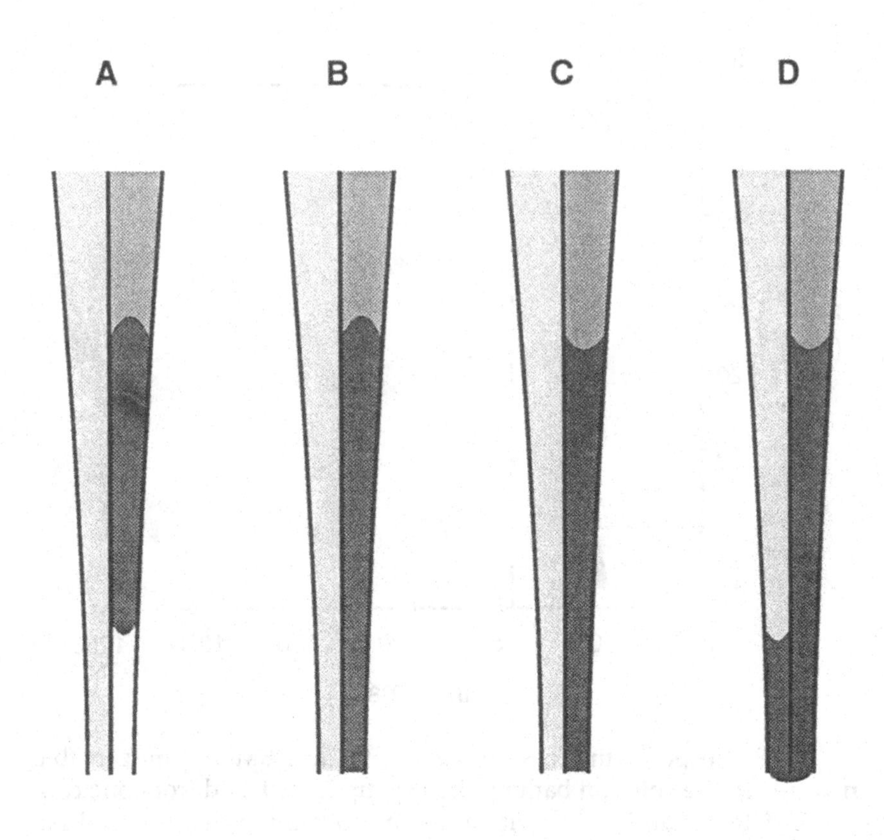

Fig. 9. Diagrammatic cross-sections of ISM tips to show the microscopic appearance of under- and oversilanized ISMs. In each case, the reference barrel is on the left, the ion-selective liquid membrane is represented as the darkest shade, the ion-standard solution is the medium shade, the reference solution is the light shade, and the external sample solution is represented as white. **A:** Undersilanized ISM, showing the convex upward meniscus between the ion-selective liquid membrane and the ion-standard solution, and showing displacement of the liquid membrane at the tip by the sample solution. **B:** Poorly silanized ISM, showing a convex liquid membrane–ion-standard solution meniscus. **C:** Ideally silanized ISM, showing a concave ion-selective liquid membrane–ion-standard solution meniscus. **D:** Oversilanized ISM, showing "crawl around" of the liquid membrane into the reference barrel.

Tripathi et al., 1985). Liquid-membrane–ion-standard solution interfaces in undersilanized ion-selective barrels are convex, with the apex of the convexity towards the stem end (Fig. 9B); the interface assumes this convex configuration because the aqueous ion-standard solution wets the glass. Silanized glass is hydrophobic and is not wetted by aqueous back-filling solutions; instead, the organic liquid ion-selective solution wets the silanized surface (Tripathi et al., 1985)—thus the concave interface profiles seen in well-silanized ISMs. The column of ion-selective solution in undersilanized ion-selective barrels can be displaced from the tip by aqueous solution—either the sample solution into which the tip is immersed or the ion-standard solution behind the liquid-membrane column (Fig. 9A). An ISM with a small (less than a few micrometers) displacement of the liquid-membrane column sometimes continues to function; usually the response time of such ISMs is considerably lengthened, because the aqueous solution trapped in the ISM tip is inhibited from free exchange with the exterior sample solution. In this situation, any changes in determinand ion activity in the sample solution will be manifested at the liquid-membrane boundary only after diffusion through the narrow micropipet orifice and through the volume of trapped aqueous solution in the displaced region. In this respect, liquid-membrane ISMs with small displacements of the sensor column from the tip resemble recessed-tip glass-membrane ISMs. Unfortunately, small displacements often go on to become severe displacements that destroy the ISM; if this should occur during an experiment, the investigator might not be cognizant of the predicament until the postexperiment calibration. The prudent tack to follow with visibly undersilanized ISMs is simply to discard them.

Grossly oversilanized ISMs sometimes will have completely plugged tips. This is especially true of ultrafine-tipped ISMs. More often, oversilanized ISMs suffer from "crawl around" of the ion-selective solution from the ion-selective barrel(s) to the reference barrel, which was made hydrophobic by excessive silanization or improper protection from silanization. Such ISMs have a distinctive microscopic appearance, exhibiting a meniscus in the reference barrel as well as in the ion-selective barrel(s) (Fig. 9D). With vigorous conditioning and/or judicious tip beveling or breakage, ISMs with a minor degree of "crawl around" can occasionally be salvaged.

4.3.2. Reference Electrode Considerations

A great deal of experience has been amassed in using the traditional reference microelectrode (Purves, 1981). Despite the familiarity and relative simplicity of reference microelectrodes, two potential pitfalls continue to bedevil the unwary: tip potentials and liquid-junction potentials. Both of these potentials affect aqueous phase reference barrels or microelectrodes, but do not affect or affect to only a minor degree ion-selective barrels or microelectrodes. Tip and junction potentials are specific to the reference side and are not common mode signals; thus, they can affect the differential signal representing ion activity (Ammann, 1986).

Junction potentials are fairly well understood as being caused by the unbalanced interdiffusion of ions across the interface between adjoining dissimilar solutions (*see* section 2.2.5.; Silver, 1985; Goodisman, 1987). Junction potentials are of especial concern to electrophysiologists when experimental solutions are changed, because changes of solution can change the junction potential at the microelectrode and/or at the preparation ground. It is also worth noting that junction potentials are sometimes larger in fine-tipped micropipets than in larger-tipped ones (Snell, 1969). The precautions necessary to minimize junction potentials are well described in Purves (1981). In brief, to minimize junction potentials, microelectrode and bath electrode filling solutions should contain high concentrations (0.5–3*M*) of cation–anion pairs of nearly equal mobility, i.e., K^+ and Cl^-, Li^+ and $acetate^-$, or NH_4^+ and NO_3^- (Purves, 1981; Ammann, 1986). Of course, it is imperative that leakage of such high concentrations of ions into the experimental preparation be prevented (*see* section 4.3.6.). Bulk flow can be restricted by sequestering the bath electrode solution in an agar matrix and by making the tips of reference microelectrodes very small (however, even the smallest of tips cannot prevent some degree of filling-solution leak). Fromm and Schultz (1981) suggest that 0.5*M* KCl be used for a reference solution because, relative to 3*M* KCl, it reduces KCl leakage from the tip fivefold without increasing junction or tip potentials (*see* section 4.3.6.).

Tip potentials are far less well understood. Operationally, tip potentials are defined as the difference in the measured EMF of a microelectrode circuit before and after the breaking back of the microelectrode tip; tip potential magnitudes can be as great as several tens of millivolts and can change considerably as the solution bathing the tip is varied (Agin, 1969; Purves, 1981). Tip potentials are separate and distinct from junction potentials, and occur even in systems in which junction potentials are minimized by the use of high-concentration solutions of equitransferent electrolytes (Lavallée and Szabo, 1969; Okada and Inouye, 1976). It has been suggested that tip potentials are ultimately the results of fixed negative charges along the interior glass surface of the micropipet tip (Lavallée and Szabo, 1969). The anionic fixed charges participate in the formation of an electrical double layer at the micropipet tip (*see* section 2.2.4.); this double layer acts as a cation-selective surface conductance in parallel with the conductance from the bulk electrolyte at the tip. According to this hypothesis, tip potentials are proportional to the ratio of surface conductance to bulk electrolyte conductance; this ratio, in turn, depends on the tip orifice diameter, the ionic strength of the filling and sample solutions, and the intrinsic glass surface charge (Lavallée and Szabo, 1969).

Tip potentials are especially troublesome in intracellular work, where the effect of the intracellular solution on the tip potential is impossible to determine *a priori*. To reduce tip potentials, the tip orifice can be made larger, the ionic strength of the filling and sample solutions can be increased, or the glass surface charge can be decreased. The latter option is usually preferable. The addition of small amounts of thorium to the filling solutions (to give a final Th^{4+} concentration ranging from 10–100 μM) will reduce tip potentials, probably by complexing with and thereby neutralizing the negatively charged fixed sites on the glass surface (Agin, 1969). Okada and Inouye (1976) found that different methods of filling reference microelectrodes resulted in tip potentials of substantially different magnitudes; glass fiber methods were associated with the smallest tip potentials.

Reference microelectrode measurements in suspensions are complicated by yet another factor: the Pallman effect (Kortya and Stulík, 1983; Ammann, 1986). Reference microelectrodes measure a different potential in a colloidal suspension solution than in an otherwise identical nonsuspension solution. The Pallman effect certainly operates in biological solutions, especially in intracellular solutions with their many proteins and small particles in suspension (Ammann, 1986). Three physicochemical hypotheses to explain the Pallman effect are current. One postulates that suspension effects are caused by anomalous liquid-junction potentials, a second hypothesis maintains that they result from Donnan potential, and the third hypothesis posits that the Pallman effect is the result of a diffuse double layer potential (Ammann, 1986). Whatever the cause, there is no easy method of avoiding suspension potentials using conventional reference microelectrodes; indeed, suspension potentials probably contribute to most intracellular measurements of potential.

In an attempt to avoid certain of the difficulties inherent in conventional aqueous reference solutions, Thomas and Cohen (1981) developed an organic reference solution consisting of 2% potassium tetrakis (*p*-chlorophenyl) borate in 1-octanol (KTpClPB in octanol). Organic reference solutions have a number of advantages. They avoid the problem posed by the difficult-to-measure alterations in liquid-junction, tip, and suspension potentials occurring when reference microelectrodes experience a change in the ambient fluid—as occurs during impalement of a cell, for example. Leakage of aqueous-phase electrolytes from the microelectrode into the cell is decreased because of the immiscibility of the organic liquid membrane and the aqueous cytosol (*but see* Oesch et al., 1987). Silanization of double- and multi-barreled ISMs is simplified when using organic reference solutions, because the reference barrel does not need to be protected against silanization; rather, it can be silanized at the same time that the ion-selective barrel(s) are. Finally, recordings made with KTpClPB in octanol as the reference solution are more stable than those made with KCl (Thomas and Cohen, 1981). There are two limitations on using KTpClPB in octanol as a reference solution: one, its relatively high resistance makes it unsuitable for recording fast transients (Thomas and Cohen, 1981); two, because KTpClPB is cation-selective, but cannot discriminate between K^+ and Na^+, it is only useful as a reference solution under conditions in which the total activity of Na^+ + K^+ remains constant (Thomas and Cohen, 1981). Fortunately, this total is usually the same outside cells as it is inside cells (Thomas and Cohen, 1981; Aicken and Brading, 1982).

4.3.3. Response Speed—How to Increase It

In many neurophysiological applications, it is important that the response speed of an ISM be reasonably fast. However, as was pointed out in section 4.2.4., the response speed of most ISMs is limited by their high resistances and resultant large RC time constants. Thus, most strategems to decrease response time rely on reducing the ISM resistance; this approach has the additional benefit of reducing noise. ISM resistance can be decreased by decreasing the specific resistance of the ion-selective membrane (Ammann, 1986), by decreasing the thickness of the ion-selective membrane (Pucacco and Carter, 1978; Ujec et al., 1979), or by increasing the area of ion-selective membrane exposed to the sample solution (Pucacco and Carter, 1978). Recessed-tip glass-membrane ISMs are a special case; their response speed is limited by the time taken for determinand ions to equilibrate between the trapped volume in the recessed tip and the free sample volume.

The specific resistance of liquid ion-selective membranes can be decreased by selecting a low-resistance solvent, by increasing the ion carrier concentration, and by adding lipophilic salts (Ammann, 1986). Of course, none of these expedients can be achieved in isolation. The liquid-membrane solvent, for example, must allow adequate solubilization of ion carriers and lipophilic salts, it must be sufficiently lipophilic to be retained in the hydrophobic ISM tip, and it must not have too high a viscosity; otherwise it will not fill the ISM tip (Ammann, 1986). Many of the requirements placed upon liquid-membrane solvents are in opposition to the requirement for low solvent resistance; increasing the ion carrier concentration or adding lipophilic salts to the liquid membrane cannot be done without consideration of the effect these manipulations will have on ISM performance parameters other than response speed. The specific resistance of glass ion-selective membranes can also be decreased—but again, not without affecting their other properties, such as selectivity (Hebert, 1969).

ISM response time can be reduced by reducing the thickness of the ion-selective membrane and thereby decreasing the length (and therefore resistance) of the low-conductance ion-transport pathway through the membrane. The thickness of liquid ion-selective membranes can be decreased by using front-filling methods to draw up only a short column of ion-selective solution (*see* section 3.2.3.; Lux and Neher, 1973) or by inserting deep into the liquid membrane a coaxial inner micropipet filled with ion-standard solution (*see* section 3.2.4.; Ujec et al., 1979, 1981). Note that both of these techniques have technical limitations (*see* sections 3.2.3. and 3.2.4.). The thickness of glass ion-selective membranes can be decreased by using techniques to stretch and thin the membrane (*see* section 3.3.3.; Pucacco and Carter, 1976, 1978; Pucacco et al., 1986); unfortunately, these methods are technically forbidding.

ISM response time can be decreased by increasing the area (hence increasing the conductance) of ion-selective membrane exposed to the sample solution. The area of exposed membrane in liquid-membrane ISMs can be increased by increasing the tip orifice diameter; obviously, this approach is incompatible with applications requiring small tip sizes. The area of exposed liquid membrane can also be increased by changing the geometry of the liquid membrane. Instead of having only a small disk of membrane exposed, a much larger area of liquid membrane can be exposed by causing it to coat the exterior of the micropipet near the tip. Plasticizers (such as polyvinylchlorine) mixed into the liquid ion-selective solution create a sol-phase liquid membrane that adheres to the outside of a silanized micropipet dipped into it; note, however, that most plasticizers increase the specific resistance of liquid membranes (Ammann, 1986). The area of exposed glass membrane can be increased by lengthening the active portion of spear-design ISMs (Hinke, 1987), by increasing the exposed area of the inner, active micropipet in recessed-tip ISMs (Thomas, 1978; Pucacco et al., 1986), or by stretching an ion-selective glass membrane over a broad, beveled opening of a micropipet made of inactive glass (Pucacco and Carter, 1976, 1978). The exterior coating method for liquid-membrane ISMs and the long active portion method for spear-design ISMs present obvious difficulties for investigators interested in making measurements in restricted spaces (i.e., intracellular measurements), because there is no assurance that the long stretch of active membrane will be contained entirely within the sample volume (Ammann, 1986). The area of exposed active membrane in recessed-tip ISMs can be increased without increasing the tip size of the external insulating micropipet but the trade-off is that the volume of sample solution "trapped" in the recessed tip is increased; increasing the trapped volume without increasing the inlet orifice diameter increases the time taken for determinand ions to equilibrate between the trapped sample volume and the exterior, free sample volume.

Finally, ISM response time can be decreased in many instances by conditioning the ISMs (*see* section 4.2.1.). It is not clear whether conditioning improves ISM response speed by reducing the specific resistance of the ion-selective membrane, by removing obstructions (such as polysiloxane plugs in liquid-membrane ISMs) at the ISM tip, or by some other mechanism(s). Whatever the explanation for its efficacy, ISM conditioning is the single most useful expedient for increasing response speed.

4.3.4. Thermal Effects

In section 4.2.2., it was pointed out that the relationship between temperature and ISM slope response is complex. In view of the complexity of the relationship between temperature and ISM slope response, it is best to avoid temperature fluctuations of the ISM or sample solution during experiments. If the experiment requires variation of temperature, careful calibrations of the ISMs over the experimental temperature range are recommended to determine empirical correction factors; given the known differences in electrode characteristics between ion-selective macro- and microelectrodes (*see* section 4.3.5.; Ammann, 1986), calculated adjustments based on theoretical models of ion-selective macroelectrode behavior cannot be relied upon to yield accurate temperature correction factors.

Liquid-membrane ISMS are also sensitive to temperature gradients along the length of their ion-selective columns (Vaughan-Jones and Kaila, 1986). In heated solutions, there may be a temperature gradient along the length of the ISM, with the immersed portions of the micropipet being warmer than the unimmersed portions. If the ISM has an ion-selective solution column spanning the analysate-solution–air boundary, then the ion-standard-solution–liquid-membrane interface will be at a different temperature than the analysate-solution–liquid-membrane interface; this difference in interface temperatures alters the ISM response (Vaughan-Jones and Kaila, 1986). Moreover, if the depth of ISM immersion in a heated sample solution varies, as can occur for example when perfusion rates are changed then not only is there a temperature gradient along the length of the ISM, but the temperature gradient and resulting effects on ISM response are time varying. To avoid these problems, the column of ion-selective solution can be made short enough that the entire column is immersed in the analysate solution, thus putting both interfaces of the liquid membrane at the same temperature (Vaughan-Jones and Kaila, 1986).

4.3.5. Tip Size Effects

The current understanding of ISM function and properties rests mainly upon experimental and theoretical work done on ion-selective microelectrodes (Morf, 1981; Ammann, 1986). It is becoming increasingly clear, however, that a number of ISM properties are critically dependent on their tip size—a parameter that is of far less importance to ion-selective macroelectrodes and therefore has been relatively neglected until recently (Armstrong and Garcia-Diaz, 1980). Among the tip-size-dependent properties of ISMs are selectivity (Orkand et al., 1984; Carlini and Ransom, 1987), response speed (*see* section 4.3.3.; Ammann, 1986), and signal fidelity (*see* section 4.3.6.; Ransom et al., 1987). Double- and multi-barreled ISMs with reference barrels are also subject to reference barrel tip-size-dependent effects (*see* section 4.3.2.), such as tip potentials (Purves, 1981) and junction potentials (Snell, 1969).

Liquid-ion-exchanger-based K^+-ISMs have selectivity profiles that depend on their tip size; the smaller the orifice diameter, the less these ISMs sense quarternary amines (Orkand et al., 1984; Carlini and Ransom, 1987). Indeed, K^+-ISMs with tip sizes less than about 1.5 μm exhibit "anomalous" selectivity, that is, $K^{pot}_{K^+,TMA^+}$ changes 10,000-fold, going from about 100–0.001, when the exterior tip diameter falls below about 1.5 μm. Carlini and Ransom (1987) discuss four possible explanations for the anomalous selectivity effect, and argue that a combination of shunting through the thin, hydrated, microcracked glass walls near the microelectrode tip (Armstrong and Garcia-Diaz, 1980; Lewis and Wills, 1980; Greaves et al., 1981), shunting along the inner surface of the glass near the tip (Armstrong and Garcia-Diaz, 1980; Munoz et al., 1983), and electrical double-layer effects near the tip (Lavallée and Szabo, 1969; Morf, 1981) might account for the reversal in selectivity that is seen as the ISM tip size is reduced. In brief, if at the ISM tip, the resistance through the glass wall, R_g, and the shunt resistance through the Gouy layer along the inner surface of the glass wall, R_s, are comparable to the resistance through the membrane, R_e (Fig. 10), then the ratio of the total shunt conductance to the ion-selective membrane conductance is inversely proportional to tip diameter. Thus, the finer the ISM tip, the less is the slope-response and the poorer is the selectivity (Armstrong and Garcia-Diaz, 1980). Moreover, the glass walls themselves might be ion-selective (Agin, 1969; Lewis and Wills, 1980), making

R_g a function not only of tip size, but also of the composition of the sample and filling solutions. The extent to which ISM selectivity depends on tip size in other types of liquid-membrane ISMs is unknown. Nevertheless, the observation of anomalous selectivity in ion-exchanger-based K^+-ISMs (Carlini and Ransom, 1987), the inverse relationship between tip size and ISM resistivity and response speed (Ammann, 1986), the variable effects of tip size on reference barrel tip and junction potentials (Snell, 1969; Lavallée and Szabo, 1969; Purves, 1981; Ammann, 1986), the inverse relationship between tip size and ISM component leakage (*see* section 4.3.6.), and the effects of tip size on signal fidelity (*see* section 4.3.6.; Ransom et al., 1987) all suggest that precautions should be taken to keep the tip size of ISMs used in experiments relatively constant (Carlini and Ransom, 1987; Ransom et al., 1987).

4.3.6. Impalement Damage

Neurophysiological applications of ISMs generally necessitate that they be inserted into tissues or cells. Insertion of an ISM into a cell or a tissue is usually associated with both biological damage to the specimen and physicochemical damage to the ISM.

Impalement damage to tissue and cells is of two forms: mechanical disruption of the tissue or cell by the insertion of the ISM, and chemical damage to the cell or tissue as a result of leakage of components from the ISM. These two forms of impalement damage will be discussed in turn. Measurements of extracellular ion activities in the mammalian central nervous system are affected by the extent of tissue disruption; fine-tipped ISMs cause less damage to the relatively restricted extracellular space than do larger-tipped ISMs, hence, they provide more faithful measurements of extracellular ion activities (Newman and Odette, 1984; Ransom et al., 1987). The physical damage done to cell membranes by microelectrode impalement is familiar to most neurophysiologists (Purves, 1981); because the tips of ISMs are usually bigger than those of conventional microelectrodes, impalement damage to the cell membrane is of even more concern to users of ISMs (Ammann, 1986). Microelectrode impalements of cells result in the rending and tearing of the plasma membrane. Cell contents can leak through the torn membrane, or extracellular contents can leak into the cell, resulting in various degrees of cell compromise or destruction; the extent of damage done to the cell is not always

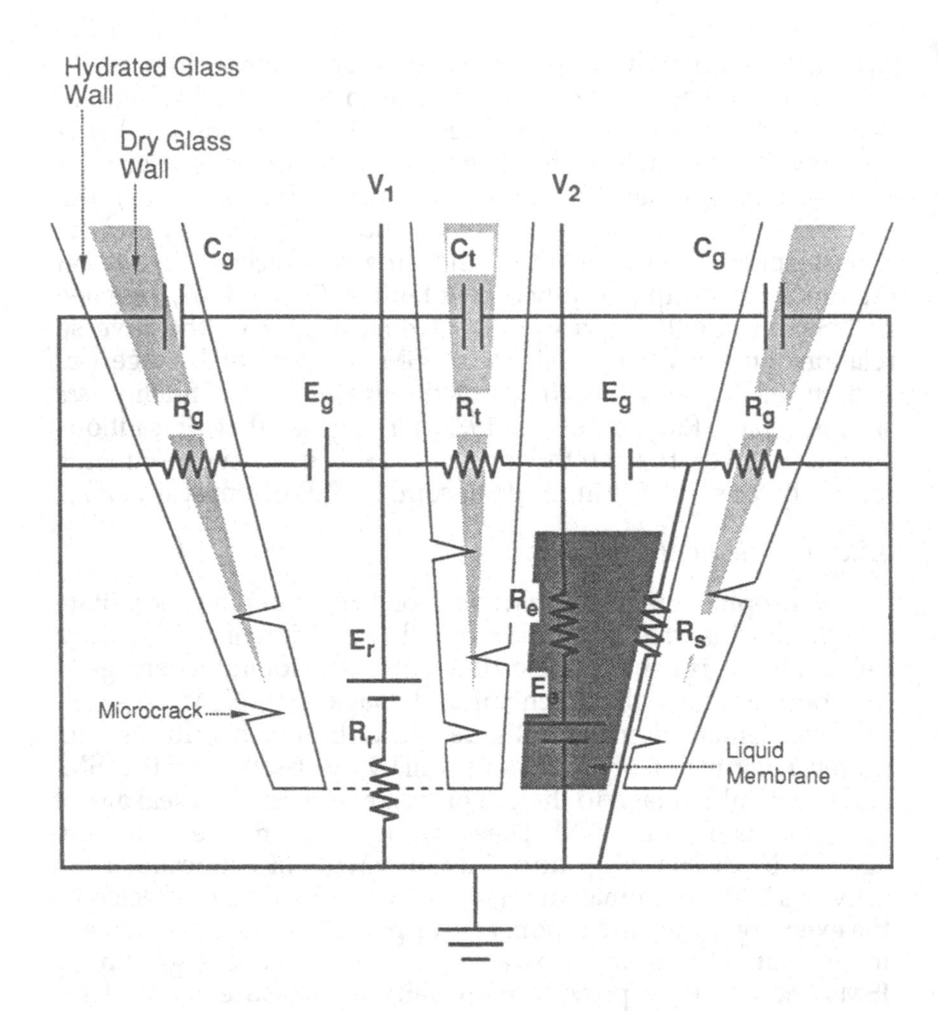

Fig. 10. Diagram of equivalent circuit at the tip of a theta glass, double-barreled ISM. V_1 and V_2 represent the voltage signal from the ion-selective and reference barrels, respectively. C_t and R_t are the interbarrel (transseptal) capacitance and coupling resistance, respectively. R_s is the shunt resistance along the aqueous surface between the ion-selective membrane and the microelectrode glass wall. R_e is the transmembrane resistance, E_e is the transmembrane potential, R_r is the tip resistance of the reference barrel, and E_r is the tip + junction potential of the reference barrel. Finally, C_g, R_g, and E_g are, respectively, the capacitance across the glass wall, the resistance across the glass wall, and the

apparent to the investigator, especially if low-impedance cells are being studied (Taylor and Thomas, 1984; Ammann, 1986). Taylor and Thomas (1984) were able to quantify the extent of impalement-induced membrane leak by monitoring the change in intracellular sodium activity in crab muscle fibers; their results warn against indiscriminate acceptance of intracellular ISM measurements. Fortunately, if cells are carefully impaled using sufficiently fine-tipped microelectrodes, the plasma membrane reseals around the microelectrode, limiting the loss of cell contents; the details of the glass–plasma-membrane interaction responsible for this resealing phenomenon are incompletely understood (Corey and Stevens, 1983). At this juncture, it is worth noting that exterior silanization of micropipets sometimes improves cell impalements (Aicken and Brading, 1982), and that side-by-side-design double- and multi-barreled micropipets may cause more cell leakage than micropipets of smoother cross section, because of the difficulty of sealing the plasma membrane around the invaginations between the micropipet barrels. The general criteria for distinguishing between suboptimal and acceptable cell impalements are well-known to most electrophysiologists and are summarized in Ammann (1986).

The other form of ISM-induced damage to cells or tissues is a result of leakage of components of the ISM into the sample solution (Ammann, 1986; Oesch et al., 1987). Leakage of aqueous electrolytes from reference microelectrodes into the cytosol can significantly alter the ionic milieu of the cell (Thomas, 1978: Fromm and Schultz, 1981; Ammann, 1986). The extent of reference electrolyte leakage depends on the concentration of the electrolyte in the microelectrode and the microelectrode resistance; experimentally determined KCl leakage rates from typical intracellular reference microelectrodes range from 1–10 fmol s^{-1} (Fromm and Schultz, 1981). These rates of electrolyte leakage are quite high, and would have serious consequences for all but the largest cells, were it not for the presence of transport mechanisms and cell-to-cell junctions

potential across the glass wall; note that, although these variables are depicted with the same symbols on both the reference and ion-selective barrels (for simplicity), in reality, their values would differ between the two barrels.

that allow the excess ion(s) to be dissipated (Fromm and Schultz, 1981).

Oesch et al. (1987) modeled the transfer of ion carriers from liquid-membrane ISMs into the cytosol and plasma membrane of the cell. Many ion carriers used in liquid-membrane ISMs can act as ionophores; if a significant concentration of these ion carriers builds up in cellular membranes, the transport properties of the cell can be affected (Ammann, 1986; Oesch et al., 1987). In the model, a number of factors influence the extent and kinetics of cytosol and plasma membrane contamination by ion carriers leached from the ISM's liquid membrane. Depending on the values chosen for the adjustable parameters of the model, the equilibrium concentration of ion carriers in the cell membrane, for example, can range anywhere from being essentially equal to the concentration of ion carriers in the ion-selective liquid membrane to being only 1% of that concentration, the estimated time over which equilibrium is achieved ranges from seconds to days (Oesch et al., 1987). Leakage of components from the ion-selective liquid membrane into the cell can be reduced by decreasing the concentrations of the components in the liquid membrane, by reducing the size of the ISM tip orifice, and by selecting a liquid membrane solvent offering favorable partition coefficients for the components vis-à-vis the cytosol or plasma membrane.

ISMs are subject to physical and chemical damage during use. Insertion of an ISM into tissue can result in tip breakage or plugging. Gross plugging or breaking of the ISM tip is readily detected and is thus not problematic. Microscopic tip breakage may not be readily apparent to the investigator, but the resultant changes in tip size could significantly alter the characteristics of the ISM (*see* section 4.3.5.). Partial obstruction of the ISM tip orifice by tissue can also change the effective tip size and thereby affect ISM performance.

Ion-selective sensors in liquid-membrane ISMs can be poisoned by various compounds, many of which are present in commonly studied cells and tissues or are compounds frequently employed for investigational purposes (Ammann, 1986). Surface contamination of both liquid- and glass-membrane ISMs by proteins and other biological species can compromise ISM function (Ammann, 1986). The utility of ISMs in certain applications is limited by selectivity considerations (Meier et al., 1982; Ammann, 1986). Often, the relevant parameters for estimating tolerable selectivity factors are unknown; this is especially true of intracellular solutions (Ammann, 1986). In some cases, ISMs designed to measure one ion are actually more selective for another (usually less common) ion that may be present in the sample solution; in these instances, even low activities of the "interferent" ion can substantially affect measurements of the ion of interest (Achenbach, 1985; Deisz and Lux, 1985).

4.4. Applications of ISMs to Neuroscience

The primary application of ISMs, of course, is the measurement of ion activities. The resting extracellular and intracellular activities of virtually all biologically important ions have been measured in a number of different nervous system structures (*see* references in Ammann, 1986). Extra- and intracellular ion activities have also been dynamically monitored, so as to elucidate the role of ion movement in a broad spectrum of nervous system functions (Ammann, 1986). By means of a variety of experimental manipulations, the ability of ISMs to measure ion activities has been taken advantage of to measure other biologically interesting parameters, such as endolymph flow (Sykova et al., 1987) and stimulation-induced volume changes of the extracellular space (Phillips and Nicholson, 1979; Hansen and Olsen, 1980; Connors et al., 1982).

The most common application of ISMs in the neurosciences is the potentiometric determination of intracellular ion activities (Walker and Brown, 1977; Tsien, 1983; Ammann, 1986). The ISM technique, being a natural extension of the electrophysiologist's usual repertoire, is technically compatible with other microelectrode methods. Transmembrane voltages are readily monitored and/or manipulated using the reference barrel of double- and multi-barreled ISMs, thus permitting investigators to record simultaneously ion activities and membrane potential (e.g., Fig. 11). The ability to record a number of interrelated electrophysiologically important variables at once, with reasonable temporal and good spatial resolution, has proved to be powerful; by studying relationships between these variables, some of the mechanisms underlying cellular ion homeostasis have been worked out (Ammann, 1986).

The breadth and depth of intracellular ion studies utilizing ISMs can be inferred from the following examples: Thomas (1972,

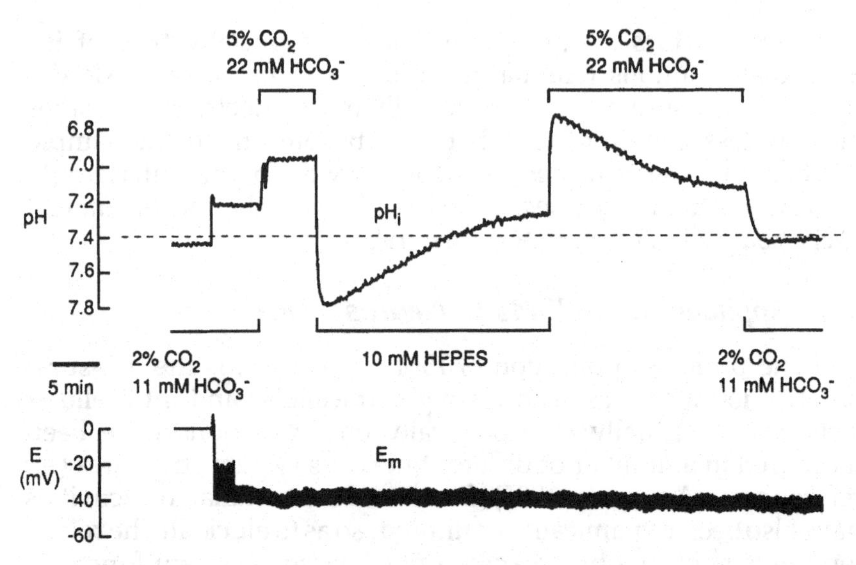

Fig. 11. Recording of intracellular pH (pH_i) and membrane potential (E_m) in a Retzius neuron during exposure to different perfusate solutions and gases. Brackets represent the periods during which solutions and gases of the specified composition were applied to the preparation. HEPES = *N*-2-Hydroxyethylpiperazine-*N*'-2-ethanesulfonic acid. The apparent noise in the records is caused by spontaneous action potentials truncated by the limited frequency response of the pen recorder. (Modified from Schlue and Deitman, 1986.)

1974, 1984), Thomas and Meech (1982), and Meech and Thomas (1987) undertook a series of intracellular studies in snail neurones that has significantly elucidated the mechanisms responsible for neuronal H^+ and Na^+ dynamics, Deitmer and Schlue (1987) have extended our knowledge of invertebrate intracellular H^+ regulation to leech glial cells and neurones; Boron (1985) studied H^+ regulation in the squid giant axon. Other workers, taking advantage of the increasing sophistication of ISM techniques, have successfully investigated intracellular ion homeostasis in the less accessible vertebrate nervous system: intracellular pH has been studied in lamprey neurones (Chesler, 1986), rat astrocytes (Chesler and Kraig, 1987), and cultured mammalian oligodendrocytes (Kettenman and Shlue, 1988). Deisz and Lux (1982) studied intracellular Cl^- in crayfish stretch receptors, Grafe et al. (1982) studied intracellular Na^+ and K^+ in frog motor neurones, Kettenman et al. (1983) analyzed intracellular K^+ in cultured oligodendrocytes, and Borelli et al (1986) studied intracellular Cl^- and K^+ in neuroblastoma cells.

Another common application of ISMs is the measurement of extracellular ion activities. Experiments with ISMs have been instrumental in advancing understanding of the complex ionic microenvironment of the nervous system (Somjen, 1979; Nicholson and Phillips, 1981, 1982; Syková, 1983; Nicholson, 1985; Ransom et al., 1986). Extracellular K^+ activity, especially, has been measured in a number of systems, including the honey bee retina (Coles and Tsacopouolos, 1979), the rat optic nerve (Connors and Ransom, 1984; Yamate and Ransom, 1985), and the rat cerebellum (Malenka et al., 1981; Kocsis et al., 1983). Extracellular pH has also been the object of intense study, having been investigated in the brain of the cat (Cragg et al., 1977; Urbanics et al., 1978; Mutch and Hansen, 1984), the rat cerebellar cortex (Kraig et al., 1983; Nicholson et al., 1985), and the rat neocortex (Kraig et al., 1985). Many of these studies pursued the relationship between extracellular pH and brain pathophysiology. A variety of other extracellular ions have also been studied both in physiological and pathological conditions: Ca^{2+} in the cat cerebellar cortex (Nicholson et al., 1978) and cat neocortex (Heinemann and Pumian, 1980), Na^+ and Cl^- in the cat neocortex (Dietzel et al., 1982), and HCO_3^- and NH_4^+ in the vertebrate CNS (Kraig and Cooper, 1986).

In the mammalian central nervous system, the extracellular space (ECS) is not a fixed volume, but rather, actively changes in size in response to neural activity and pathological conditions, such as anoxia (Dietzel et al., 1980; Ransom et al., 1985a,b, 1986). Changes in ECS volume can be measured with ISMs by measuring the activity of impermeant extracellular marker ions. Exogenously applied quarternary amines are frequently used as marker ions, because they cross cell membranes reluctantly, they are easily detected with ion-exchanger based ISMs, and at the concentrations in which they are applied, they do not have marked physiological effects (Phillips and Nicholson, 1979; Dietzel et al., 1980; Connors et al., 1982; Ransom et al., 1985b). Because quarternary amines are plasma-membrane impermeant, changes in ECS volume are reflected as changes in quarternary amine activity detected by the extracellular ISM. Quarternary amines have been exogenously applied to the experimental preparation in two different fashions: as brief, small puffs ejected from nearby micropipets (Dietzel et al.,

1980), or added directly to the perfusate solution (Hansen and Olsen, 1980; Connors et al., 1982); each of these techniques has its particular drawbacks and advantages (Connors et al., 1982). Nicholson and Phillips (1981) have cleverly employed extracellular marker ions and ISMs to determine absolute extracellular volume fraction and extracellular path tortuosity.

Similarly, flow rates of in vivo solutions can be measured using ISMs by tagging flowing solutions with a brief pulse of marker ion and then measuring the time taken for the marker ion to be detected by an ISM a known distance downstream from the point of marker ion introduction. Applying this method, Syková et al. (1987) used quarternary amines as markers and ion-exchanger-based K^+-ISMs as detectors to measure the flow of endolymph in the guinea-pig cochlea.

As ISM technology continues to evolve, further applications for ISMs in neuroscience will undoubtably be found. The development of improved neutral carrier anion-selective liquid membranes (Ammann et al., 1987) will spur experimentation that should give better understanding of Cl^- and HCO_3^- homeostasis in the nervous system. New sensors selective for inorganic (Pungor and Tóth, 1978) or organic ions (Meyerhoff and Opdycke, 1986; Noble and Way, 1987) will allow investigators to apply high-resolution potentiometric techniques to the difficult problem of real-time microanalysis of biochemical species. Novel techniques of ion-selective electrode manufacture and implementation (Tamura et al., 1982; Noble and Way, 1987; Rubinstein et al., 1988) will widen the experimental domain, and heretofore daunting experiments (such as measuring ion activities in subcellular organelles) may become relatively routine.

References

Achenbach C. (1985) Effects of thallous ions on the measurement of intracellular ion activities, in *Ion Measurements in Physiology and Medicine* (Kessler M., Harrison D. K., and Höper J., eds.) Springer Verlag, New York, pp. 256–263.

Agin D. P. (1969) Electrochemical properties of glass microelectrodes, in *Glass Microelectrodes* (Lavallée M., Schanne O., and Hebert N. C., eds.) Wiley, New York, pp. 62–75.

Aicken C. C. and Brading A. F. (1982) Measurement of intracellular chloride in guinea-pig vas deferens by ion analysis, ^{36}chloride efflux and microelectrodes. *J. Physiol.* **326,** 139–154.

Ammann D. (1986) *Ion-Selective Microelectrodes* (Springer Verlag, New York).

Ammann D., Oesch U., Bührer T., and Simon W. (1987) Designs of ionophores for ion-selective microsensors. *Can. J. Physiol. Pharmacol.* **65,** 879–884.

Ammann D., Lanter F., Steiner R. A., Schulthess P., Shijo Y., and Simon W. (1981) Neutral carrier based ion-selective microelectrode for extra- and intracellular studies. *Anal. Chem.* **53,** 2267–2269.

Armstrong W. McD. and Garcia-Diaz J. F. (1980) Ion-selective microelectrodes: Theory and technique. *Fed. Proc.* **39,** 2851–2859.

Bailey P. L. (1980) *Analysis with Ion-Selective Electrodes* (Heyden, Philadelphia).

Bartsch R. A., Charewicz W. A., Kang S. I., and Walkowiak W. (1987) Proton-coupled transport of alkali metal cations across liquid membranes by ionizable crown ethers, in *ACS Symposium Series,* Vol. 347: *Liquid Membranes Theory and Applications* (Noble R. D. and Way J. D., eds.) American Chemical Society, Washington, DC, pp. 86–97.

Bates R. G. (1969) Inner reference electrodes and their characteristics, in *Glass Microelectrodes* (Lavallée M., Schanne O., and Hebert N. C., eds.) Wiley, New York, pp. 1–24.

Bates R. G. and Robinson R. A. (1978) Trends in the standardization of ion-selective electrodes, in *Ion-Selective Electrodes* (Pungor E. and Buzás I., eds.) Elsevier Scientific, New York, pp. 3–19.

Bauer H. (1972) *Electrodics* (George Thieme, Stuttgart).

Boron W. F. (1985) Intracellular pH-regulating mechanism of the squid axon. *J. Gen. Physiol.* **85,** 325–245.

Borrelli M. J., Carlini W. G., Dewey W. C., and Ransom B. R. (1985) A simple method for making ion-selective microelectrodes suitable for intracellular recording in vertebrate cells. *J. Neurosci. Methods* **15,** 141–154.

Borrelli M. J., Carlini W. G., Dewey W. C., and Ransom B. R. (1986) Ion-sensitive microelectrode measurements of free-intracellular chloride and potassium concentrations in hyperthermia-treated neuroblastoma cells. *J. Cell. Physiol.* **129,** 175–184.

Briano R. A., Jr. (1983) A reproducible technique for breaking glass micropipettes over a wide range of tip diameters. *J. Neurosci. Methods* **9,** 31–34.

Brown K. T. and Flaming D. G. (1977) New microelectrode techniques for intracellular work in small cells. *Neurosci.* **2,** 813.

Brown K. T. and Flaming D. G. (1979) Technique for precision beveling of relatively large micropipettes. *J. Neurosci. Methods* **1,** 25–34.

Buck R. P. (1978) Theory and principles of membrane electrodes, in *Ion-selective Electrodes in Analytical Chemistry* (Freiser H., ed.) Plenum Press, New York, pp. 1–141.

Caldwell P. C. (1954) An investigation of the intracellular pH of crab muscle fibres by means of micro-glass and micro-tungsten electrodes. *J. Physiol.* **126,** 169–180.

Carlini W. G. and Ransom B. R. (1987) Tip size of ion-exchanger based K^+ selective microelectrodes. I. Effects on selectivity. *Can. J. Physiol. Pharmacol.* **65,** 889–893.

Chesler M. (1986) Regulation of intracellular pH in reticulospinal neurones of the lamprey, *Petromyzon marinus. J. Physiol.* **381,** 241–261.

Chesler M. and Kraig R. P. (1987) Intracellular pH of astrocytes increases rapidly with cortical stimulation. *Am. J. Physiol.* **253(4),** R666–R670.

Coles J. A. and Tsacopoulos M. (1977) A method of making fine double-barrelled potassium-sensitive micro-electrodes for intracellular recording. *J. Physiol.* **270,** 13–14P.

Coles J. A. and Tsacopoulos M. (1979) Potassium activity in photoreceptors, glial cells, and extracellular space in the drone retina: changes during photostimulation. *J. Physiol.* **290,** 525–549.

Coles J. A., Munoz J. L., and Deyhemi F. (1985) Surface and volume resistivity of pyrex glass used for liquid membrane ion-selective microelectrodes, in *Ion Measurements in Physiology and Medicine* (Kessler M., Harrison D. K., and Höper J., eds.) Springer Verlag, New York, pp. 67–73.

Connors B. W. and Ransom B. R. (1984) Chloride conductance and extracellular potassium concentration interact to modify the excitability of rat optic nerve fibers. *J. Physiol.* **355,** 619–633.

Connors B. W., Ransom B. R., Kunis D. M., and Gutnick M. J. (1982) Activity-dependent K^+ accumulation in the developing rat optic nerve. *Science* **216,** 1341–1343.

Corey D. P. and Stevens C. F. (1983) Science and technology of patch-recording electrodes, in *Single-Channel Recording* (Sakmann B. and Neher E., eds.) Plenum Press, New York, pp. 53–68.

Cragg P., Patterson L., and Purves D. (1977) The pH of brain extracellular fluid in the cat. *J. Physiol.* **272,** 137–166.

Deisz R. A. and Lux H. D. (1982) The role of intracellular chloride in hyperpolarizing post-synaptic inhibition of crayfish stretch receptor neurones. *J. Physiol.* **326,** 123–138.

Deisz R. A. and Lux H. D. (1985) Thiocyanate interference at chloride-selective microelectrodes in crayfish stretch receptor neurons: Evidence for a non-passive thiocyanate distribution, in *Ion Measurements in Physiology and Medicine* (Kessler M., Harrison D. K., and Höper J., eds.) Springer Verlag, New York, pp. 158–165.

Deitmer J. M. and Schlue W. R. (1987) The regulation of intracellular pH by identified glial cells and neurones in the central nervous system of the leech. *J. Physiol.* **388,** 261–283.

Deyhimi F. and Coles J. A. (1982) Rapid silylation of a glass surface: Choice of reagent and effect of experimental parameters on hydrophobicity. *Helv. Chir. Acta* **65,** 1752–1759.

Dietzel I., Heinemann U., Hofmeier G., and Lux H. D. (1980) Transient changes in the size of the extracellular space in the sensorimotor cortex of the cat in relation to stimulation-induced changes in potassium concentration. *Exp. Brain Res.* **40,** 432–439.

Dietzel I., Heinemann U., Hofmeier G., and Lux H. D. (1982) Stimulus-induced changes in extracellular Na^+ and Cl^- concentration in relation to changes in the size of the extracellular space. *Exp. Brain Res.* **46,** 73–84.

Durst R. A. (1978) Sources of error in ion-selective electrode potentiometry, in *Ion-selective Electrodes in Analytical Chemistry* (Freiser H., ed.) Plenum Press, New York, pp. 311–338.

Eisenman G. (1969) The ion-exchange characteristics of the hydrated surface of Na^+ selective glass electrodes, in *Glass Microelectrodes* (Lavallée M., Schanne O., and Hebert N. C., eds.) Wiley, New York, pp. 32–61.

Ellermann A., Höper J., Brunner M., and Kessler M. (1985) Computer-assisted processing of ion-selective electrode measurements, in *Ion Measurements in Physiology and Medicine* (Kessler M., Harrison D. K., and Höper J., eds.) Springer Verlag, New York, pp. 90–95.

Elmer T. H. (1980) Glass surfaces, in *Silyated Surfaces* (Leyden D. and Collins W., eds.) Gordon and Breach Science Publishers, London, pp. 1–30.

Finkel A. S. (1987) *Axoprobe-1A Microelectrode Amplifier Operator's and Service Manual.* Axon Instruments, Inc., Burlingame, California.

Finkel A. S. and Redman S. (1983) A shielded microelectrode suitable for single-electrode voltage clamping of neurones in the CNS. *J. Neurosci. Methods* **9,** 23–29.

Flaming D. G. and Brown K. T. (1982) Micropipet puller design—form of the heating filament and effects of filament width on tip length and diameter. *J. Neurosci. Methods* **6,** 91–102.

Fromm M. and Schultz S. G. (1981) Some properties of KCl-filled microelectrodes: Correlation of potassium "leakage" with tip resistance. *J. Membr. Biol.* **62,** 239–244.

Fujimoto M. and Honda M. (1980) A triple-barreled microelectrode for simultaneous measurements of intracellular Na^+ and K^+ activities and membrane potential in biological cells. *Jpn. J. Physiol.* **30,** 859–875.

Garcia-Diaz J. F. and Armstrong W. McD. (1980) The steady-state relationship between sodium and chloride transmembrane electrochemical potential differences in *Necturus* gallbladder. *J. Membr. Biol.* **55,** 213–222.

Goodisman J. (1987) *Electrochemistry: Theoretical Foundations.* Wiley, New York, 374 pp.

Grafe P., Rimpel J., Reddy M. M. and Ten Bruggencate G. (1982) Changes of intracellular sodium and potassium ion concentrations in frog spinal motoneurones induced by repetitive synaptic stimulation. *Neurosci.* **7,** 3213–3220.

Greaves G. N., Fontaine A., Lagarde P., Raoux D., and Gurman S. J. (1981) Local structure of silicate glasses. *Nature* **293,** 611–615.

Güggi M., Oehme M., Pretsch E., and Simon W. (1976) Neutraler Ionophor für Flüssigmembranelektroden mit hoher Selektivität für Natrium—gegenüber Kalium-Ionen. *Helv. Chir. Acta* **59,** 2417–2420.

Hamer W. J. (1968) *Theoretical Mean Activity Coefficients of Strong Electrolytes in Aqueous Solutions from 0 to 100°C.* National Bureau of Standards, Washington, DC.

Hansen A. J., and Olsen C. E. (1980) Brain extracellular space during spreading depression and ischemia. *Acta Physiol. Scand.* **108,** 355–365.

Harris R. J. and Symon L. (1984) Extracellular pH, potassium, and calcium activities in progressive ischaemia of rat cortex. *J. Cereb. Blood Flow Metab.* **4,** 178–186.

Hebert N. C. (1969) Properties of microelectrode glasses, in *Glass Microelectrodes* (Lavallée M., Schanne O., and Hebert N. C., eds.) Wiley, New York, pp. 25–31.

Heinemann U. and Pumian R. (1980) Extracellular calcium activity changes in cat sensorimotor cortex induced by iontophoretic application of aminoacids. *Exp. Brain. Res.* **40,** 247–250.

Hille B. (1984) *Ionic Channels of Excitable Membranes* (Sinauer Associates Inc. Sunderland, Massachusetts).

Hinke J. A. M. (1967) Cation-selective microelectrodes suitable for intracellular use, in *Glass Electrodes for Hydrogen and Other Cations* (Eisenman E., ed.) Marcel Dekker, New York, pp. 464–477.

Hinke J. A. M. (1987) Thirty years of ion-selective microelectrodes: Disappointments and successes. *Can. J. Physiol. Pharmacol.* **65,** 873–878.

Horvath A. L. (1985) *Handbook of Aqueous Electrolyte Solutions.* Wiley, New York, 631 pp.

Kaila K. and Voipio J. (1985) A simple method for dry-bevelling of micropipettes used in the construction of ion-selective microelectrodes. *J. Physiol.* **369,** 8P.

Kessler M., Hajek K., and Simon W. (1976) Four-barreled microelectrode for the measurement of potassium, sodium, and calcium ion activity, in *Ion and Enzyme Electrodes in Biology and Medicine* (Kessler M., Clark L. C., Lubbers D. W., Silver A., and Simon W., eds.) Urban & Schwarzenberg, Munich, pp. 136–140.

Kettenman H. and Schlue W. R. (1988) Intracellular pH regulation in cultured oligododendrocytes. *J. Physiol.* **406,** 147–162.

Kettenman H., Sonnhof U. and Schachner M. (1983) Exclusive potassium dependence of the membrane potential in cultured mouse oligodendrocytes. *J. Neurosci.* **3,** 500–505.

Kirk-Othmer D. (1984) *Kirk-Othmer Encyclopedia of Chemical Technology,* 3rd Ed., vol. 24, *Water Properties,* (Wiley-Interscience, New York).

Kocsis J. D., Malenka R. C., and Waxman S. G. (1983) Effects of extracellular potassium concentration on the excitability of the parallel fibers of the rat cerebellum. *J. Physiol.* **334,** 225–244.

Koryta J. (1982) *Ions, Electrodes, and Membranes,* (Wiley, New York).

Koryta J. and Stulík, K. (1983) *Ion-selective Electrodes* (Cambridge University Press, Cambridge, UK).

Koryta J., Dvorák J., and Bohácková V. (1970) *Electrochemistry* (Methuen, London).

Kraig R. P. and Cooper A. J. L. (1986) Bicarbonate and ammonia changes in brain during spreading depression. *ISM Symposium Abstracts* **98.**

Kraig R. P., Ferreira-Filho C. R. and Nicholson C. (1983) Alkaline and acid transients in cerebellar microenvironment. *J. Neurophysiol.* **49,** 831–850.

Kraig R. P., Pulsinelli W. A., and Plum F. (1985) Hydrogen ion buffering during complete brain ischemia. *Brain Res.* **342,** 281–290.

Kriz N. and Sykova E. (1981) Sensitivity of K^+-selective microelectrodes to pH and some biologically active substances, in *Ion-Selective Microelectrodes and Their Use in Excitable Tissues* (Syková E., Hník P., and Vyklicky L., eds.) Plenum Press, New York, pp. 25–39.

Lanford W. A. (1977) Glass hydration: A method of dating glass objects. *Science* **186,** 975–976.

Lavallée M. and Szabo G. (1969) The effect of glass surface conductivity phenomena on the tip potential of glass micropipette electrodes, in *Glass Microelectrodes* (Lavallée M., Schanne O., and Hebert N. C., eds.) Wiley, New York, pp. 95–110.

Lee C. O. (1981) Determination of selectivity coefficients of ion-selective microelectrodes, in *Ion-Selective Microelectrodes and their Use in Excitable Tissues* (Syková E., Hník P., and Vyklicky L., eds.) Plenum Press, New York, pp. 47–52.

Lev A. A. (1964) Determination of activity and activity coefficients of potassium and sodium ions in frog muscle fibres. *Nature* **201,** 1132–1134.

Levy S., Tillem L., and Tillotson D. L. (1985) Ion selective microelectrodes: Computer-controlled calibration, plotting, and data analysis. *J. Neurosci. Methods* **15,** 253–261.

Lewis S. A. and Wills N. K. (1980) Resistive artifacts in liquid ion-exchanger microelectrode estimates of Na^+ activity in epithelial cells. *Biophys. J.* **31,** 127–138.

Lux H. D. and Neher E. (1973) The equilibration time course of $[K^+]o$ in cat cortex. *Exp. Brain Res.* **17,** 190–205.

Malenka R. C., Kocsis J. D., Ransom B. R., and Waxman S. G. (1981) Modulation of parallel fiber excitability by postsynaptically mediated changes in extracellular potassium. *Science* **214,** 339–341.

Meech R. W. and Thomas R. C. (1987) Voltage-dependent intracellular pH in *Helix aspera* neurones. *J. Physiol.* **390,** 433–452.

Meier P. C., Lanter F., Ammann D., Steiner R. A., and Simon W. (1982) Applicability of available ion-selective liquid-membrane microelectrodes to intracellular ion-activity measurements. *Pflügers Arch.* **393,** 23–30.

Meyer G., Rossetti C., Bottá G., and Cremaschi D. (1985) Construction of K^+- and Na^+-sensitive theta-microelectrodes with fine tips: an easy method with high yield. *Pflügers Arch.* **404,** 378–381.

Meyerhoff M. E. and Opdycke W. N. (1986) Ion-selective electrodes. *Adv. Clin. Chem.* **25,** 1–47.

Morf W. E. (1981) *The Principles of Ion-Selective Electrodes and of Membrane Transport* (Elsevier, Oxford, UK).

Munoz J.-L. and Coles J. A. (1987) Quartz micropipettes for intracellular voltage microelectrodes and ion-selective microelectrodes. *J. Neurosci. Methods* **22,** 57–64.

Munoz J.-L., Deyhimi F., and Coles J. A. (1983) Silanization of glass in the making of ion sensitive microelectrodes. *J. Neurosci. Methods* **8,** 231–247.

Mutch W. A. C. and Hansen A. J. (1984) Extracellular pH changes during spreading depression and cerebral ischemia: Mechanisms of brain pH regulation. *J. Cereb. Blood Flow Metab.* **4,** 17–27.

Newman E. A. and Odette L. L. (1984) Model of electroretinogram B-wave generation–A test of the K^+ hypothesis. *J. Neurophysiol.* **51,** 164–182.

Nicholson C. (1985) Diffusion from an injected volume of a substance in brain tissue with arbitrary volume fraction and tortuosity. *Brain Res.* **333,** 325–329.

Nicholson C. and Phillips J. M. (1981) Ion diffusion modified by tortuosity and volume fraction in the extracellular microenvironment of the rat cerebellum. *J. Physiol.* **321,** 225–257.

Nicholson C. and Phillips J. M. (1982) Diffusion in the brain cell microenvironment. *Lectures on Mathematics in the Life Sciences* **15,** 103–122.

Nicholson C., Kraig R. P., Ferreira-Filho C. R., and Thompson P. (1985) Hydrogen ion variations and their interpretation in the microenvironment of the vertebrate brain, in *Ion Measurements in Physiology and Medicine* (Kessler M., Harrison D. K. and Höper J., eds.) Springer Verlag, New York, pp. 206–213.

Nicholson C., Ten Bruggencate G., Stöckle H., and Steinberg R. (1978) Calcium and potassium changes in extracellular microenvironment of cat cerebellar cortex. *J. Neurophysiol.* **41,** 1026–1039.

Noble R. D. and Way J. D. (1987) Liquid membrane technology: An overview, in *ACS Symposium Series,* Vol. 347: *Liquid Membranes Theory and Applications* (Noble R. D. and Way J. D., eds.) American Chemical Society, Washington, DC, pp. 1–27.

Oehme M., Kessler M., and Simon W. (1976) Neutral carrier Ca^{2+}-microelectrode. *Chimia.* **30,** 204–206.

Oesch U., Ammann D., and Simon W. (1987) Cell contamination due to the use of carrier-based microelectrodes. *Can. J. Physiol. Pharmacol.* **65,** 885–888.

Oesch U., Dinten O., Ammann D., and Simon W. (1985) Lifetime of neutral carrier based membranes in aqueous systems and blood serum, in *Ion Measurements in Physiology and Medicine* (Kessler M., Harrison D. K., and Höper J., eds.), Springer Verlag, New York, pp. 42–47.

Okada Y. and Inouye A. (1976) Studies on origin of tip potential of glass microelectrode. *Biophys. Struct. Mech.* **2,** 31–42.

Orkand R. K., Dietzel I., and Coles J. A. (1984) Light induced changes in extracellular volume in the retina of the drone, *Apis Mellifera*. *Neurosci. Lett.* **45**, 273–278.

Orme F. W. (1969) Liquid ion-exchanger microelectrodes, in *Glass Microelectrodes* (Lavallée M., Schanne O., and Hebert N. C., eds.) Wiley, New York, pp. 376–395.

Phillips J. M. and Nicholson C. (1979) Anion permeability in spreading depression investigated with ion-sensitive microelectrodes. *Brain Res.* **173**, 567–571.

Plueddemann E. P. (1980) Chemistry of silane coupling agents, in *Silyated Surfaces* (Leyden D. and Collins W. eds.) Gordon and Breach Science Publishers, London, pp. 1–30.

Pretsch E., Wegmann D., Ammann D., Bezegh A., Dinten O., Läubli M. W., Morf W. E., Oesch U., Sugahara K., Weiss H., and Simon W. (1985) Effects of lipophilic charged sites on the electromotive behavior of ligand membrane electrodes, in *Ion Measurements in Physiology and Medicine* (Kessler M., Harrison D. K., and Höper J., eds.) Springer Verlag, New York, pp. 11–16.

Pucacco L. R. and Carter N. W. (1976) A glass-membrane pH microelectrode. *Anal. Biochem.* **73**, 501–512.

Pucacco L. R. and Carter N. W. (1978) A submicrometer glass-membrane pH microelectrode. *Anal. Biochem.* **89**, 151–161.

Pucacco L. R., Corona S. K., Jacobson H. R., and Carter N. W. (1986) pH microelectrode: Modified Thomas recessed-tip configuration. *Anal. Biochem.* **153**, 251–261.

Pungor E. and Tóth K. (1978) Precipitate-based ion-selective electrodes, in *Ion-selective Electrodes in Analytical Chemistry* (Freiser H., ed.) Plenum Press, New York, pp. 143–210.

Purves R. D. (1981) *Biological Techniques Series*, vol. 6: *Microelectrode Methods for Intracellular Recording and Ionophoresis* (Academic, New York).

Ransom B. R., Carlini W. G., and Connors, B. W. (1986) Brain extracellular space: Developmental studies in rat optic nerve. *Ann. NY Acad. Sci.* **481**, 87–105.

Ransom B. R., Carlini W. G., and Yamate C. L. (1987) Tip size of ion-exchanger based K^+-selective microelectrodes. II. Effects on measurement of evoked $[K^+]_o$ transients. *Can J. Physiol. Pharmacol.* **65**, 894–897.

Ransom B. R., Yamate C. L., and Connors B. W. (1985a) Developmental studies on brain extracellular space: Activity-dependent K^+ accumulation and shrinkage, in *Ion Measurements in Physiology and*

Medicine (Kessler M., Harrison D. K. and Höper, J. eds.) Springer Verlag, New York, pp. 206–213.

Ransom B. R., Yamate C. L., and Connors B. W. (1985b) Activity-dependent shrinkage of extracellular space in the rat optic nerve: A developmental study. *J. Neurosci.* **5,** 532–535.

Raynauld J.-P. and Laviolette J. R. (1987) The silver-silver chloride electrode: A possible generator of offset voltages and currents. *J. Neurosci. Methods* **19,** 249–255.p

Rubinstein I., Steinberg S., Tor Y., Shanzer A., and Sagiv J. (1988) Ionic recognition and selective response in self-assembling monolayer membranes on electrodes. *Nature* **332,** 426–429.

Sakmann B. and Neher E. (1983) Geometric parameters of pipettes and membrane patches, in *Single-Channel Recording* (Sakmann, B. and Neher, E., eds.) Plenum Press, New York, pp. 37–51.

Schlue W. R. and Deitmer J. W. (1986) Direct measurement of intracellular pH in identified glial cells and neurons of the leech central nervous system. *Can. J. Physiol. Pharmacol.* **65,** 978–985.

Schlue W. R. and Thomas R. C. (1985) A dual mechanism for intracellular pH regulation by leech neurones. *J. Physiol.* **364,** 327–338.

Schlue W. R. and Wuttke W. (1983) Potassium activity in leech neuropile glial cells changes with external potassium concentration. *Brain Res.* **270,** 368–372.

Schwartz T. L. (1971) The thermodynamic foundations of membrane physiology, in *Biophysics and Physiology of Excitable Membranes* (Adelman W. J., ed.) Van Nostrand Reinhold Co., New York, pp. 47–95.

Siebens A. W. and Boron W. F. (1987) Effect of electroneutral luminal and basolateral lactate transport on intracellular pH in salamander proximal tubules. *J. Gen. Physiol.* **90,** 799–831.

Silver B. J. (1985) *The Physical Chemistry of Membranes* (Allen & Unwin, Boston).

Snell F. M. (1969) Some electrical properties of fine-tipped pipette microelectrodes, in *Glass Microelectrodes* (Lavallée M., Schanne O., and Hebert N. C., eds.) Wiley, New York. pp. 111–121.

Somjen G. G. (1979) Extracellular potassium in the mammalian central nervous system. *Ann. Rev. Physiol.* **41,** 159–177.

Starzak M. E. (1984) *The Physical Chemistry of Membranes* (Academic, San Francisco).

Syková E. (1983) Extracellular K^+ accumulation in the central nervous system. *Prog. Biophys. Mol. Biol.* **42,** 135–189.

Syková E., Syka J., Johnstone B. M., and Yates G. K. (1987) Longitudinal flow of endolymph measured by distribution of tetraethylammonium and choline in scala media. *Hear. Res.* **28,** 161–171.

Tamura H., Kimura K., and Shono T. (1982) Coated wire sodium- and potassium-selective electrodes based on bis(crown ether) compounds. *Anal. Chem.* **54,** 1224–1227.

Taylor P. S. and Thomas R. C. (1984) The effect of leakage on microelectrode measurements of intracellular sodium activity in crab muscle fibers. *J. Physiol.* **352,** 539–550.

Thomas R. C. (1972) Intracellular sodium activity and the sodium pump in snail neurones. *J. Physiol.* **220,** 55–71.

Thomas R. C. (1974) Intracellular pH of snail neurones measured with a new pH-sensitive glass microelectrode. *J. Physiol.* **238,** 159–180.

Thomas R. C. (1978) *Ion-Sensitive Microelectrodes.* (Academic, London).

Thomas R. C. (1984) Experimental displacement of intracellular pH and the mechanism of its subsequent recovery. *J. Physiol.* **354,** 3P–22P.

Thomas R. C. and Cohen C. J. (1981) A liquid ion-exchanger alternative to KCl for filling intracellular reference microelectrodes. *Pflügers Arch.* **390,** 96–98.

Thomas R. C. and Meech R. W. (1982) Hydrogen ion currents and intracellular pH in depolarized voltage-clamped snail neurones. *Nature* **299,** 826–828.

Tinoco J., Sauer K., and Wang J. C. (1978) *Physical Chemistry: Principles and Applications in Biological Sciences* (Prentice Hall, Englewood Cliffs, New Jersey).

Tripathi S., Mirgunov N., and Boulpaep E. L. (1985) Submicron tip breakage and silanization control improve ion-selective microelectrodes. *Am. J. Physiol.* **249(5),** C514–C521.

Tsien R. Y. (1983) Intracellular measurements of ion activities. Ann. Rev. *Biophys. Bioeng.* **12,** 91–116.

Tsien R. Y. and Rink T. J. (1980) Neutral carrier ion-selective microelectrodes for measurement of intracellular free calcium. *Biochim. Biophys. Acta* **599,** 623–638.

Tsien R. Y. and Rink T. J. (1981) Ca^{2+}-selective electrodes: A novel pvc-gelled neutral carrier mixture compared with other currently available sensors. *J. Neurosci. Methods* **4,** 73–86.

Ujec E., Keller O., Machek J., and Pavlik V. (1979) Low impedance coaxial K^+ selective microelectrodes. *Pflügers Arch.* **382,** 189–192.

Ujec E., Keller O., Kríz N., Pavlik V., and Machek J. (1981) Double-barrel ion selective [K^+, Ca^{2+}, Cl^-] coaxial microelectrodes (ISCM) for measurements of small and rapid changes in ion activities, in *Ion-*

Selective Microelectrodes and their Use in Excitable Tissues (Syková E., Hník P., and Vyklicky L., eds.) Plenum Press, New York, pp. 41–46.

Urbanics R., Leniger-Follert E., and Lübbers D. W. (1978) Time course of changes of extracellular H^+ and K^+ activities during and after direct electrical stimulation of the brain cortex. *Pflügers Arch.* **378,** 47–53.

Vaughan-Jones R. D. and Kaila K. (1986) The sensitivity of liquid sensor, ion-selective microelectrodes to changes in temperature and solution level. *Pflügers Arch.* **406,** 641–644.

Walden J., Lehmenkühler A., Speckman E.-J., and Witte O. W. (1984) Continuous measurement of pentylenetetrazol concentration by a liquid ion exchanger microelectrode. *J. Neurosci. Methods* **11,** 187–192.

Walker J. L. (1971) Ion specific liquid ion exchanger microelectrodes. *Anal. Chem.* **43,** 89A–93A.

Walker J. L. and Brown H. M. (1977) Intracellular ionic activity measurements in nerve and muscle. *Physiol. Rev.* **57,** 729–778.

Yamate C. L. and Ransom B. R. (1985) Effects of altered gliogenesis on activity-dependent K^+ accumulation in the developing rat optic nerve. *Dev. Brain Res.* **21,** 167–173.

Zeuthen T. (1981) How to make and use double-barreled ion-selective microelectrodes. *Current Topics in Membranes and Transport* **13,** 31–47.

In Vivo Voltammetry

The Use of Carbon-Fiber Electrodes to Monitor Amines and Their Metabolites

Nigel T. Maidment, Keith F. Martin,
Anthony P. D. W. Ford, and Charles A. Marsden

1. General Introduction

One of the major challenges for the neuropharmacologist is to relate functions to the numerous ligand binding sites identified in brain tissue; there are, for example, at least seven postulated 5-hydroxytryptamine (5-HT) receptors and associated subtypes. Such studies involve biochemical measurement of changes in second messenger systems, behavioral tests of receptor function or measurement of physiological responses produced by receptor activation, such as release. Most studies measuring release have used radiolabeled transmitters and their release from slices in vitro. A more direct approach would be to measure changes in the release of endogenous transmitter in response to selective agonists and antagonists. In recent years, two methods for measuring endogenous transmitter metabolism and release in vivo have attracted considerable attention. The first involves intracerebral perfusion, using microdialysis tubes in which the perfusate samples are collected and assayed for their transmitter and metabolite content, using specific and highly sensitive assay methods (Ungerstedt, 1984; Marsden, 1985). The advantage of this method is the positive analytical identification of the substances within the perfusates. Disadvantages include the relatively large size of the dialysis probe (300–1000 μm) (which limits the application to larger brain areas) and the relatively slow time resolution resulting from the long collection time between each sample, although the recent advent of microbore high-performance liquid chromatography (HPLC) has allowed dopamine (DA) release to be measured in

3-min dialysates of the striatum "on line" (Church and Justice, 1987).

The second approach attempts to monitor changes in amine release and metabolism *in situ* electrochemically, using carbon-fiber microelectrodes that both oxidize the relevant amine or its metabolite and measure the resultant current. This method overcomes the problem of probe size, since the electrodes can have a tip diameter of 8 μm or less and a resolution time that can be as little as 25 ms when electrically stimulated DA release is monitored. In vivo electrochemical techniques are, however, performed without positive external identification of the transmitters and their metabolites. This chapter will consider some general theoretical aspects and practical applications of in vivo voltammetry, and subsequently will describe the development and application of carbon-fiber electrodes to the monitoring of DA, 5-HT, and their metabolites, 3,4-dihydroxyphenylacetic acid (DOPAC) and 5-hydroxyindoleacetic acid (5-HIAA), respectively. The status of in vivo voltammetry with regard to the different types of electrodes has been the subject of a recent critical commentary (Marsden et al., 1988).

2. Electrochemical Principles

Electrochemical detection (ECD) coupled with HPLC has become the method of choice for the assay of trace amounts of amine neurotransmitters and their metabolites from brain tissue since Adams and his colleagues first made neuroscientists aware of its capabilities (Adams, 1969; Kissinger et al., 1973a; Adams and Marsden, 1982). The in vivo voltammetric technique essentially involves implanting into the brain a miniaturized version of the HPLC electrochemical detector. Electrochemical detection exploits a well-known property of catechol- and indole-based substances—their susceptibility to undergo oxidation (Fig. 1). The ability of such compounds as oxygen to act as oxidizing agents for this reaction is a result of the energies of their unfilled electron orbitals lying sufficiently below those of the filled orbitals of the oxidizable moieties of the catechols/indoles so that electron transfer can occur with a net decrease in free energy. This situation can also be expressed in terms of the electron environment of the oxidants being "more positive" than that of the reductants, so that electrons

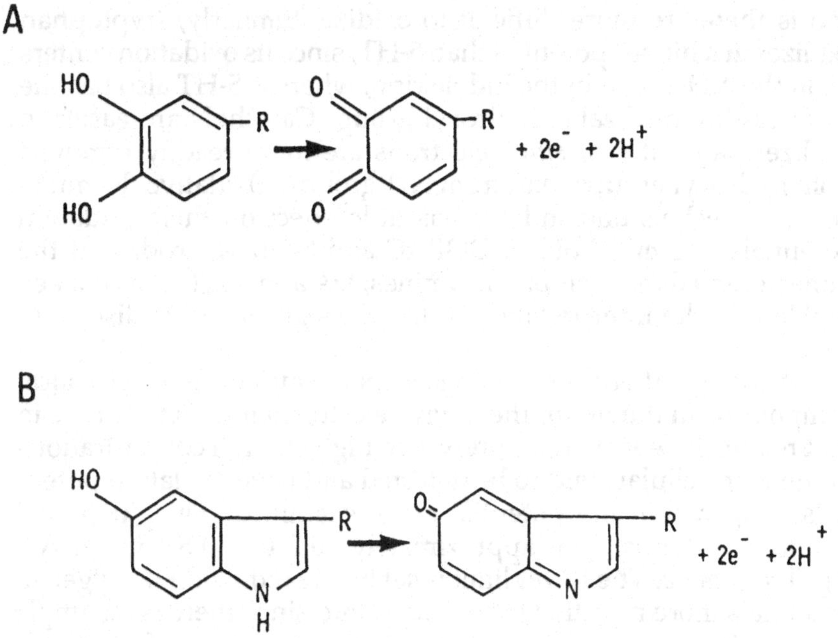

Fig. 1. Schematic representation of the oxidation of catecholamines (A) and indoleamines (B).

readily pass to the oxidants. The electrochemical detection system uses as its oxidizing agent a carbon-based electrode, at the surface of which is applied a positive potential of a "strength" sufficient to remove electrons from the oxidizable species concerned. The resultant flow of electrons can be measured in the form of an electrical current, the amplitude of which is directly proportional to the amount of material oxidized.

A clear difference between the use of electrochemical detection for HPLC and in vivo voltammetry is the absence of a chromatographic separation procedure in the latter, which instead relies upon the individual oxidation potentials of the compounds of interest to provide resolution. Some electroactive species will undergo oxidation with greater ease than others, and by slowly increasing the strength of the potential applied to the electrode, individual substrates in the neuronal matrix can be sequentially oxidized. For instance, DA will oxidize at a lower potential than tyrosine, since the latter compound has only one hydroxyl group and is therefore more difficult to oxidize. Similarly, tryptophan oxidizes at a higher potential than 5-HT, since its oxidation centers upon the NH-group in the indole ring, whereas 5-HT also has the more readily oxidizable hydroxyl group. Catechols are easier to oxidize than indoles, since electrons are more readily removed from hydroxyl groups than from NH-groups. Oxidative deamination of catechols and indoles has little effect on their oxidation potentials; the metabolites, DOPAC and 5-HIAA, oxidize at the same potentials as their parent amines, DA and 5-HT. This is a key problem in the interpretation of in vivo signals and is discussed later.

Although there are obviously many hundreds of chemical compounds in the brain, the in vivo electrochemist is fortunate in that relatively few are both present in high enough concentrations in the extracellular fluid to be detected and have oxidation potentials lying within the potential range examined. This "potential window" extends from approximately –0.2 to +0.8V vs an Ag/AgCl reference. The lower limit is set by the reduction if oxygen at potentials more negative than –0.2 V, and since there is an ample supply of oxygen in the brain, an excessive amount of current is generated, overshadowing any other redox process. Similarly, the upper potential limit is determined by the oxidation of water, the supply of which, again, is inexhaustible. Figure 2 shows the approximate redox potentials of compounds of neurochemical interest that are oxidizable within this potential window. Ranges rather than specific values are given, since the precise potential will vary with such parameters as the type of working electrode used and method of application of potential. Attention should be drawn to the fact that several drugs used to manipulate catechol and indole neurotransmission are electroactive; this must be considered when designing experiments. The inclusion in Fig. 2 of two compounds not readily associated with neurotransmission—ascorbic acid and uric acid—deserves comment at this point. The presence of these compounds in high concentrations in brain extracellular fluid (Zetterstrom et al., 1983) has caused particular problems for the in vivo electrochemist, since they have oxidation potentials very similar to catechols and indoles, respectively. This problem is discussed fully in later sections.

The degree to which individual oxidizable species can be resolved in the brain with in vivo voltammetry is largely dependent

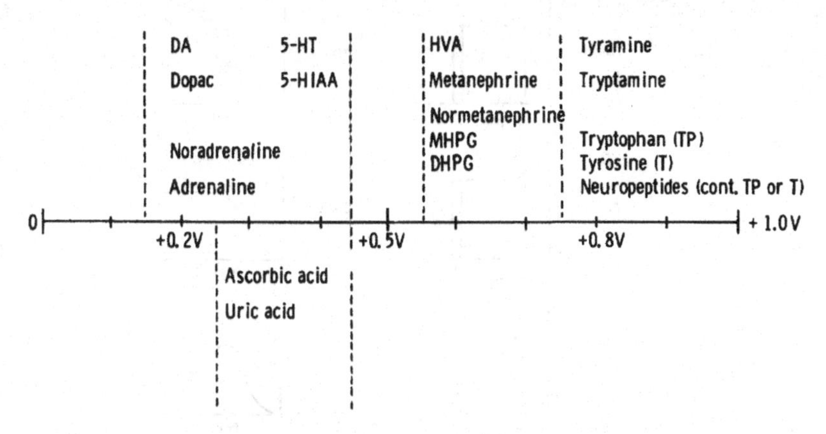

Fig. 2. Diagram demonstrating the variety of compounds capable of undergoing oxidation at the surface of carbon-based electrodes implanted in the brain. The list is not comprehensive, and precise oxidation potentials are not given since these vary with the type of electrodes employed.

upon two factors: (a) the type of electrode employed and (b) the measurement technique—that is, the way in which the potential is applied to the electrode. The latter of these factors is covered in the following section.

3. Measurement Techniques

3.1. Chronoamperometry

This is the simplest method of applying a potential to the working electrode and involves instantaneous switching of the electrode from open circuit to the desired applied potential, E_{app}, which is generally slightly higher than the oxidation potential of the compound to be oxidized. The potential is held constant for a fixed time (between 50 ms and 1 s) and then switched back to open circuit. Also, the current is measured at a set time within this period (Fig. 3A1). Figure 3A2 shows the current–time curve resulting from the application of a potential to an electrode in a stirred solution. Clearly, apart from initial charging effects, the current

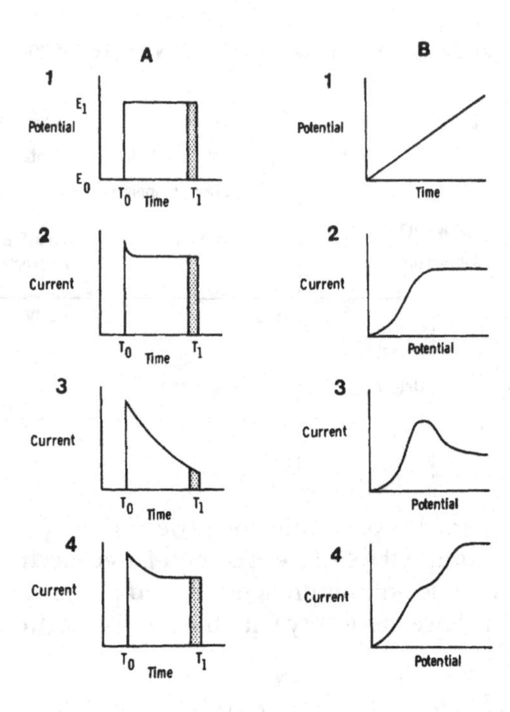

Fig. 3. (A) (1) Chronoamperometry involves instantaneous switching of the electrode from open circuit (E_0) to a predetermined potential (E_1). The electrode is held at this potential for a fixed time (T_1) and then switched back to open circuit. (2) Diagrammatic representation of a current–time curve obtained from a chronoamperometric measurement shown in (1), using a standard-sized electrode (>300 µm) in a stirred solution. Apart from an initial non-faradaic charging current, the response is time-independent, since fresh oxidizable material is constantly being presented to the electrode surface. The current is measured at a set time after application of the potential (shaded portion). (3) As (1), but in an unstirred solution. The current recorded diminishes with time, since the process of diffusion of oxidizable species to the electrode surface is unable to keep pace with oxidative depletion. (4) The current–time response of a microelectrode in an unstirred solution. The small size of such electrodes increases the effectiveness of replenishment of the electrode surface by diffusion, such that steady-state conditions are rapidly attained and the response is largely time-dependent.

(B) (1) Linear sweep voltammetry involves increasing with time the amplitude of the potential applied to the electrode, in the form of a linear generated is independent of time, since fresh, oxidizable material is continuously being presented to the electrode surface; this is similar to the situation with HPLC-ECD. However, electrodes implanted into the brain can be considered to be lying in a pool of extracellular fluid that bathes, and thus provides contact with, the neuronal matrix composed of synapses, cell bodies, axons, and glial cells. (The size of the electrodes employed—8–300 μm—precludes implantation into individual synapses.) From an electrochemical viewpoint, this is an unstirred solution and, as such, differs markedly from the HPLC-ECD situation. When considering the dynamics of the electrochemical oxidation process, it is important to realize that electrons generated as a result of chemical oxidation cannot be transferred over distances greater than a few molecular diameters. Therefore, electroactive species must be very close to the surface of the electrode in order to be detected. In the unstirred situation, replenishment of oxidizable material at the electrode surface is totally dependent upon diffusion down a concentration gradient. Thus, as soon as the potential is applied to an electrode surface in an unstirred solution, the resulting electrolysis depletes the immediate surroundings of oxidizable material, setting up a concentration gradient between the electrode surface and the bulk of the solution. The process of diffusion down this concentration gradient is unable to keep pace with the oxidative depletion, so the current falls off with time (Fig. 3A3). The current–time relationship is described by the Cottrell equation:

$$i = \frac{nFAD \quad C^{b}}{t} \tag{1}$$

---

ramp. (2) Oxidation of a single species in a stirred solution at the surface of a standard-sized electrode using linear sweep voltammetry produces a limiting current. (3) The situation pertaining in an unstirred environment using a standard-sized electrode. Peak-shaped voltammograms are obtained, since the process of diffusion is unable to replenish the surface of the electrode with oxidizable species as rapidly as they are being oxidized. (4) Microelectrodes placed in an unstirred solution (e.g., in vivo) produce voltammograms with limiting currents as a result of enhanced diffusion effects; thus, it becomes difficult to resolve the oxidation of two compounds with similar oxidation potentials.

where i is current, n is the number of electrons transferred/mol of species oxidized, F is the Faraday constant, A is the area of the electrode in cm^2, D is the diffusion coefficient of the oxidizable species in cm^2/s, C is the bulk concentration in mol/mL of the species, and t is time in seconds. If the current is taken at $t = 1$ s, then the above equation can be simplified to show that current is directly proportional to bulk concentration:

$$i = kC^b \tag{2}$$

However, the Cottrell equation is valid only when replenishment of a planar electrode surface takes place only by perpendicular linear diffusion and, as such, the equation is inadequate for describing the situation with the small electrodes used in vivo, especially the cylindrical carbon-fiber electrodes used extensively as described below. With such electrodes, perpendicular linear diffusion is supplemented by so-called edge contributions, so that the current response of these electrodes is largely time-independent and rapidly becomes steady-state (Fig. 3A4). This can be accounted for by the following modification to the Cottrell equation:

$$i = \frac{nFAD \quad C}{t} + 1/r\ nFADC \tag{3}$$

where r is the radius of a hypothetical sphere having the same area as the electrode tip. As the size of the electrode gets very small or the time of electrolysis increases, the steady-state component dominates the response.

However, it should be emphasized that the linear relationship between current measurement at a fixed potential and bulk concentration still holds with these small electrodes. These modifications become important only when using voltammetry to study diffusion coefficients of substances in the brain (McCreery et al., 1974a; Dayton et al., 1983). When, as in most in vivo voltammetric applications, the only information sought is regarding changes in the extracellular concentration of an oxidizable species, the precise current–time relationship can be ignored. As long as the moment at which the current is sampled is kept constant throughout the experiment, the current measured will be directly proportional to the concentration of the compound undergoing oxidation, and pre- or postcalibration of the electrode in vitro will provide an *estimate* of the extracellular concentration.

Chronoamperometry was the technique used in the original in vivo voltammetric studies and has been used in several laboratories (Conti et al., 1978; Marsden et al., 1979; Schenk et al., 1983; Schenk and Adams, 1984; Nagy et al., 1985). The major advantages of this system are its capability for relatively high-frequency, repetitive sampling of the extracellular environment (up to 1 measurement/s) and the ease with which it can be automated using relatively unsophisticated equipment (*see* Schenk and Adams, 1984).

3.2. Linear Sweep Voltammetry

With this technique, the potential applied to the electrode is increased progressively over a period of many seconds in the form of a linear ramp (Fig. 3B1) (usually at a rate of 5–20 mV/s), producing a current-vs-voltage plot, or voltammogram. In this way, it is possible to sequentially oxidize species and to identify their oxidation potentials. As with chronoamperometry, there are fundamental differences between the unstirred and stirred situation, as well as between in vitro and in vivo measurements with small electrodes. As the voltage applied to a standard sized electrode (>300 μm) in stirred solution is increased, the current generated rises to a plateau, after which further increases in voltage produce no further increase in current (Fig. 3B2). This is because the potential has reached a value capable of oxidizing all the electroactive material reaching the electrode surface. The value of the current at this plateau is termed the limiting current, i_L, and the value of E_{app} at $i = i_{L/2}$ is termed the half-wave potential, and is closely related to the standard redox potential. However, if this electrode is placed in an unstirred solution of the same oxidizable species, the voltammogram will not show a limiting plateau (Fig. 3B3). Rather, peak-shaped voltammograms are obtained as the current reaches a maximum and then decays. This is because diffusion is not sufficient to replenish the electrode surface of oxidizable species as rapidly as it is being oxidized. The value of E_{app} at the peak is used to identify the compound undergoing oxidation, and the amplitude of the current at this peak is directly proportional to the bulk concentration of the compound.

As mentioned previously, the small electrodes that are used in vivo exhibit different diffusion characteristics, such as tending to behave as though they are in a stirred solution. This is particularly evident when using linear sweep, since the long time of the scan increases the relative contribution from the steady-state factor in equation 3, so that voltammograms in unstirred solutions (e.g., in vivo) tend to exhibit limiting plateaus rather than peaks. Consequently, it becomes difficult to measure individual contributions to the current generated by the sequential oxidation of two or more compounds in a mixed solution (Fig. 3B4). Effectively, two compounds with oxidation potentials less than 150 mV apart cannot be resolved with linear sweep voltammetry and small in vivo electrodes. However, this technique has been used effectively in some laboratories (Kennett and Joseph, 1982; O'Neill et al., 1982, 1983; O'Neill and Fillenz, 1985), and it should be noted that peak-shaped voltammograms are obtained with linear sweep from chronically implanted electrodes. This results from the formation of a restricted compartment owing to gliosis around the electrode, so that the process of diffusion of oxidizable species to the electrode again becomes the limiting factor (Albery et al., 1983).

3.3. Differential Pulse Voltammetry

In order to aid the process of resolving compounds with similar oxidation potentials, the linear sweep technique can be modified to produce peak-shaped voltammograms. This technique, differential pulse voltammetry, has been used extensively and involves the application of a linear ramp potential as above, but superimposed on this ramp are regular step potentials of constant amplitude and duration (Fig. 4). The current is sampled immediately before the pulse is applied (i_a) and immediately before the pulse is switched off (i_b), and the difference ($i_b - i_a$) is plotted against applied potential. Peak-shaped voltammograms are obtained, since it is the rate of increase of current with respect to time (and therefore E_{app}) that is measured. The sensitivity and resolving power can be manipulated by changing the characteristics of the pulses superimposed on the ramp. Increasing the amplitude of the pulses (modulation amplitude) increases the sensitivity and decreases the resolution between peaks. A modulation amplitude of 50 mV is generally used. Varying the scan rate also affects sensitivity and resolution. Slower scan rates improve both of these parameters, but obviously have the disadvantage of producing unacceptably long sampling times.

3.4. Linear Sweep with Semidifferentiation

This technique essentially involves electronic manipulation of the output from a normal linear sweep polarograph to obtain semidifferentiated voltammograms similar to those produced by differential pulse techniques (Lane et al., 1979).

3.5. Normal Pulse Voltammetry

This technique was developed from chronoamperometry. It involves applying a series of pulses of increasing amplitude to the electrode, with the potential returning to baseline between pulses, unlike differential pulse voltammetry (Ewing et al., 1982).

3.6. Differential Normal Pulse Voltammetry

This is an interesting modification of differential pulse and normal pulse voltammetry in which potential pulses of increasing amplitude are applied to the working electrode, the potential returning to baseline between pulses as with normal pulse voltammetry. However, at the end of each pulse is a step potential allowing for differential measurement. An advantage of this technique over differential pulse voltammetry is that the electrode is submitted to positive potentials for an overall shorter period of time, since the potential returns to zero between pulses. This may reduce the amount of filming of the electrode by oxidized species and increase electrode life as a result (Gonon et al., 1984a; Gonon and Buda, 1985).

3.7. Cyclic Voltammetry

This is an adjunct to linear sweep voltammetry involving the application of a triangular wave potential to the electrode, so that the voltage is first scanned out to a predetermined limit and then scanned back to an initial value. Thus, provided that rapid sweep rates are used (approximately 500 mV/s), oxidizable species are first oxidized and then reduced to their original state.

The so-called high-speed cyclic voltammetry technique is a more interesting area of development, in which the electrodes are scanned repeatedly at rates of up to 300 V/s. This allows electrical-

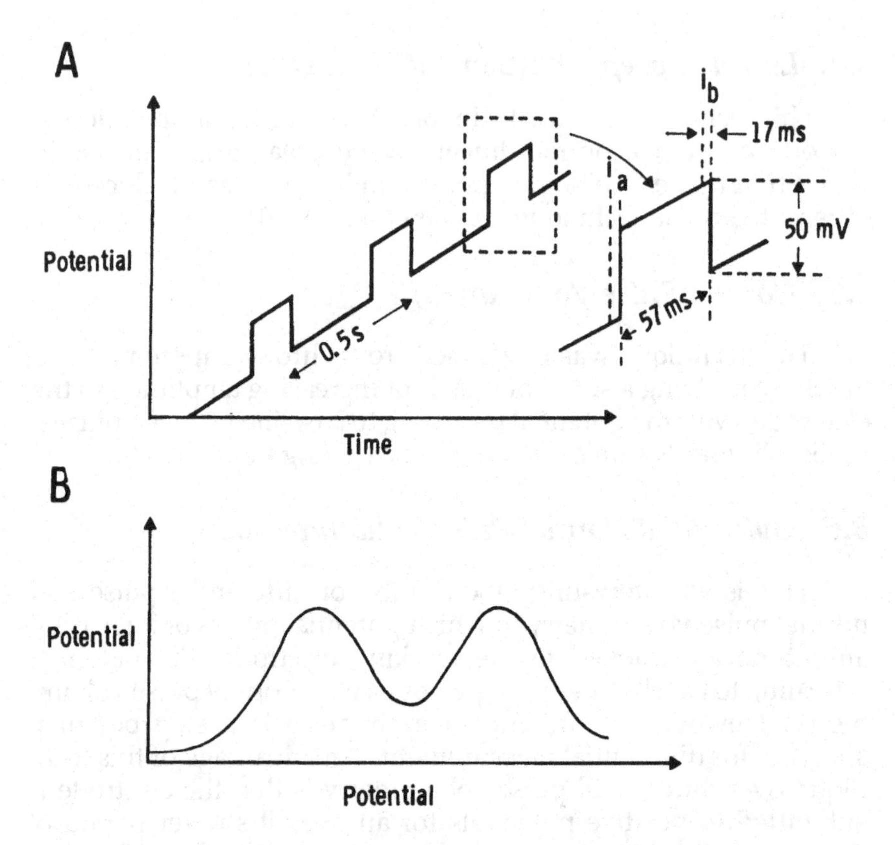

Fig. 4. Differential pulse voltammetry is a modification of linear sweep voltammetry, in which pulses of constant amplitide and duration are superimposed on the linear ramp at regular intervals (A). The current produced as a result of oxidation is sampled immediately before the application of a pulse (i_a) and immediately before it is switched off (i_b), and the difference ($i_b - i_a$) is plotted against applied potential, thus producing peak-shaped voltammograms (B).

stimulation-evoked release of neurotransmitters to be monitored on a millisecond time scale (Armstrong-James and Millar, 1984; Millar et al., 1985; *see* Future Developments in this chapter).

3.8. Differential Pulse Amperometry

This recently developed technique (Marcenac and Gonon, 1985) is similar to differential normal pulse voltammetry, except that the pulses are at a constant potential, so it resembles chronoamperometry with step potentials to allow for differentiation. The time resolution of this technique approaches, but does not attain, that of high-speed cyclic voltammetry, but its sensitivity is apparently greater.

4. Development of the Working Electrode

4.1. Requirements

Perhaps the most important factor in attaining meaningful voltammetric signals in vivo is the choice of working electrode. The type of material used for the working electrode is governed by the requirement of a low residual current. This is the current generated when a potential is applied to the electrode in the absence of electroactive species as a result of oxidation of the electrode surface. This is particularly important, since the faradaic currents generated in vivo are very small (in the nanoampere range). Although a modicum of success was achieved with the use of platinum (Lane and Hubbard, 1976; Lane et al., 1976), most studies have employed carbon-based electrodes.

Development of the electrodes has centered around the requirement for electrodes with optimally selective properties. As previously described, several compounds present in the extracellular fluid of the brain oxidize at similar potentials, and of particular importance is the presence of high concentrations of ascorbic acid, which interferes with the oxidation of the catechols. It is the purpose of this section to describe the various types of electrodes that have evolved in order to overcome these problems.

4.2. Carbon-Paste Electrodes

These electrodes were the first to be used successfully in vivo (Kissinger et al., 1973b; McCreery et al., 1974a,b; Wightman et al., 1976), and simply consist of graphite powder mixed with liquid paraffin or silicone oil, which is packed into the exposed Teflon™ sheath of a length of Teflon™-coated silver wire. Paste electrodes are therefore easy to construct (Brazell and Marsden, 1982a) and have been reported to have working lives in vivo of several weeks (O'Neill et al., 1983; O'Neill and Fillenz, 1985). However, these electrodes have several disadvantages; one of these is their large size (up to 300 μm), but more important is their inability to resolve the oxidation of ascorbic acid (AA) from that of DA or DOPAC. The structure of AA has certain features in common with DA and DOPAC, and in fact, has an absolute redox potential below that of the catechols. However, the oxidation potential of AA at carbon-paste electrodes is displaced several hundred millivolts in a positive direction, and mixtures of AA and DA produce one broad peak with differential pulse voltammetry. In addition, the oxidation of DA (or DOPAC) in the presence of AA results in chemical oxidation of AA, with concomitant regeneration of DA (or DOPAC) as follows:

$$DA \rightleftharpoons DOQ$$

$$DOQ + AA \rightleftharpoons DA + DHA \quad (4)$$

where DOQ = dopamine orthoquinone (the oxidized form of dopamine) and DHA = dehydroascorbate (the oxidized form of ascorbate). This has the effect of enhancing the oxidation signal from DA (or DOPAC) (Dayton et al., 1980). Furthermore, manipulation of the oxidation peaks obtained in vivo with these electrodes suggests that the signal that originally was attributed to catechol oxidation is, instead, largely caused by oxidation of AA. Thus, microinjection of ascorbate oxidase (an enzyme that converts AA to the electrochemically inert dehydroascorbate) in the vicinity of the electrode in vivo almost completely suppressed the apparent catechol signal (Brazell and Marsden, 1982b). These electrodes—in unmodified form—are therefore generally considered unsuitable for measuring DA/DOPAC levels in vivo. However, they have been used by some groups to monitor 5-HT neurotransmission (Joseph and Kennett, 1981, 1983; Kennett and Jospeh, 1982; Echizen and Freed, 1983, 1984) and are currently being used to record HVA peaks in vivo over periods of several weeks, although the signals generated by this compound are small and require computer-controlled subtraction of background current to produce usable results (O'Neill et al., 1983; O'Neill and Fillenz, 1985). Furthermore, it has been suggested recently that the in vivo peak recorded with these electrodes, previously attributed to 5-hydroxyindoles, is in fact entirely caused by uric acid oxidation (O'Neill et al., 1984).

4.3. Graphite-Epoxy Electrodes

These electrodes are similar to carbon-paste electrodes and, in their basic form, suffer from the same limitations, except that the incorporation of an epoxy resin produces a harder and more durable surface for chronic implantations (Conti et al., 1978; Marsden et al., 1979). Subsequently, Falat and Cheng (1982) reported that electrochemical treatment of graphite-epoxy electrodes *(see below)* produced clear separation of AA and DA/DOPAC oxidation peaks. However, such electrodes proved unstable in our hands, and their application has not been described subsequently in the literature.

4.4. Carbon-Fiber Electrodes

These electrodes essentially take two forms—cylindrical and disc-shaped. The former electrodes were first described by Gonon and coworkers (Ponchon et al., 1979; Gonon et al., 1980, 1981), and consist of one or more 8-μm od pyrolytic graphite fibers supported in a pulled glass capillary, the active surface of which is a 0.5-mm length of fiber(s) protruding from the capillary. The electrodes are capable of distinguishing between the oxidation of AA and catechols after an electrochemical pretreatment, and have been used extensively in our laboratory. A full description of their preparation and application is given in later sections.

By cutting the fiber flush with the glass capillary, one obtains a very small, disc-shaped, active surface, and such electrodes have been used for in vivo studies (Dayton et al., 1980; Ewing et al., 1982; Wightman, 1981; Kuhr et al., 1984; Ewing and Wightman, 1984). The extremely small size of these electrodes results in their exhibiting unique electrochemical properties. The catalytic reaction between AA and the oxidized form of the catechols is not a problem, since the extensive spherical diffusion contributions associated with such a small electrode result in the oxidized product diffusing away from the electrode before it can undergo reaction with AA to reform DA (Dayton et al., 1980). Furthermore, the small differences in the shapes of voltammograms obtained from AA, DA, and DOPAC as a result of the varying degrees of reversibility of their oxidation processes are accentuated at these small electrodes. DA undergoes oxidation with the greatest degree of reversibility and produces voltammograms similar to those seen with larger carbon-paste electrodes. However, the relatively poor reversibility of AA and DOPAC oxidation produces ill-defined, drawn-out voltammograms, which enables a certain degree of differentiation of DA from DOPAC and AA in vivo, on the basis of the shapes of voltammograms obtained (Dayton et al., 1981; Ewing et al., 1982). However, from such in vivo studies, it has been concluded that the major contributor to the oxidation signal under basal conditions is AA. Thus, although these electrodes, in combination with normal pulse voltammetry and chronoamperometry, have been used to monitor extracellular DA resulting from electrical or drug-induced stimulation of DA neurons (Dayton et al., 1981; Ewing et al., 1982, 1983; Kuhr et al., 1984; Ewing and Wightman, 1984), they are unable to measure basal extracellular levels of DA or, indeed, DOPAC. Similar electrodes have been used in conjunction with high-speed cyclic voltammetry to monitor electrically stimulated DA release in the striatum on a millisecond time scale (Stamford et al., 1984; Millar et al., 1985) or for quantification of iontophoretically applied amine neurotransmitters (Armstrong-James et al., 1980; Kruk et al., 1980; Millar et al., 1981). In the latter case, the same etched-fiber electrodes were used for electrophysiological unit recording.

4.5. Ascorbic Acid Eliminator Electrode

A novel approach to the problem of selectivity between AA and catechols has involved the use of the enzyme ascorbate oxidase. As mentioned above, this enzyme converts AA to the electrochemically inert dehyroascorbate. By immobilizing this enzyme on the surface of a conventional carbon-paste or graphpoxy electrode, using a dialysis membrane or glutaraldehyde cross-linking agent, Adams and coworkers produced an electrode that was essentially insensitive to AA, since this compound is converted before it reaches the active surface of the electrode (Nagy et al., 1982). Problems with long-term immobilization of the enzyme have prevented its full potential being realized.

4.6. Amine-Selective Electrodes

The various electrodes described above are capable of monitoring basal extracellular levels of the DA and 5-HT metabolites, DOPAC, HVA, and 5-HIAA, or of detecting DA released into the extracellular space following electrical or K^+-induced stimulation of DA neurones. However, measurement of basal levels of the neurotransmitters themselves has proved more problematical. This results from the facts that (a) DA and 5-HT oxidize at potentials almost identical to those of their respective metabolites and (b) the extracellular concentration of the metabolites is at least 100 × higher than that of the neurotransmitters themselves (Zetterstrom et al., 1983).

One approach to the problem of measuring basal DA release has been to remove the contribution to the oxidizing current from DOPAC by administering monoamine oxidase inhibitors. The small signal remaining after such treatment using electrically pretreated carbon-fiber electrodes *(see below)* is probably a result of oxidation of DA, and can be manipulated by pharmacological and stimulatory means (Gonon and Buda, 1985; Marcenac and Gonon, 1985). However, the functional relevance of studies incorporating such measurements after such drastic manipulation of DA neurochemistry is questionable. Inhibition of monoamine oxidase, by removing the major route of degradation, will undoubtedly produce artificially high extracellular levels of DA.

Perhaps a more promising approach has been the development of cation-selective electrodes. The first of these was described by Blaha and Lane (1983), who incorporated stearic acid into the graphite-paste electrodes previously described. Since both AA and DOPAC are anions at physiological pH, these compounds are repelled by the negatively charged carboxyl groups of the stearic acid molecules, whereas the positively charged DA is able to reach the electrode surface and undergo oxidation. (Blaha and Lane, 1984; Lane and Blaha, 1986).

The same principle of cation selection is employed in the more recently described Nafion-coated electrodes originating from Professor Adams' laboratory (Gerhardt et al., 1984; Brazell et al., 1987). Nafion is a perfluorosulfonated derivative of Teflon™ and, as such, is highly negatively charged. Similarly, it is claimed that these electrodes are relatively selective for DA in preference to AA and DOPAC, and they have been used in brain-slice preparations to monitor K^+-stimulated DA release. More recently, successful attempts have been made to coat carbon-fiber electrodes with Nafion (Kuhr and Wightman, 1986; Crespi et al., 1988; *see also* section 9.1.1.)

5. Preparation of Electrically Pretreated Carbon-Fiber Electrodes

The cylindrical carbon-fiber electrodes used in our laboratory were first described by Gonon et al. (1980). These electrodes comprise a length of carbon fiber(s) protruding from a pulled glass capillary, the suface(s) of which is modified by an electrochemical pretreatment prior to implantation into the brain. This treatment results in a transformation of the electrochemical properties of the electrode surface, such that AA and catechols exhibit separable oxidation peaks using differential pulse voltammetry. However, it should be emphasized that the oxidation of amines (i.e., DA and 5HT) is not distinguishable from that of their deaminated metabolites (i.e., DOPAC and 5-HIAA), although evidence from experiments described later in this chapter suggests that signals obtained in vivo are predominantly caused by metabolite oxidation. This section will concentrate on the manufacture and in vitro evaluation of these electrodes following various forms of electrochemical pretreatment.

5.1. Electrode Fabrication

Carbon-fiber working electrodes are manufactured in a manner similar to that described by Ponchon et al. (1979), as follows:

1. A length of glass capillary (Clark Electromedical Instruments, 1.2-mm od, 0.69-mm id, ref GC150-15) is pulled to a fine tip, using a conventional electrode puller, and cut to a length of approximately 25mm.
2. 1–3 carbon fibers (Le Carbone Lorraine, 8μm, ref AGT/F) are isolated from a bunch of fibers, threaded into the pulled capillary, and pushed to the tip.
3. The tip of the capillary is cut, using a pair of iris scissors, so that the fiber can be pushed through the end to protrude approximately 5 mm. The opposite end of the fiber is then cut flush with the capillary.
4. A carbon paste is prepared by mixing graphite power (0.18 g) (Ultra Carbon, ref UCP-1-1M), with polyester resin (200μL) (Escil ref Sody 33).

5. The capillary is inverted into the paste, which is then forced to the tip using a suitably sized plunger. Only electrodes in which the resin separated from the graphite at the tip of the capillary are used.
6. A length of Teflon™-coated silver wire (Clark Electromedical Instruments ref AG-10), stripped of its coating at each end, is pushed into the paste-filled glass capillary.
7. The electrode is cured overnight at 60°C.
8. Finally, the protruding fiber is cut to a length of 0.5 mm or 0.3 mm immediately prior to use.

A modification of the above procedure has been employed more recently, using electrically conductive paint to form an electrical contact between the fiber and the wire, as follows:

1.–3. as above.
4. One exposed end of a length of Teflon™-coated silver or plastic-coated copper wire is dipped in electrically conductive paint (Radio Spares) and placed in contact with the carbon fiber.
5. This connection is then strengthened with polyester resin (Escil ref Sody 33).
6. The tip of the electrode is sealed by the application of a drop of low-viscosity resin (Loctite Glassbond) using a piece of thin wire and a microscope, taking care to avoid fouling the exposed fiber with resin. Capillary action ensures a good seal between glass and fiber.
7. The fiber is cut to length as described above.

In the majority of studies, electrodes incorporated either two or three carbon fibers, since this increases the amplitude of the signals recorded and produces a stronger electrode tip that is less likely to break during implantation. The diameter of such electrodes is thus increased to approximately 16 μm. Electrodes in which fibers splay out are discarded. Figure 5 is a diagrammatic representation of a completed electrode.

5.2. Reference and Auxiliary Electrodes

Electrochemical detection involves the application of a potential difference to the surface of an electrode with respect to a reference

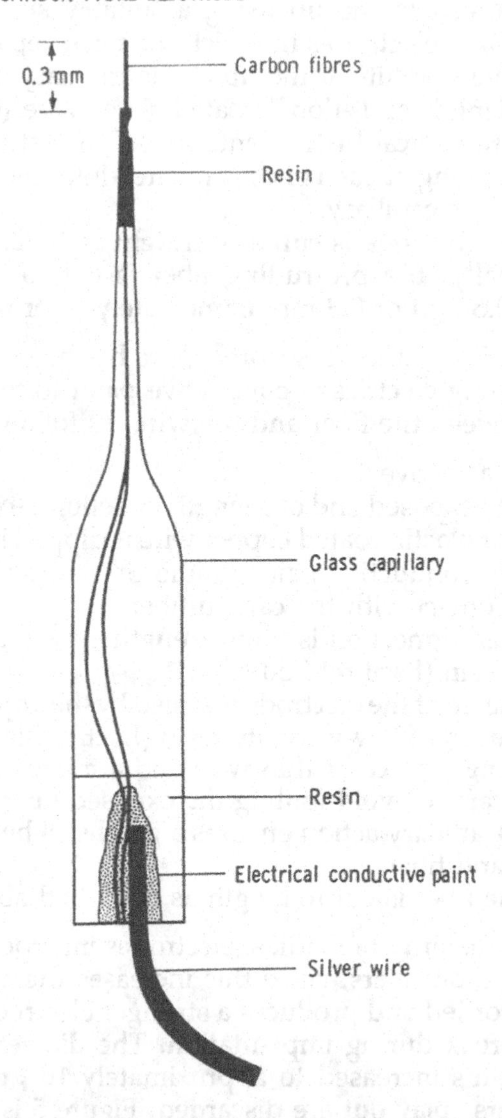

Fig. 5. Diagram of a completed carbon-fiber electrode, in which electrically conductive paint has been used to form electrical contact between fibers and silver wire. *See* text for details of preparation.

electrode, with concomitant measurement of the current generated as a result of chemical oxidation or reduction. In practice, this necessitates the use of at least two, but generally three, electrodes. The potential is applied between the oxidizing electrode (the working electrode) and a reference half-cell (normally Ag/AgCl) that completes the circuit for electron flow, and against which all applied potentials are expressed. A third, auxiliary electrode (platinum, silver, or copper) is added to provide for potentiostatic control. The potentiostat is normally composed of three operational amplifier sections: a signal portion generates the applied potential, a control amplifier maintains these potential differences between the working and reference electrodes (by passing the required current through the working and auxiliary electrode), and an output section measures the current generated as a result of the oxidation or reduction processes. There are several commercially available polarographic systems that provide the potentiostatic control (e.g., Princeton, Tacussel, Metrohm). The auxiliary electrode plays no role in the measurement; it is simply a part of the control system preventing current flow between the working and reference electrodes.

Ag/AgCl reference electrodes are made by stripping approximately 5 mm of Teflon™ from both ends of a length of Teflon™-coated silver wire (300 μm, Clark Electromedical Instruments), connecting one end to the positive pole of a regulated power supply, and immersing the other end in 1*M* HCl. A length of copper or platinum wire immersed in the acid and connected to the negative pole completes the circuit. The silver wire is anodized by applying +5 V for 1 min, causing a thin layer of AgCl to form on the exposed wire.

Auxiliary electrodes are simply a length of platinum wire for in vitro experiments or, for in vivo studies, a piece of silver or copper wire placed adjacent to the reference electrode on the surface of the brain.

5.3. Electrical Pretreatment

To obtain separable oxidation peaks for AA and catechols, it is necessary to pretreat the carbon-fiber electrode electrically (Gonon et al., 1980, 1981) by immersing it, together with a reference and auxiliary electrode, in phosphate-buffered saline and connecting the electrodes to the relevant terminals of a suitable polarographic

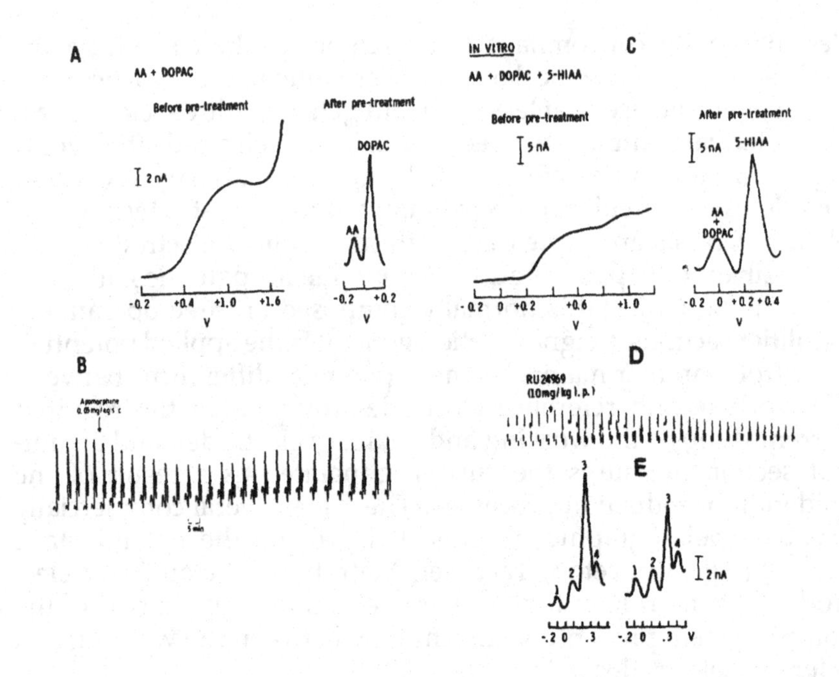

Fig. 6. (A) The effect of electrical pretreatment of carbon-fiber electrodes on the separation of ascorbic acid ($5 \times 10^{-4}M$) and DOPAC ($5 \times 10^{-5}M$) in vitro at pH 7.4 using differential pulse voltammetry. Note the absence of separation without electrical pretreatment. Electrical pretreatment: Triangular waveform 0 to +3 V, 70 Hz for 20 s; then held at +1.5 V DC for 5 s, 0.9 V DC for 3 s, and finally +1.5 V DC for 5 s.

(B) The effect of apomorphine (0.05 mg/kg s.c.) on the voltammograms obtained with an electrically pretreated carbon-fiber electrode implanted in the rat striatum. Note the decrease in the DOPAC oxidation peak followed by recovery of this peak, whereas the smaller ascorbic acid peak increases after apomorphine.

(C) The effect of electrical pretreatment of carbon-fiber electrodes on the separation of ascorbic acid ($5 \times 10^{-4}M$), DOPAC ($5 \times 10^{-5}M$), and 5-HIAA ($5 \times 10^{-5}M$) in vitro at pH 7.4 using differential pulse voltammetry. Note the absence of separation without electrical pretreatment. Electrical pretreatment: Triangular waveform 0 to +3 V, 70 Hz for 20 s; then 0 to +2 V, 70 Hz for 20 s; and finally 0 to +1 V, 70 Hz for 20 s (*see* Gonon et al., 1984b and Sharp et al., 1984 for details.

(D) The effect of the 5-HT_{1B} receptor agonist RU 24969 (10 mg/kg ip) on the 5-HIAA oxidation peak obtained with the electrode implanted in the suprachiasmatic nucleus. Note the prolonged decrease in the peak.

analyzer (e.g., Princeton Applied Research 174A or Tacussel "Biopulse"). If the Princeton Applied Research 174A is being used, the output and earth terminals of a waveform generator are connected to the PAR174A via pins 9 and 10 of plug J36, and a 0 to +3 V, 70 Hz triangular wave potential is applied for 20 s, with the PAR174A in the "SAMPLED DC" mode, initial potential, 0. The signal generator is then disconnected and the following potentials applied to the electrode, using the PAR174A in the "DC" mode: +1.5 V for 5 s, –0.9 V for 3 s, and +1.5 V for 5 s. The effect of such treatment on the performance of the electrode in solution is shown in Fig. 6A and in vivo in Fig. 6B. With the "Biopulse" machine, the electrical pretreatment parameters are preset automatically.

Optimum conditions for monitoring indoles with carbon-fiber electrodes are obtained with a slightly different pretreatment, as follows: 0 to +3 V triangular wave, 70 Hz for 20 s; 0 to +2 V, 70 Hz for 20 s; 0 to +1 V, 70 Hz for 20 s (Cespuglio et al., 1981a). Such pretreatment does not produce separation of AA and catechols, but provides well-defined indole peaks in vitro and in vivo (Fig. 6C). A further modification allows separation of catechols and indoles simultaneously, the pretreatment being as follows: 0 to +3 V triangular wave, 70 Hz for 20 s; 0 to +2.5 V, 70 Hz for 20 s; 0 to +1.5 V, 70 Hz for 20 s. Although such electrodes again display minimal differentiation between AA and catechols in vitro, these compounds are distinguishable in vivo, the electrodes producing up to four peaks when implanted in the nucleus accumbens or striatum (Crespi et al., 1984a, Fig. 6D).

It is still unclear why the electrical pretreatment increases the sensitivity and selectivity of the carbon fibers. The effects of electrical pretreatment on the chemical and structural features of the fibers are currently under study, using transmission and scanning

---

(E) Separation of ascorbic acid (1), DOPAC (2), 5-HIAA (3), and HVA (4) in vivo obtained with carbon-fiber electrodes pretreated as follows: –0.3 V at 70 Hz for 20 s, then a continuous potential of +1.5 V for 5 s. The first voltammogram obtained in the striatum was obtained in the absence of uricase, and the second voltammogram was recorded following injection of uricase (0.2 U) locally, close to the electrode. Note the decrease in the 5-HIAA peak by about 30%, indicating that uric acid may contribute to this peak as uricase converts uric acid to the electrochemically inert allantoin (*see* text for details).

electron microscopy combined with X-ray microanalysis. Initial results indicate that the pretreatment may increase the redox couples and oxides both on the surface and within the core of the fibers, thereby enhancing their interaction with electroactive compounds. As yet, there is no rationale that can be applied to produce the correct pretreatment parameters for a particular electroactive compound—serendipity is still way ahead.

6. Signal Verification: Comparison with Intracerebral Dialysis

Evidence that carbon-fiber electrodes are able to monitor changes in DOPAC (Peak 2) and 5-HIAA (Peak 3) has come from two types of experiments: first, pharmacological manipulation of the recorded peaks (*see* Gonon et al., 1984b; Cespuglio et al, 1984; Crespi et al., 1984 a,b; Marsden, 1985; summarized in Table 1) and, secondly, direct comparison of the effects of drugs on changes in the metabolites monitored simultaneously by intracerebral dialysis and in vivo voltammetry (Sharp et al., 1984). The dialysis technique (Ungerstedt, 1984) involves the stereotaxic implantation of a small-diameter (250 μm) dialysis tube (folded into a loop and supported by two steel cannulae) into the brain area of interest. The tube is then perfused with physiological saline (0.7–2 μL/min) and fractions collected into a small inverted tube placed on the outlet cannula. The fractions are then assayed by HPLC-ECD for DA, 5-HT, DOPAC, HVA, and 5-HIAA. Thus, by implanting a dialysis probe in one striatum of anaesthetized rats and a carbon-fiber electrode in the other, it is possible to compare changes in the voltammetric signals directly with changes in the extracellular levels of DOPAC and 5-HIAA measured by dialysis with HPLC—ECD (Sharp et al., 1984). Haloperidol, which increases DA release and metabolism, produces an almost identical percentage increase in both DOPAC measured by dialysis and the presumed DOPAC voltammetric oxidation peak (Peak 2). Similarly, amphetamine reduces both DOPAC levels in the dialysis samples and the height of the DOPAC voltammetric oxidation peak.

In general, the evidence that Peak 2 is solely a result of the oxidation of DOPAC is convincing. However, with regard to the 5-HIAA peak (Peak 3), there is a problem regarding the contribution of uric acid. The decrease in Peak 3 in the striatum after

Table 1
Summary of the Effects of Pharmacological Manipulation of Peak 2 (DOPAC) and Peak 3 (5-HIAA) Observed with Electrically Pretreated Carbon Fiber Electrodes[a]

Treatment	Peak 2[b]	Peak 3[b]
Catecholamine		
6-hydroxydopamine	↓	–
α-methyl-*p*-tyrosine	↓	–
DOPA/5-HTP decarboxylase inhibitors	↓	↓
MAO inhibitors	↓	↓
D-amphetamine	↓	↓
Neuroleptics	↑	↓
Reserpine	↑	↑
L-DOPA	↑	–
Indoleamine		
5,7-dihydroxytryptamine	–↑?	↓
p-chlorophenylalanine	–	↓
Stimulation of the raphé nucleus	↓?	↑
L-tryptophan	–	–↑?
5-HTP	–	↑
Other compounds		
Probenecid	↑	↑
Uric acid	–	↑
Uricase	–	↓

[a]Data from Cespuglio et al., 1984; Crespi et al., 1984a,b; Gonon et al., 1984a,b; Marsden, 1985, and references contained therein.
[b]↑ = increase, ↓ = decrease, – = no change, ? = small response in direction indicated.

5,7-dihydroxytryptamine lesions, monoamine oxidase inhibition, and 5-HTP decarboxylase inhibition is usually about 20% less than the decrease in tissue 5-HIAA (Cespuglio et al., 1984). Conversely, the increase in Peak 3 after 5-HTP (25 mg/kg) was considerably less than the increase in tissue 5-HIAA (Cespuglio et al., 1981b), whereas there was no significant increase in Peak 3 following L-tryptophan (100 mg/kg) (Cespuglio et al., 1981b), although striatal 5-HIAA levels increase.

The difference between the change in Peak 3 and the biochemical data might suggest that some other compound contributes to the oxidation peak. Another explanation is that, though the voltammetric electrodes measure extracellular 5-HIAA, the tissue measurements include intra- and extracellular metabolite, and the drugs have a greater effect on intra- than extracellular metabolite levels. This latter view is supported by the finding that, when intracerebral dialysis and voltammetry were performed simultaneously in the striatum of the same rat, there was a very close correlation between the change in Peak 3 and the decrease in 5-HIAA in the perfusates after administration of the monoamine oxidase inhibitor tranylcypromine (10 mg/kg), with maximal decreases of 58% and 54%, respectively (Sharp et al., 1984). In contrast, however, are the results obtained following 5-HTP (25 mg/kg) administration. In this case, although both Peak 3 and 5-HIAA in the perfusates increased, there was a large difference in the maximal response (+97% in Peak 3 and +447% 5-HIAA) (Sharp et al., 1984).

The failure of Peak 3 to keep pace with the increased 5-HIAA levels may, in part, relate to the sensitivity of the electrodes to such high concentrations of 5-HIAA, since the electrodes show a linear response only over a concentration range of 5–100 μM, and above this, the response flattens. Following 5-HTP administration, the extracellular 5-HIAA concentration is probably well above 100 μM. Alternatively, the pool of 5-HIAA sampled by dialysis may differ from that sampled by voltammetry, with damaged tissue pools making a significant contribution to the former. However, there still remains the possibility that some other compound contributes to Peak 3, and the main candidate is uric acid, which is found in high concentrations in the brain extracellular space (Zetterstrom et al., 1983). Uric acid and 5-HIAA have similar oxidation potentials, and uricase, which converts uric acid to the electroactively inert allantoin, abolishes the uric acid oxidation peak in vitro (Crespi et al., 1983).

Intrastriatal injection of uric acid (10 μg) increases the height of the striatal Peak 3, whereas injection of uricase decreases Peak 3 by about 30%, and when this is followed by injection of the monoamine oxidase inhibitor pargyline (75 mg/kg), the peak is abolished within the subsequent 3 h. These results suggest that uric acid contributes about 30% to Peak 3 recorded with electrochemically pretreated fiber electrodes implanted in the striatum, and so this needs to be considered when interpreting changes in this peak. The uric acid problem may be greater in acute than in chronic preparations and relates to uric acid in blood being brought down the electrode tip during implantation rather than to uric acid in brain. We find that small-diameter (16–20 μm) electrodes placed carefully, without causing loss of blood, into the suprachiasmatic nucleus (SCN) appear not to detect uric acid.

7. Some Problems To Be Considered with In Vivo Studies

7.1. Electrode Calibration

Several attempts have been made to estimate the basal extracellular concentrations of both DA and DOPAC, as well as 5-HIAA, using various forms of the carbon-fiber voltammetric electrode (Sharp et al., 1984; Ewing et al., 1982). Comparison of the values obtained with such electrodes to those obtained using intracerebral dialysis indicate that the voltammetric electrodes produce higher values than does dialysis (Sharp et al., 1984). These results highlight one of the major problems with the voltammetric electrodes—namely calibration, especially with respect to the relationship between the sensitivity of an electrode in vitro using standard solutions (artificial CSF, pH 7.4) and the sensitivity in vivo. Recalibration after use in vivo indicates that the sensitivity may alter in vivo compared to the original in vitro value, so a method that permits an in vivo calibration is required. For example, the injection of an electroactive compound not normally present in brain may be employed, but there are problems in interpreting values obtained with this approach. First, the substances injected may have different distribution and metabolism characteristics from the endogenous compound. Second, the properties of the electrode regarding the oxidation of the substance injected may differ from those with respect to the amines and their metabolites. Indeed, there are differences in the sensitivity of carbon-fiber electrodes for DA and DOPAC, with greater sensitivity for the amine than for the metabolite, and the sensitivities of these electrodes for DOPAC and 5-HIAA also differ. Such variations in sensitivity make it unlikely that a single substance can act as a reference for the range of compounds detected by the carbon-fiber electrodes in vivo, but may provide a general indication as to sensitivity in vivo. In summary, it is difficult to calibrate the carbon-fiber electrodes for in vivo use, but it is essential to check in vitro the separation characteristics and relative sensitivities towards the compounds to be detected prior to electrode use in vivo. At present, the carbon-fiber electrodes provide reliable *qualitative* and interesting semiquantitative information about changes in DOPAC and 5-HIAA.

7.2. Effects of Anaesthetics on the Signals

Several earlier studies have indicated that the basal levels of the DOPAC and 5-HIAA signals (Peaks 2 and 3) in vivo can be affected by both the type of anaesthetic used and the level of anaesthesia. We have recently studied the effects of both of these, as well as the responses obtained by various drugs on the striatal DOPAC-oxidation peak in rats under various anaesthetics (halothane, α-chloralose, pentobarbitone, and chloral hydrate). The results show that the choice of anaesthetic needs to be considered with care (Ford and Marsden, 1986).

The results indicate that, for routine use, halothane (2–3% in 50:50 O_2/N_2O [1 L/h]) produces stable baseline DOPAC levels, upon which significant changes induced by DA agonists and antagonists are clearly seen (Fig. 7). In practical terms, halothane is very convenient to administer, and allows rapid induction and constant anaesthesia. The steady Peak 2 baseline observed with this anaesthetic may reflect the well-balanced nature of the anaesthesia induced by halothane. This contrasts with the situation observed with pentobarbitone or chloral hydrate (Fig. 7). Although α-chloralose produces a stable baseline and large drug-induced responses (Fig. 7), the physical characteristics of the anaesthesia induced (i.e., certain reflex irritability retained) coupled with the inconvenient administration of the drug raise doubts as to its suitability as a general anaesthetic for rats. The mechanisms involved in the effects of the different anaesthetics warrant further investigation, but chloral hydrate and pentobarbitonone should obviously be avoided in experiments monitoring amine neurotransmission.

7.3. Electrode Life

Carbon-fiber electrodes show a limited useful life in vivo, since the separation between the amine metabolites and AA achieved by the electrical pretreatment is retained for a limited time. For AA

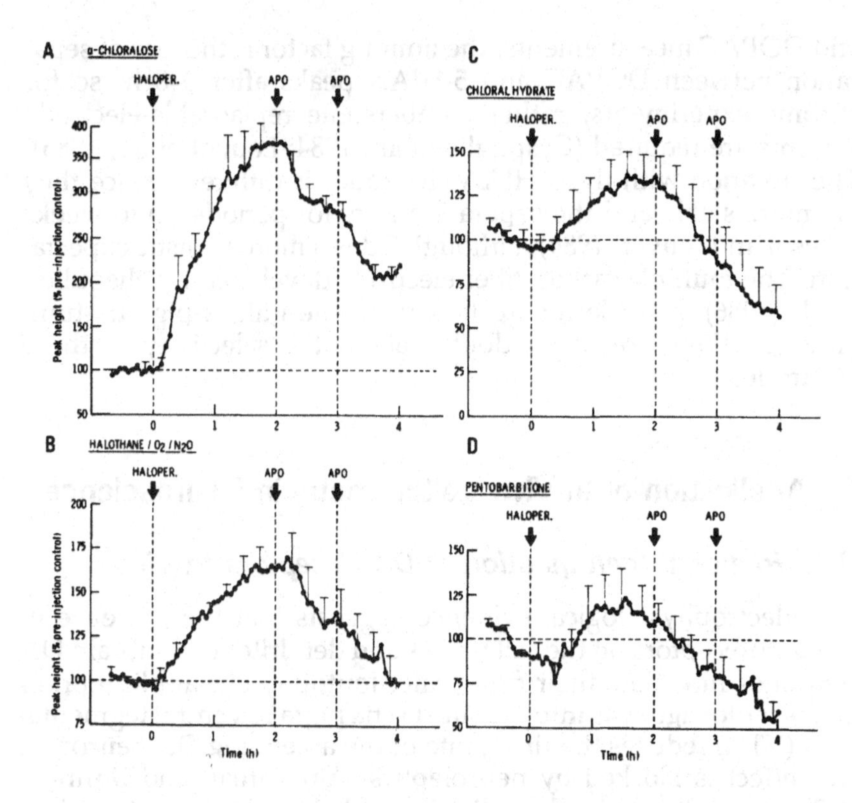

Fig. 7. Effect of haloperidol (0.3 mg/kg ip) at time 0, followed by apomorphine (0.5 mg/kg s.c.) administration at times 2 and 3, on the DOPAC oxidation peak in the striatum of rats anaesthetized with α-chloralose, halothane, chloral hydrate, or pentobarbitone. Note the different scale for peak height with α-chloralose data. The results are given as a percentage of preinjection control values. Note the stable preinjection baseline obtained with α-chloralose and halothane compared to the falling baseline with pentobarbitone. Haloperidol significantly increased the DOPAC peak under α-chloralose, halothane, and chloral hydrate ($P < 0.01$), but not under pentobarbitone anaesthesia. Each dose of apomorphine clearly decreased the haloperidol effect under α-chloralose and halothane, whereas under chloral hydrate and pentobarbitone anaesthesia, it is hard to distinguish the effect of apomorphine from the declining baseline. The maximal peak height values seen under α-chloralose after haloperidol, and the changes seen after apomorphine, are significantly different ($P < 0.01$) from those obtained under the three other anaesthetics. Reproduced with permission from Ford and Marsden (1986).

and DOPAC measurements, the limiting factor is the loss of separation between DOPAC and 5-HIAA peaks after 5–8 h, so for chronic experiments, rather cumbersome replaceable-electrode systems are required (Cespuglio et al., 1984; Louilot et al., 1985). The situation with the 5-HIAA electrodes is different, since they are more stable and the separation lasts for periods up to weeks (Crespi and Jouvet, 1984), although 2 d is a more realistic expectation. The multiple carbon-fiber electrode developed by Nieoullon et al. (1986) is stable for up to 1 yr in the guinea-pig striatum, although there are many doubts about the selectivity of these electrodes.

8. Application of In Vivo Voltammetry in Neuroscience

8.1. Regional Identification of DA Receptors In Vivo

Electrophysiological evidence suggests that the presence of DA autoreceptors on the cell bodies and dendrites of midbrain DA neurones modulate their firing rate. Iontophoretic application of DA receptor agonists into the substantia nigra or ventral tegmental area (VTA) reduces the firing rate of the ascending DA neurones; this effect is blocked by neuroleptics (Aghajanian and Bunney, 1977). In addition, local application of haloperidol into the nigra increases DA neuronal activity, suggesting the presence of a tonic self-inhibition of nigro-striatal DA neurones (Groves et al., 1975). Furthermore, low doses of peripherally administered apomorphine produce hypolocomotion in rodents and are reported to be beneficial in the treatment of schizophrenia. It is possible that such effects result from an action on somatodendritic autoreceptors in the midbrain, since these have been shown to be more sensitive than postsynaptic DA receptors (Skirboll et al., 1979).

We have investigated the role of mesolimbic somatodendritic autoreceptors in controlling the turnover of DA in a terminal region in vivo. Using differential pulse voltammetry with carbon-fiber electrodes implanted in the nucleus accumbens of chloral hydrate-anaesthetized rats, it was possible to monitor changes in extracellular DOPAC levels following infusion of DA or haloperidol into the ipsilateral VTA (Maidment and Marsden, 1985). Figure 8 represents a typical series of voltammograms recorded in the nu-

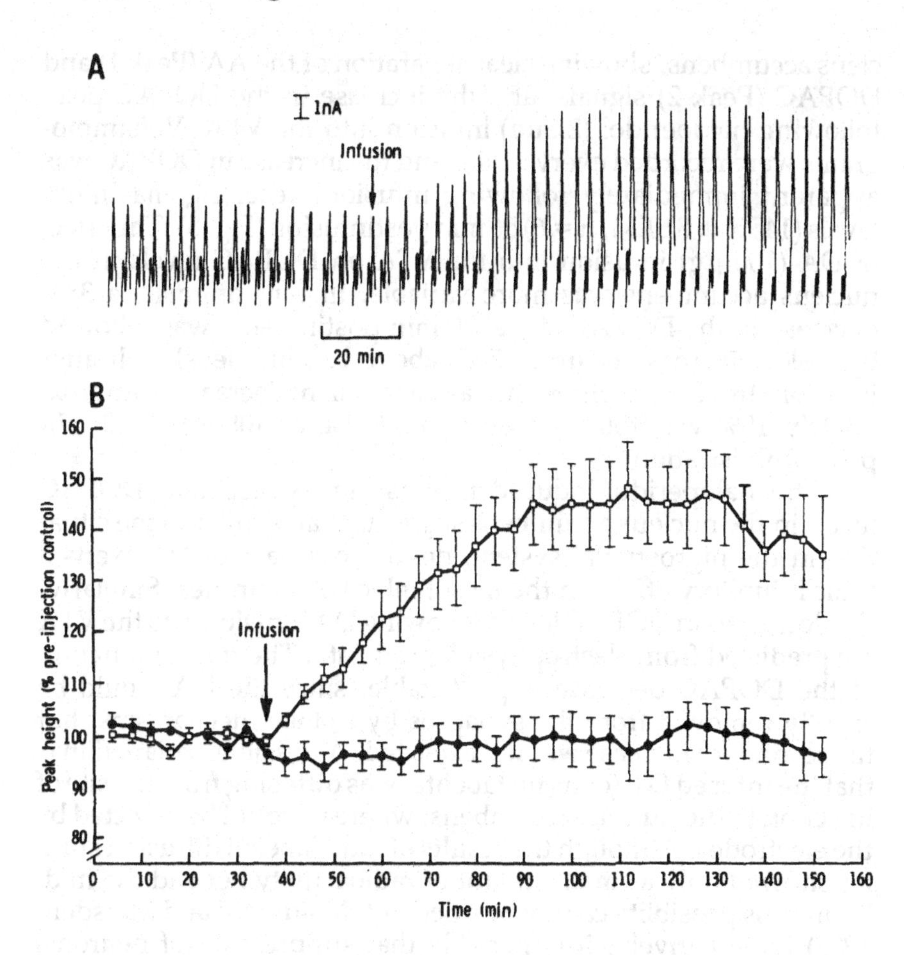

Fig. 8. (A) A typical series of voltammograms recorded every 4 min from the nucleus accumbens of a chloral hydrate-anaesthetized rat. Peak 1 is the result of ascorbic acid oxidation and Peak 2 of DOPAC oxidation. The increase in the DOPAC peak following haloperidol infusion (2.5 μg/0.5 μL) into the ipsilateral ventral tegmental area is clearly shown. (B) Graph showing the mean effect (n = 6) of haloperidol, given as described above, on the DOPAC peak in the nucleus accumbens. The results are expressed as a percentage with the mean of 10 preinfusion values taken as 100%. The error bars are the standard error of the mean. (Adapted from Maidment and Marsden, 1985.)

cleus accumbens, showing clear separation of the AA (Peak 1) and DOPAC (Peak 2) signals, and the increase in the DOPAC peak following haloperidol (2.5μg) infusion into the VTA. Voltammograms were recorded every 4 min, and the increase in DOPAC was apparent immediately following infusion, reaching maximum levels (145% ± 8 SEM, n = 6) 56 min postinfusion (Fig. 8). The effect of DA (100 μg) infusion into the VTA on DOPAC levels in the nucleus accumbens was more variable. In four animals, a 38% decrease in the DOPAC signal 20 min postinfusion was followed by a slow increase of up to 76% above baseline levels 1 h after infusion. In a further three animals, a transient decrease of approximately 15% was followed by a rapid rise of 108–208% 40 min postadministration.

The haloperidol-induced increase in extracellular DOPAC levels in the nucleus accumbens suggests that, similar to the situation in the nigrostriatal system, dendritic release of DA exerts a tonic inhibitory effect on the mesolimbic DA neurones. Similarly, the decreases in DOPAC levels following DA infusion into the VTA are predicted from electrophysiological data. The transient nature of the DOPAC decrease is predictable, since the DA would be rapidly removed from the synapses by uptake mechanisms, but the secondary increase was not expected. A possible explanation is that the infused DA (or its metabolite) was diffusing from the site of injection to the nucleus accumbens, where it would be detected by the electrodes. Although the results of radiolabeled diffusion studies showed only a small amount of radioactivity beyond the midbrain, this possiblity cannot be ruled out (Maidment and Marsden, 1985). Alternatively, it is possible that suppression of neuronal firing, while reducing DA release, also caused an increase in intraneuronal DA turnover, resulting in increased extracellular DOPAC levels. Similarly, it is possible that the increased DOPAC levels in the nucleus accumbens following haloperidol infusion into the VTA reflects an increase in DA turnover alone, not necessarily involving DA release. This possibility was investigated using the well-established behavioral model of hyperlocomotion, which results from DA release in the nucleus accumbens. Bilateral infusion of haloperidol into the VTA of freely moving rats produced a short period of hyperlocomotion, peaking 28–36 min postinfusion, with normal activity returning at approximately 48 min. Since the voltammetric data comes from anaesthetized animals, the two experiments are not directly comparable. It is, nonetheless, interesting to note that extracellular DOPAC levels followed a different time course, rising gradually to reach a plateau at 56 min and remaining high after 116 min. This could be interpreted as suggesting that increased DA release is transient, although the associated increase in DA turnover continues for longer periods. An alternative explanation is that DA release remains elevated, whereas the behavioral consequences are prevented by compensatory mechanisms distal to the postsynaptic DA receptor (Maidment and Marsden, 1985).

Which of the above explanations is correct remains unclear. However, the voltammetric and behavioral data, taken together, provide clear in vivo evidence for the presence of somatodendritic autoreceptors in the VTA that mediate changes in DA release and metabolism in the nucleus accumbens. The possibility that the mesolimbic and nigrostriatal DA systems may be differentially affected by "atypical" neuroleptics provided the impetus for the experiments described in section 8.3.

8.2. Identification of Receptors Controlling 5-HT Release and Metabolism

Voltammetry can also be used to monitor extracellular 5-HIAA, and we have used the technique to study the 5-HT receptor subtypes involved in controlling 5-HT release in the suprachiasmatic nucleus (SCN) and the frontal cortex. Several different recognition sites for the neurotransmitter serotonin (5HT) have been identified on neurones within the central nervous system. The initial binding studies of Peroutka and Snyder (1979) revealed two sites, designated 5-HT_1 and 5-HT_2 receptors. More recent evidence has suggested that the 5-HT_1 binding site can be further subdivided into 5-HT_{1A} and the 5-HT_{1B} subtypes (Pedigo et al., 1981; Schnellman et al., 1984) and that certain 5-HT agonists bind preferentially to a particular subtype. For instance, the tetralin derivative 8-hydroxy-2-(*n*-dipropylamino) tetralin (DPAT) is reported to be a selective ligand for the 5-HT_{1A} binding site (Middlemiss and Fozard, 1983), and the piperidinyl indole derivative 5-methoxy-3-[1,2,3,6-tetrahydro-4-pyridinyl]-1H indole (RU 24969) has been proposed as a selective ligand for the 5-HT_{1B} site (Sills et al., 1984; Doods et al., 1985). It is accepted, on the basis of in vitro studies, that the prejunctional 5-HT autoreceptor is a 5-HT_1 receptor (for review, *see* Moret, 1985; Middlemiss, 1985). However, there has been controversy as to whether it is a 5-HT_{1A} or 5-HT_{1B} receptor. The present experiments were carried out to identify and characterize these receptors in vivo.

8.2.1. Frontal Cortex

Using the techniques of brain dialysis and differential pulse voltammetry with carbon-fiber electrodes, the effect of stimulation and blockade of the 5-HT autoreceptors in the rat frontal cortex was studied. Initially, dialysis loops were implanted in the right frontal cortex, and used to monitor 5-HIAA and 5-HT levels before and after the administration of the putative $5HT_1$ receptor agonist RU 24969. Furthermore, the changes in extracellular 5-HIAA were compared with the changes in Peak 3 recorded using a carbon-fiber working electrode implanted in the contralateral brain region (Brazell et al., 1985; Marsden et al., 1986).

Extracellular 5-HIAA levels decreased after administration of RU 24969 (10 mg/kg ip) as measured by simultaneous voltammetry and dialysis. The decrease followed approximately the same time course with both techniques (Fig. 9). Levels of 5-HT (dialysis) also decreased following RU 24969 administration; there was, however, a 20-min delay in the 5-HT response to drug injection as compared to that of 5-HIAA (Brazell et al., 1985). The fact that levels of extracelluar 5-HIAA decreased approximately 20 min before 5-HT implies that extracellular 5-HIAA is not an index of 5-HT release, but only of 5-HT turnover. However, this time lag could be explained if RU 24969 had effects in addition to inhibiting 5-HT release. The agonist could (a) decrease 5-HIAA turnover and (b) inhibit 5-HT uptake and/or metabolism; both of these could decrease extracellular 5-HIAA before a decrease in extracellular 5-HT was observed. This is supported by reports that RU 24969 does have inhibitory effects in addition to a 5-HT_1 agonist action (Euvard and Boissier, 1980). The decrease in 5-HIAA measured by dialysis after RU 24969 was also observed in freely moving animals, but was more pronounced, suggesting that the anaesthetic may itself decrease 5-HT release (*see* section 7.2.).

The decrease in extracellular 5-HIAA (Peak 3) following RU 24969 was prevented by pretreatment with the nonselective 5-HT antagonist metergoline (2 mg/kg ip) (Fuxe et al., 1975; Gothert, 1980). The antagonist on its own produced a transient small (+10%) increase in Peak 3. In contrast, the selective 5-HT_2 receptor antagonist ritanserin (Janssen Pharmaceuticals) (1 mg/kg) failed to

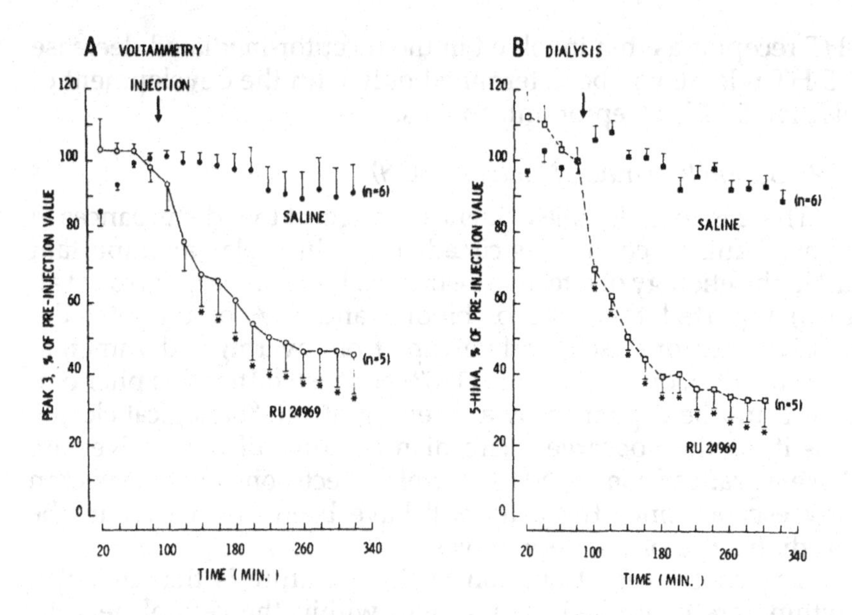

Fig. 9. The effect of RU-24969 (10 mg/kg) (O) on extracellular 5-HIAA in the frontal cortex measured by voltammetry (A) and dialysis (B) in anaesthetized rats and in saline-injected control animals (●). Injections (↓) were made 100 min after the start of the experiment. Note the similar decrease in 5-HIAA (dialysis) and Peak 3 (voltammetry) following administration of RU-24969. $*P < 0.01$. Results are mean values, with vertical lines showing the sem. From Brazell et al. (1985) with permission.

attenuate the decrease produced by RU 24969 (10 mg/kg) or have any significant effect of its own on the height of Peak 3.

The fall in extracellular 5-HIAA and 5-HT in vivo following RU 24969 and the blockade of the response by pretreatment with the 5-HT antagonist metergoline, when compared with the failure of a 5-HT_2 antagonist to affect the response, support the view from in vitro studies (Middlemiss, 1985; Brazell et al., 1985) that 5-HT_1 receptors are involved in the autoregulation of 5-HT release and metabolism. The small increase in Peak 3 in the frontal cortex produced by metergoline on its own was similar to that observed in the hippocampus by Baumann and Waldmeier (1984). In vitro studies have also shown that nonselective 5-HT receptor antagonists enhance K^+-stimulated release of [^{3}H 5-HT (Engel et al., 1983; Middlemiss, 1984). These results suggest that antagonism of the 5-HT autoreceptor increases release. Whether it is the same

5-HT receptor as that involved in the receptor-mediated decrease in 5-HT release will be determined only with the development of selective 5-HT_1 receptor antagonists.

8.2.2. Suprachiasmatic Nucleus (SCN)

There is considerable evidence to suggest that disturbances in the mechanisms controlling circadian rhythms play an important part in the etiology of affective disorders. For instance, Carroll et al. (1976) reported that 88% of bipolar and 50% of unipolar depressives demonstrated early escape from overnight dexamethasone suppression. Wehr et al. (1983) suggested that this phenomenon could be explained by a resetting of the "biological clock." Thus if, as they observed, circadian rhythms of depressives are phase-advanced compared to control subjects, one might expect an early escape, since the drug will have been given later in the circadian cycle of the depressives.

The consensus of opinion in the literature is that circadian rhythms are controlled by a system within the central nervous system, the suprachiasmatic nucleus (SCN) of the hypothalamus being the major site (for review, *see* Moore, 1983). Unfortunately, little evidence is available to suggest the mechanisms involved within the SCN by which control is exerted. The SCN receives a rich serotoninergic innervation that may play a part in the control of circadian rhythms (Wirz-Justice et al., 1982). We have investigated the regulation of 5-HT release and metabolism in the SCN using carbon-fiber microelectrodes implanted in the SCN, and studied the effects of selective 5-HT receptor agonists and antagonists on the height of the recorded indole peak. Our carbon-fiber microelectrodes are particularly suited for use in small nuclei like the SCN, since the tip diameter is only 20 μm.

In common with other workers (Faradji et al., 1983), we have found that the size of the oxidation peak recorded at approximately +0.3 V (Peak 3) increased as the electrode was lowered towards the SCN. However, there was a noticeable decrease upon entry, a phenomenon that we have used as an indicator of electrode position prior to histological verification (Martin and Marsden, 1986a,b; Fig. 10).

Intraperitoneal injection of the 5-HT_1 receptor agonist RU 24969 was associated with a transient rise in the size of Peak 3 (5-HIAA), followed by a prolonged and marked decrease 2 h after

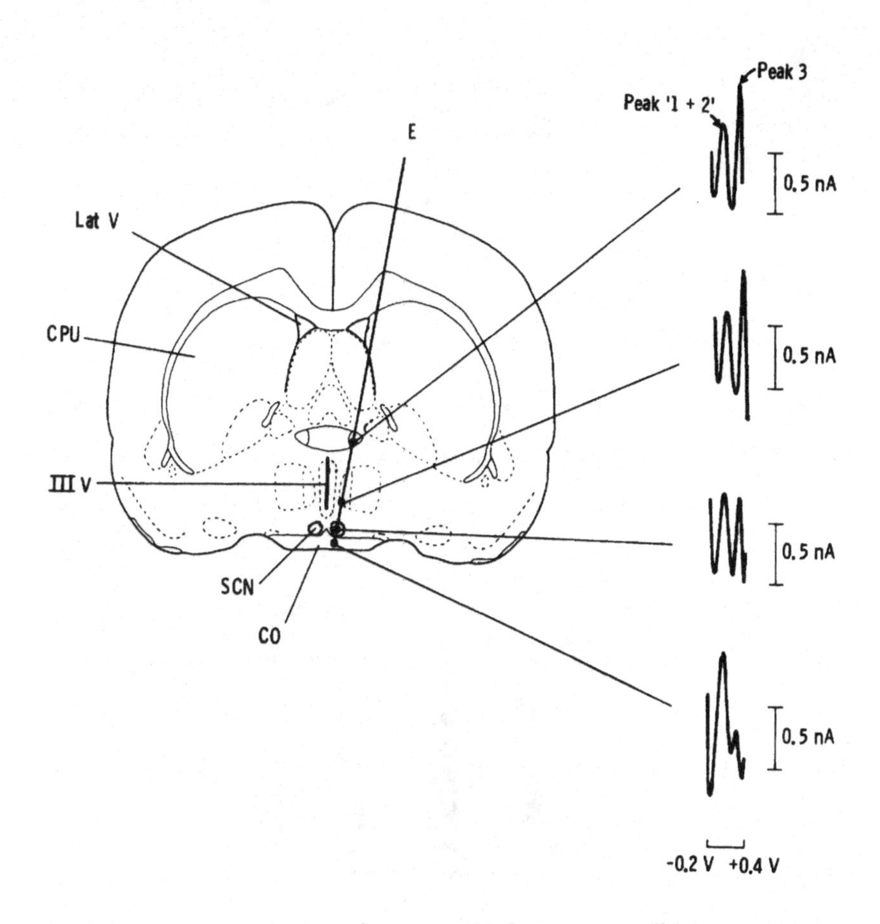

Fig. 10. Coronal section of the rat brain illustrating the path of implantation of the working electrode (E) to the suprachiasmatic nucleus (SCN). Representative voltammograms from a single animal are shown on the right. Each voltammogram was obtained when the electrode tip was in the approximate region of the indicated black dot on the line. The position of the electrode in the SCN was verified histologically. Note the decrease in the size of the oxidation peak at +300 mV (Peak 3) when the electrode tip was in the SCN. Abbreviations used: Lat V, lateral ventricle; CPU, caudate putamen; IIIV, third ventricle; CO, optic chiasm. From Marsden and Martin (1986) with permission.

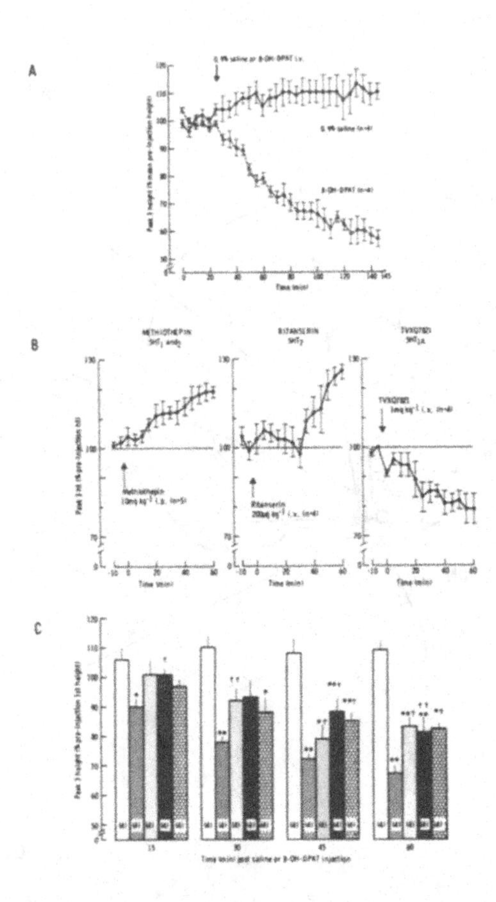

Fig. 11. (A) The effect of 0.9% w/v NaCl solution (saline 1 mL/kg iv, ●, $n = 4$) 8-hydroxy-2-(n-dipropylamino) tetralin (8-OH-DPAT, 0.1 mg/kg iv O, $n = 4$) on the height of Peak 3 recorded in the suprachiasmatic nucleus (SCN). Each point represents the mean, and the SEM is illustrated by the vertical lines. (B) The effect of methiothepin (10 mg/kg ip, $n = 5$), ritanserin (0.2 mg/kg iv, $n = 4$), and TVX Q 7821 (1 mg/kg iv, $n = 4$) on the height of Peak 3 recorded in the SCN. Each point represents the mean, and the SEM is also shown. Note the decrease in Peak 3 height following injection of TVX Q 7821. (C) The effect of 0.9% w/v NaCl solution (saline 1 ml/kg^{-1} iv, open columns) or 8-hydroxy-2-(n-dipropylamino) tetralin (8-OH-DPAT, 0.1 mg/kg iv, hatched columns) on the height of Peak 3 recorded in the SCN 15, 30, 45, and 60 min after their injection. The effect of administration of methiothepin (1 mg/kg iv, stippled columns), ritanserin (0.2 mg/kg, solid columns), or TVX Q 7821 (1

administration. This decrease lasted for up to 6 h and was similar to the results described in the frontal cortex. The decrease in the size of Peak 3 was probably the result of decreased 5-HT release and metabolism to 5-HIAA following stimulation of the 5-HT_1 autoreceptor that is thought to regulate 5-HT release (Middlemiss, 1985; Ennis, 1986). The nonselective 5-HT receptor antagonist methiothepin again prevented the RU 24969-mediated decrease in Peak 3, whereas on its own, the antagonist increased the size of Peak 3 by 20 ± 2% 75 min after administration. These results are similar to those obtained in the frontal cortex with RU 24969 and the 5-HT antagonist metergoline described earler, and support the view that 5-HT release in the SCN is under the control of a 5-HT_1 autoreceptor.

DPAT, when given intravenously, also decreased 5-HT metabolism in the SCN such that 60 min after DPAT administration, Peak 3 height had decreased by 33 ± 3% (n = 4) (Fig. 11). This finding is in general agreement with that of Hjorth et al. (1982), who reported that DPAT dose-dependently decreased tissue levels of 5-HIAA. In order to determine the 5-HT receptor involved in this response to DPAT, we administered methiothepin 5 min before DPAT. At a dose that would block the response to RU 24969 (Martin and Marsden, 1986a), we were able only to attenuate the response to DPAT. The selective 5-HT_2 receptor antagonist ritanserin also appeared to attenuate the response to DPAT; however, when its own effect on Peak 3 height was taken into consideration (20% increase 60 min postadministration, it became doubtful whether 5-HT_2 receptors were involved in the response to DPAT.

Isapirone (TVXQ7821) has been reported to be a selective antagonist ligand for the 5-HT_{1A} binding site (Dompert et al., 1985; Schumann et al., 1984; Martin et al., 1986; Reynolds et al., 1986). In our experimental protocol, we observed that isapirone was able to block the effects of DPAT on Peak 3 height recorded in the SCN and

mg/kg iv, honeycombed columns) 5 min before an injection of 8-OH-DPAT (0.1 mg/kg iv) is also shown. Each column represents the mean of four observations, and the standard error of the mean is shown by the vertical lines. *P < 0.05, **P < 0.01 compared to saline. P < 0.05, P < 0.01 compared to 8-OH-DPAT. Compiled from data in Marsden and Martin (1986) with permission.

had some agonist effects (Fig. 11). Thus, we were able to conclude that the effects of DPAT involved 5-HT_{1A} receptors, and also that isapirone is a partial agonist at 5-HT_{1A} receptors.

The cardiovascular responses to peripheral administration of DPAT have been suggested to involve α_2 adrenoceptors (Fozard and McDermott, 1985). To determine whether α_2 adrenoceptors were involved in the neurochemical responses to DPAT, we attempted to block the effects of DPAT with the selective α_2 adrenoceptor antagonist idazoxan (Doxey et al., 1983). Interestingly, at a dose that had no effect on Peak 3 height alone, idazoxan completely abolished the response to DPAT (Fig. 12). These data indicate that the effects of iv DPAT involved not only 5-HT_{1A} receptors, but also α_2 adrenoceptors (Marsden and Martin, 1986).

We next needed to determine where these 5-HT_1 receptor agonists were acting to decrease 5-HT release and metabolism in the SCN, i.e., in the nerve terminal region (the SCN), or in the cell bodies (the dorsal raphé nucleus [DRN]), so we administered RU 24969 and DPAT directly into both of these nuclei while recording from the SCN. RU 24969 administered directly into the dorsal raphé in doses of up to 10 μg in 1 μL of saline had no effect on the size of the indole oxidation peak (Marsden and Martin, 1985a). However, RU 24969 injected into the SCN contralateral to the one containing the recording electrode was associated with a dose-dependent decrease in the height of Peak 3. The peak height decreased by 26 ± 7% 1 h after 5 μg, and by 81 ± 7% 1 h after a further 5 μg (Fig. 13). These results strongly suggest that the 5-HT_1 receptor regulating 5-HT release and metabolism in the SCN is located on the terminals, and not on the cell bodies within the DRN.

The results obtained following local injection of DPAT into the SCN or DRN were in marked contrast to those obtained following local injection of RU 24969. Infusion of DPAT (4 or 8 μg) into the SCN had no effect on the height of Peak 3 recorded in the SCN (Marsden and Martin, 1985b). This observation led us to conclude that the presynaptic autoreceptor in the rat was in fact a 5-HT_{1B} receptor, a finding that is in agreement with the previously quoted in vitro data (Middlemiss, 1984).

Several groups have now reported that 5-HT_{1A} recepter agonists, such as DPAT, inhibit the firing of 5-HT neurones in the DRN

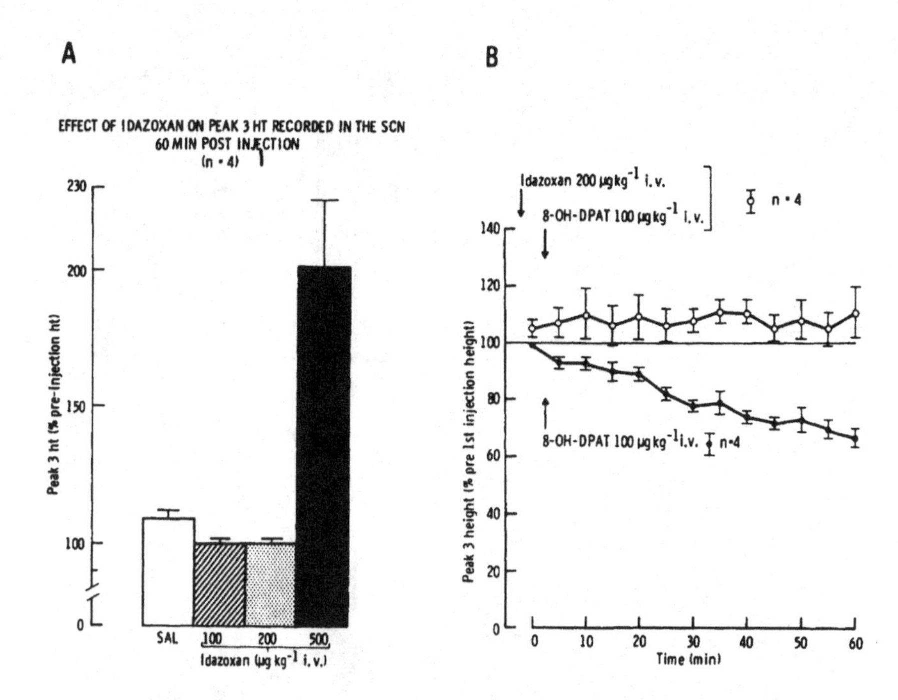

Fig. 12. (A) The effect of 0.9% w/v NaCl solution (saline 1 ml/kg^{-1}, open columns) and idazoxan (0.1 mg/kg iv, hatched columns; 0.2 mg/kg iv, stippled columns; and 0.5 mg/kg iv, solid columns) on Peak 3 height recorded in the suprachiasmatic nucleus (SCN). Each column represents the mean, and the standard error of the mean is shown by the vertical lines. (B) The effect of idazoxan (0.2 mg/kg iv, (0), n = 4) 5 min before 6-hydroxy-2-(*n*-dipropylamino) tetralin (8-OH-DPAT, 0.1 mg/kg iv) and the effect of 8-OH-DPAT (0.1 mg/kg iv, (●), *n* = 4) on Peak 3 height recorded in the SCN of the chloral hydrate anaesthetized rat. Each point represents the mean, and the SEM is shown by the vertical lines. Compiled from data in Marsden and Martin (1986) with permission.

(e.g., De Montigny et al., 1984). On the basis of these findings and our own following iv injection of DPAT (described above), one might expect local infusion of DPAT into the DRN to be associated with a decrease in the height of Peak 3 recorded in the SCN. However, we observed the exact oposite of this predicted effect. The lowest dose that we used (1 μg) had no significant effects,

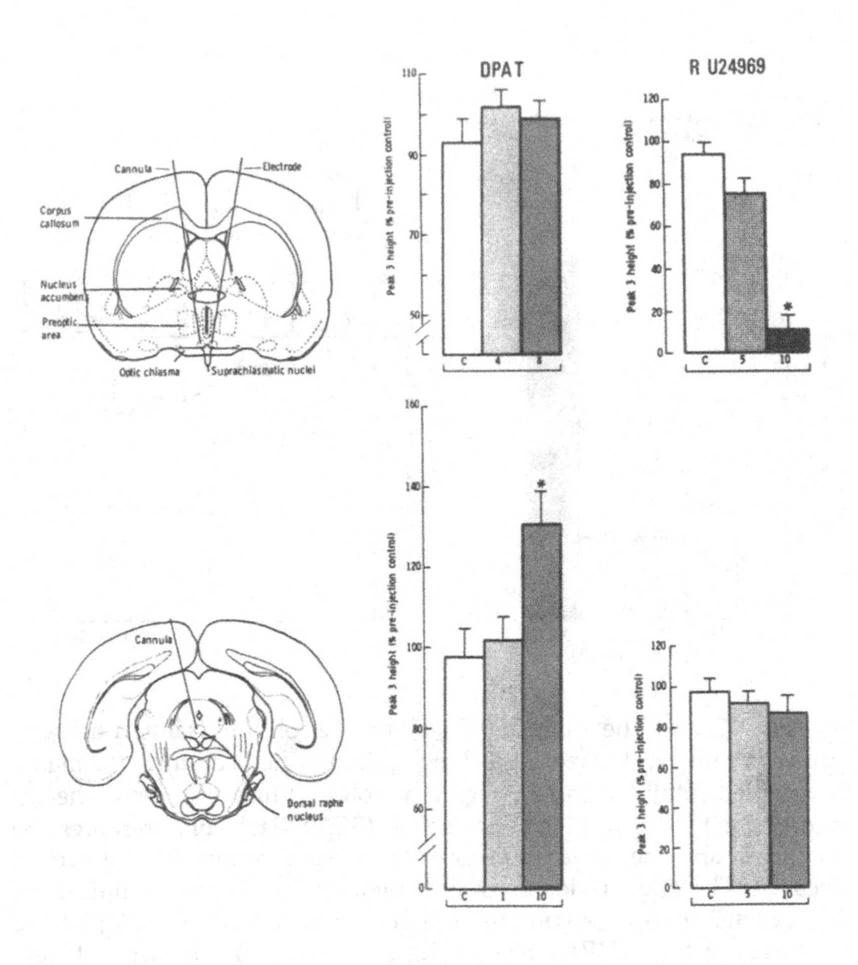

Fig. 13. The effect of infusion of DPAT or RU 24969 into the SCN (upper panels) or the DRN (lower panels) on Peak 3 height recorded in the SCN 1 h after infusion. The open histobars (C) represent the height of Peak 3 after infusion of saline. The numbers at the bottom of each histobar indicate the quantity of drug infused (μg) in 0.5 μL of saline over 1 min. Each histobar represents the mean ±SEM of at least four observations. *$P < 0.05$ compared to appropriate saline control (Student's t-test).

whereas the highest dose was actually associated with a significant increase in Peak 3 height (31 ± 8%, n = 5). Thus, although decreased serotonergic unit activity is probably associated with a decrease in 5-HT release, quite clearly this is not reflected by a similar change in 5-HT metabolism. Thus, the conditions under which neurone firing is coupled to 5-HT metabolism remain to be determined.

8.3. Comparison of Acute and Chronic Drug Action

A major problem with the clinical use of most existing antischizophrenic (neuroleptic) drugs is the occurrence of extrapyramidal side effects, especially tardive dyskinesia (Marsden et al., 1975). However, the degree of these side effects does not always correlate with the clinical efficacy of the drugs, which has led to the suggestion that the site of beneficial action of the neuroleptics lies outside the striatum, possibly in the mesolimbic DA system (Crow et al., 1976). The effort to identify neuroleptic compounds with selective effects on the mesolimbic DA system, in an attempt to reduce unwanted side effects, has been based on such assumptions. For example, thioridazine and clozapine have been reported to activate DA cell firing selectively in the ventral tegmental area (VTA), the DA cell body region of the mesolimbic system, without affecting those in the substantia nigra (the nigrostriatal DA system) following acute administration (White and Wang, 1983). Following chronic treatment, however, DA cell firing is selectively decreased in the VTA, explained by the phenomenon of depolarization inactivation (Chiodo and Bunney, 1983; White and Wang, 1983). This pharmacological selectivity could relate to the fewer extrapyramidal side effects reported with these drugs. A feature of typical neuroleptics is their ability to increase DA metabolism in the nucleus accumbens and striatum measured postmortem following acute administration, and the development of tolerance to this effect after chronic treatment (e.g., Clow et al., 1980; Saller and Salama, 1985; Scatton, 1977). Evidence in the literature is conflicting as to whether atypical neuroleptics selectively increase mesolimbic DA metabolism after acute administration (*see* Maidment and Marsden, 1987a).

Using voltammetry, we have monitored the acute and chronic effects of thioridazine and clozapine (Maidment and Marsden,

1987a,b) on DA metabolism in vivo, in an attempt to determine the selectivity of these compounds on the mesolimbic and striatal DA systems and to compare them with reported electrophysiological studies. The use of a custom-made automation unit linked to the PAR 174A polarographic analyzer allowed continual measurements to be taken alternately from each of two electrodes placed in the striatum and nucleus accumbens of halothane/N_2O-anaesthetized rats.

Thioridazine (20 mg/kg subcutaneous) given to rats 1 d after 21 daily doses of saline (2 mL/kg) increased extracellular DOPAC above prechallenge control levels in both the nucleus accumbens (+66%) and the striatum (+91%) 2 h postinjection. However, in rats given thioridazine for 21 d and then a challenge dose on d 22, there was no significant increase in extracellular DOPAC above prechallenge levels in either the striatum or the nucleus accumbens (Fig. 14), suggesting that tolerance to the acute effect of thioridazine on DA turnover occurred in both regions. However, expressing the results as percentage changes from preinjection control values masks possible differences in the prechallenge DOPAC levels between animals treated with saline and those treated with thioridazine. To assess the effects of chronic thioridazine administration on extracellular DOPAC levels 24 h after the last dose (d 21) of neuroleptic prior to subsequent challenge on d 22, the absolute DOPAC peak heights (nA) (corrected for in vitro sensitivity) were determined. These show that the mean preinjection striatal DOPAC peak heights on d 22 of animals treated for 21 d with thioridazine tended to be elevated compared to those of rats given saline chronically, although this difference failed to attain statistical significance, as a result of the wide variation in the absolute peak height values obtained from chronic thioridazine-treated rats. However, the small increase produced by subsequent challenge with the drug was sufficient to attain values statistically different from those of prechallenge saline-treated animals. Figure 15 demonstrates the clearly different pattern of DOPAC changes between the nucleus accumbens and striatum, and provides some indication that absolute tolerance may have occurred only in the nucleus accumbens.

Therefore, to investigate further the possible link between tolerance to DA turnover and the development of depolarization inactivation produced by chronic thioridazine treatment (Chiodo and Bunney, 1983; White and Wang, 1983), apomorphine, which

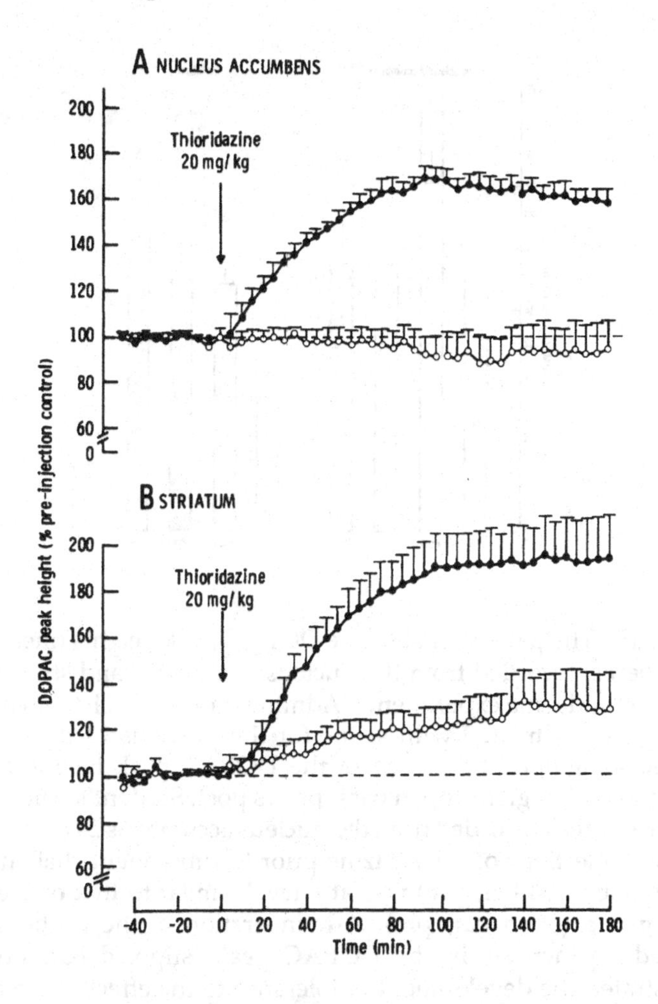

Fig. 14. The time course of changes in the height of the DOPAC peak produced by a challenge dose of thioridazine (20 mg/kg s.c.) on d 22 recorded every 5 min from the nucleus accumbens (A) and striatum (B) of rats. Rats were treated with saline (closed circles) or thioridazine (20 mg/kg s.c./d, open circles) for the previous 21 d. Results are expressed as percentage changes, taking the mean of the last 10 preinjection peak heights as 100%. Although no increase in extracellular DOPAC was apparent in the nucleus accumbens of rats to whom thioridazine was repeatedly administered, a small but not significant increase was observed in the striatum of these animals. (Student's paired t-test, $n = 5$. Error bars represent the SEM.) From Maidment and Marsden (1987b) with permission.

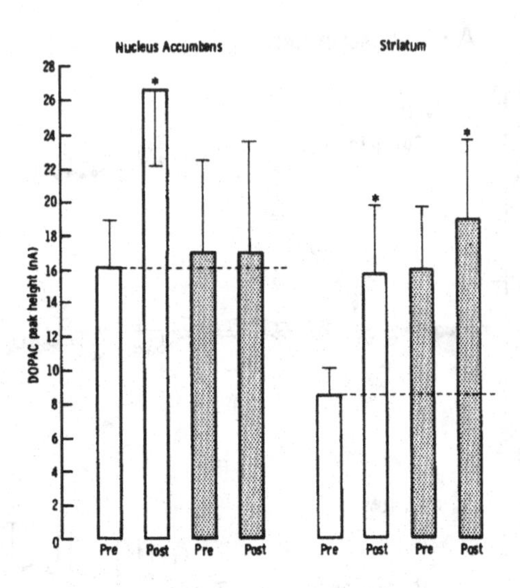

Fig. 15. The pattern of absolute DOPAC peak heights (measured in nanoamperes) recorded from the nucleus accumbens and striatum on d 22 of the thioridazine experiments. Administration of a challenge dose of this drug to rats treated with saline for the previous 21 d (open bars) produced a significant increase in the DOPAC peak height 2 h after injection in both regions (open bars, pre vs post; Student's paired t-test). The DOPAC peak recorded from the nucleus accumbens 24 h after the last of 21 daily injections of thioridazine prior to subsequent challenge with this drug (stippled bar, pre) was at a level similar to that of the saline-treated group (open bar, pre). Administration of the challenge dose produced no increase in the DOPAC peak (stippled bar, post) and demonstrated the development of tolerance to the effect of the drug on dopamine matabolism in this region. Conversely, DOPAC levels in the striatum of rats treated with thioridazine for 21 d prior to subsequent challenge (stippled bars, pre) appeared to remain elevated above, but were not significantly different from, those of animals treated with saline prior to drug challenge (open bars, pre; Student's unpaired t-test). However, the challenge dose of thioridazine increased the DOPAC peak sufficiently to attain values (stippled bars, post) significantly above those recorded from prechallenge saline-treated animals (open bars, pre; Student's unpaired t-test), suggesting that absolute tolerance may not have developed in this region. (*$p < 0.05$, error bars represent the SEM of five animals). From Maidment and Marsden (1987b) with permission.

stimulates DA autoreceptors (and so acutely decreases DA release and metabolism [Zetterstrom et al., 1986; Ford and Marsden, 1986]), was administered to rats previously treated chronically with thioridazine. Such treatment has been shown to reverse the apparent state of depolarization inactivation of mesolimbic DA neurones (Bunny and Grace, 1978; White and Wang, 1983). Thus, if the tolerance to increased DA metabolism in the nucleus accumbens following repeated thioridazine was a direct consequence of depolarization inactivation, apomorphine might be expected to reverse the tolerance phenomenon and produce an increase in extracellular DOPAC in this region, instead of the normal decrease seen acutely. Administration of increasing doses of apomorphine (0.05, 0.1, 0.25 mg/kg s.c.) at 30-min intervals 1 h after administration of thioridazine (20 mg/kg s.c.) on d 22 to rats treated with the neuroleptic for the previous 21 d similarly decreased extracellular DOPAC in both regions.

The above results clearly demonstrate that thioridazine increases DA metabolism in both the nucleus accumbens and the striatum after acute administration, and appears to induce tolerance to this effect in both regions after chronic treatment (although analysis of absolute peak heights suggested a possible selective action). Similar experiments with clozapine also failed to provide evidence of a selective action after acute or chronic administration (Maidment and Marsden, 1987a,b). Furthermore, the experiments with apomorphine also indicate that the phenomena of tolerance to increased DA metabolism and depolarization inactivation of DA neurones are not directly linked, but are most likely produced by separate mechanisms.

Alternative approaches to the development of neuroleptic compounds with fewer extrapyramidal side effects have been considered. The neuropeptide neurotensin is found in high concentrations in DA-rich regions together with high-affinity binding sites (*see* review by Jolicoeur et al., 1985). Initial electrophysiological results indicate that neurotensin inhibits neurones in the nucleus accumbens, and excites neurones in the substantia nigra and ventral tegmental area (Andrade and Aghajanian, 1981; McCarthy et al., 1979). Furthermore, the profile of neurotensin neuropharmacology shares many features associated with the neuroleptics. In particular, neurotensin attenuates DA-induced hyperactivity in the accumbens (but not amphetamine-induced stereotypy

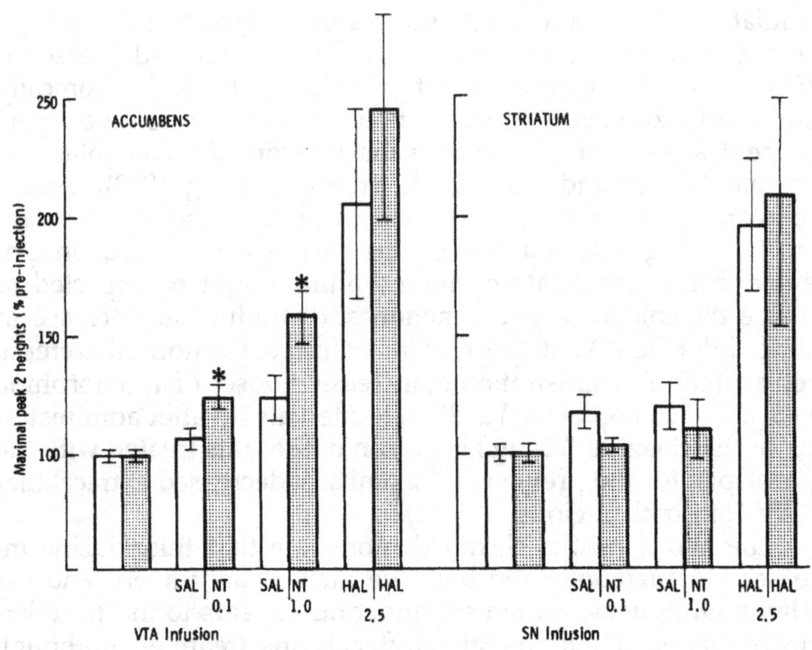

Fig. 16. The effect of 0.9% saline (SAL 0.5 μL), neurotensin (NT, 100 ng and 1 μg in 1 μL) and haloperidol (HAL, 2.5 μg in 1 μL) injected into the ventral tegmental area (VTA infusion) or the substantia nigra (SN infusion) on the DOPAC oxidation peak in the nucleus accumbens and striatum, respectively. Note that haloperidol infusion increased the peak in both regions, whereas neurotensin increased the DOPAC peak in the accumbens only following its infusion into the VTA (*) $n = 6$ for each group; *$P < 0.05$ compared to saline infusion.

mediated by the striatum [Jolicoeur et al., 1985]) and increases DA metabolism (Widerlov and Breese, 1982). In many respects, the pharmacology of neurotensin resembles the profile of "atypical" neuroleptics, such as thioridazine, rather than classical neuroleptics (e.g., haloperidol, chlorpromazine) (Nemeroff and Cain, 1985; Kalivas et al., 1984).

Although the results with "atypical" neuroleptics failed to show biochemical selectivity between their effects on the mesolimbic and striatal DA systems, initial data with neurotensin indicate that this peptide selectively increases DA metabolism in the mesolimbic system (Ford and Marsden, 1987). Thus, injection of saline (0.5 μL) or neurotensin (0.1 or 1.0 μg in 0.5 μL) into the substantia nigra produced no changes in the DOPAC oxidation peak in the ipsilateral striatum, whereas injection of haloperidol (2.5 μg in 0.5 μL) produced an immediate and significant increase of about 100% above preinjection values (Fig. 16). However, injection of neurotensin (0.1 and 1 μg) into the VTA significantly increased the DOPAC signal in the ipsilateral nucleus accumbens (Fig. 16), with subsequent haloperidol injection producing a further 100% increase.

The present results suggest that this action does not only depend on effects occurring within the terminal region (nucleus accumbens), but also involves neurotensin receptors in the ventral tegmental area (VTA), where the DA cell bodies are located. The development of neurotensin-like compounds able to recognize neurotensin high-affinity binding sites may result in new anti-schizophrenic compounds with fewer side effects. At present, we are further investigating the regional effects of neurotensin and stable analogs of neurotensin on DA metabolism in vivo. Voltammetric techniques will have an important role in such studies, since the small size of the electrodes permit amine metabolism to be monitored in regions too small to be accessible to such other techniques as intracerebral dialysis.

9. Future Developments

9.1. *Measurement of Dopamine Release*

9.1.1. Basal Dopamine Release

In section 4.6., we briefly described the approaches that have been adopted to develop DA-selective electrodes; in this section we extend the discussion to cover some of the problems encountered with such work, in the hope of stimulating new ideas and approaches. The major difficulty to overcome is the fact that DA and DOPAC oxidize at the same potential, but DOPAC is present in the extracellular space at a concentration about 100 × greater than that of DA. The first amine-selective electrode was described by Blaha and Lane (1983), and involved incorporation of stearic acid into a standard carbon-paste electrode, the resultant surface of which is highly negatively charged and thus effectively repels DOPAC and AA while allowing DA to be oxidized. These electrodes have been used with chronoamperometry or linear sweep voltammetry with semidifferentiation, and the pharmacological evidence supports their being able to measure basal DA release (Blaha and Lane, 1984). However, an exception is the observation that D-amphetamine (1mg/kg) produced only a 70% increase in the proposed DA signal recorded in the rat striatum (Gazzara and Howard-Butcher, 1986), whereas a similar dose of this drug (2 mg/kg) produced a 14-fold increase in extracellular DA measured by dialysis (Zetterstrom et al., 1983). Additionally, an anomaly exists with regard to estimation of absolute levels. Dialysis studies and carbon-fiber electrode results described above suggest extracellular DA levels to be in the 20–50 n*M* range, whereas the stearate-modified paste electrodes provide values of approximately 0.5 μ*M*. It is possible that these electrodes are overestimating DA levels because their negatively charged nature actually attracts DA to its surface. Adsorption of DA to the electrode surface is certainly a problem for this reason. The other major theoretical problem associated with these electrodes is that of electro-catalytic regeneration of DA in the presence of AA, another factor that would tend to produce artificially high estimates of DA levels but, more importantly, could lead to false interpretation if AA levels were to change during the course of an experiment.

A similar approach has been employed in the development of Nafion-coated electrodes. This compound is a perfluorosulfonated derivative of Teflon™ that has been used to coat the surface of graphite-epoxy or, more recently, carbon-fiber electrodes, imparting a negative charge (Gerhardt et al., 1984). The early Nafion-coated electrodes certainly went a long way towards discriminating between DA and its metabolite, DOPAC, and AA, with ratios of sensitivity of 1000:1 for DA:AA and 100:1 for DA:DOPAC (Brazell et al., 1987, and personal communication). However, these figures must be considered in the light of similar estimated ratios for the basal concentrations of these substances in the extracellular space. Since basal signals recorded with such electrodes are likely to consist of approximately equal components of all three compounds, an increase in sensitivity was required. In the laboratory at Nottingham, we have now shown that chemically and electrically pretreated carbon-fiber electrodes, electro-coated with Nafion and combined with differential pulse voltammetry (DPV), are capable of measuring basal levels of 5-HT in the dorsal raphé and frontal cortex with no interference from 5-HIAA or uric acid (Crespi et al., 1988).

Finally, there are the electrically pretreated, multiple carbon-fiber electrodes described by Nieoullon and coworkers (Nieoullon et al., 1986). These consist of approximately 10,000 fibers bonded together with an epoxy resin and sharpened to a tip diameter of 50 μm. These electrodes, when used in vitro with differential pulse voltammetry, apparently exhibit selectivity for DA over DOPAC and AA with sensitivity ratios of 1000:1 and 10,000:1, respectively; the reason for this is unclear, and no explanation for such a surprising profile has been offered. As far as the in vivo signal is concerned, the pharmacological data look convincing (apart from the amphetamine response, which is small compared to dialysis data), but also in common with the stearate paste electrode, the estimated basal concentration is high (approximately 1 μ*M*; Nieoullon et al., 1986). The other reported feature of these electrodes is their stability in vivo, proposed DA signals being recorded for up to a year in the guinea-pig striatum. There is clearly a need for further investigation of these electrodes.

9.1.2. Measurement of Stimulated Release

One way around the problem of measuring the apparently low levels of DA in the extracellular space is to raise the concentration artificially by electrical or K^+-induced stimulation of the DA neurones. Voltammetry in combination with a fast sampling technique is well suited to this approach, and carbon-fiber electrodes have been used with chronoamperometry (Schenk et al., 1983) and electrically pretreated carbon-fibers with differential pulse amperometry in pargyline-treated rats (Marcenac and Gonon, 1985). The other type of electrode that has been used successfully is the simple untreated single carbon fiber, which either may be cut flush to provide a disc-shaped active surface (Wightman, 1981) or an approximately 50-μm tip length may be used (Armstrong-James et al., 1980). The small size of these electrodes results in their exhibiting unique electrochemical properties (*see* section 4.4), and their use with fast cyclic voltammetry to monitor electrically stimulated DA release on a millisecond time scale is now established

(Stamford et al., 1986; Kuhr and Wightman, 1986), but at the present time, this technique is not sensitive enough to measure basal or drug-induced increases in DA release.

9.2. Conclusions

This chapter has described the use of electrically pretreated carbon-fiber electrodes to monitor changes in DA and 5-HT metabolites in order to provide in vivo information about amine metabolism. A question that remains is what relevance the measurement of metabolites offers to required information about amine release. Further studies are required, using both intracerebral dialysis and voltammetry, to obtain some answers. Indeed, the nature of the relationship between neurotransmitter release and metabolism and neuronal activity remains to be successfully investigated. To obtain this information, studies are required in which these three factors are measured simultaneously, preferably in freely moving (unanaesthetized) animals.

There is also a need to develop further amine-selective electrodes with greater sensitivity, so that basal amine levels can be routinely monitored using voltammetry. A further challenge would be to investigate the possibility of using in vivo voltammetry to monitor nonamine compounds, such as amino acids (tyrosine and tryptophan) and neuropeptides containing these amino acids.

Finally, there are certain situations in which in vivo voltammetry may have particular relevance. These include monitoring in small brain nuclei, such as the SCN, and situations in which rapid resolution of the changes in metabolism and release are required. In the future, voltammetry and such techniques as intracerebral dialysis should be considered as complementary methods, used to unravel problems relating to the physiological and pharmacological aspects of neurotransmitter function and the mechanisms of drugs acting in the brain.

Acknowledgments

The work described in this chapter from the Nottingham laboratory was supported financially by The Wellcome Trust, Medical Research Council, and The Nottingham University Research Fund.

References

Adams R. N. (1969) Application of modern electroanalytical techniques to pharmaceutical chemistry. *J. Pharm. Sci.* **58,** 1171–1178.

Adams R. N. and Marsden C. A. (1982) Electrochemical detection methods for monoamine measurements in vitro and in vivo, in *Handbook of Psychopharmacology,* vol. 15 (Iversen L. L., Iversen S. D. and Snyder S. H., eds.), Academic, New York, pp. 1–74.

Aghajanian G. K. and Bunney B. S. (1977) Dopamine autoreceptors: pharmacological characterisation by microiontophoretic single cell recording studies. *Naunyn Schmiedeberg's Arch. Pharmacol.* **297,** 1–7.

Albery W. J., Fillenz M., and O'Neill R. D. (1983) The compartment model for chronically implanted voltammetric electrodes in the rat brain. *Neurosci. Lett.* **38,** 175–180.

Andrade R. and Aghajanian G. K. (1981) Neurotensin selectively activates dopaminergic neurons of the substantia nigra. *Soc. Neurosci. Abstr.* **7,** 573.

Armstrong-James M. and Millar J. (1984) High-speed cyclic voltammetry and unit recording with carbon fibre microelectrodes, in *Measurement of Neurotransmitter Release In Vivo* (Marsden C. A., ed.), Wiley, Chichester, pp. 209–224.

Armstrong-James M., Millar J., and Kruk Z. L. (1980) Quantification of noradrenaline iontophoresis. *Nature* **288,** 181–183.

Baumann P. A. and Waldmeier P. C. (1984) Negative feedback control of serotonin release in vivo: comparison of 5-hydroxyindoleacetic acid levels measured by voltammetry in conscious rats and by biochemical techniques. *Neuroscience* **11,** 195–204.

Blaha C. D. and Lane R. F. (1983) Chemically modified electrode for in vivo monitoring of brain catecholamines. *Brain Res. Bull.* **10,** 861–864.

Blaha C. D. and Lane R. F. (1984) Direct in vivo electrochemical monitoring of dopamine release in response to neuroleptic drugs. *Eur. J. Pharmacol.* **98,** 113–117.

Brazell M. P. and Marsden C. A. (1982a) Differential pulse voltammetry in the anaesthetised rat. Identification of ascorbic acid, catechol and indoleamine oxidation peaks in the striatum and frontal cortex. *Br. J. Pharmacol.* **75,** 539–547.

Brazell M. P. and Marsden C. A. (1982b) Intracerebral injection of ascorbate oxidase—effect on in vivo electrochemical recordings. *Brain Res.* **249,** 167–172.

Brazell M. P., Marsden C. A., Nisbet A. P., and Routledge C. (1985) The 5HT receptor agonist RU24969 decreases 5-hydroxytryptamine

(5HT) release and metabolism in the rat frontal cortex in vitro and in vivo. *Br. J. Pharmacol.* **86,** 209–218.

Brazell M. P., Kasser A. J., Renner K. J., Feng J., Moghaddam B., and Adams R. N. (1987) Electrocoating carbon fibre microelectrodes with Nafion improves selectivity for electroactive neurotransmitters. *J. Neurosci. Methods* **22,** 167–172.

Bunney B.S. and Grace A. A. (1978) Acute and chronic haloperidol treatment: comparison of effects on nigral dopaminergic cell activity. *Life Sci.* **23,** 1715–1728.

Carroll B. J., Curtis G. C., and Mendels J. (1976) Neuroendocrine regulation in depression, II. Discrimination of depressed from non-depressed patients. *Arch. Gen. Psychiatry* **33,** 1951–1958.

Cespuglio R., Faradji H., Hahn Z., and Jouvet M. (1984) Voltammetric detection of brain 5-hydroxyindoleamines by means of electrochemically treated carbon fibre electrodes: chronic recordings for up to one month with movable cerebral electrodes in the sleeping or waking rat, in *Measurement of Neurotransmitter Release In Vivo* (Marsden C. A., ed.), Wiley, Chichester, pp. 173–191.

Cespuglio R., Faradji H., Ponchon J. L., Riou F., Buda M., Gonon F., Pujol J. F., and Jouvet M. (1981a) Differential pulse voltammetry in brain tissue: I: Detection of 5-hydroxyindoles in the rat striatum. *Brain Res.* **223,** 187–298.

Cespuglio R., Faradji H., Riou F., Buda M., Gonon F., Pujol J. F., and Jouvet M. (1981b) Differential pulse voltammetry in brain tissue: II: Detection of 5-hydroxyindoleacetic acid in the rat striatum. *Brain Res.* **223,** 299–311.

Chiodo L. A. and Bunney B. S. (1983) Typical and atypical neuroleptics: differential effects of chronic administration on the activity of A9 and A10 midbrain dopaminergic neurones. *J. Neurosci.* **3,** 1607–1619.

Church W. H. and Justice J. B. (1987) Rapid sampling of extracellular dopamine In Vivo. *Anal. Chem.* **59,** 712–716.

Clow A., Theodorou A., Jenner P., and Marsden C. D. (1980) Changes in rat striatal dopamine turnover and receptor activity during one year's neuroleptic administration. *Eur. J. Pharmacol.* **63,** 135–144.

Conti J., Strope E., Adams R. N., and Marsden C. A. (1978) Voltammetry in brain tissue: chronic recording of stimulated dopamine and 5-hydroxytryptamine release. *Life Sci.* **23,** 2705–2716.

Crespi F. and Jouvet M. (1984) Differential pulse voltammetric determination of 5-hydroxyindoles in four raphe nuclei of chronic freely moving rats simultaneously recorded by polygraphic technique. Physiological changes with vigilance states. *Brain Res.* **299,** 113–119.

Crespi F., Martin K. F., and Marsden C. A. (1988) Measurement of extracellular basal levels of 5HT in vivo using Nafion coated carbon fibre electrodes combined with differential pulse voltammetry. *Neuroscience,* **27**, 885–896

Crespi F., Sharp T., Maidment N. T., and Marsden C. A. (1983) Differential pulse voltammetry in vivo: evidence that uric acid contributes to the indole oxidation peak. *Neurosci. Lett.* **43,** 203–207.

Crespi F., Sharp T., Maidment N. T., and Marsden C. A. (1984a) Differential pulse voltammetry: simultaneous in vivo measurement of ascorbic acid, catechols and 5-hydroxyindoles in the rat striatum. *Brain Res.* **322,** 135–138.

Crespi F., Paret J., Keane P. E., and Moore M. (1984b). An improved differential pulse voltammetry technique allows the simultaneous analysis of dopaminergic and serotonergic activities in vivo with a single carbon fibre electrode. *Neurosci. Lett.* **52,** 159–64.

Crow T. J., Johnstone E. C., and McClelland H. A. (1976) The coincidence of schizophrenia and Parkinsonism: some neurochemical implications. *Psychol. Med.* **6,** 227–233.

Dayton M. A., Ewing A. G., and Wightman R. M. (1980) Response of microvoltammetric electrodes to homogeneous catalytic and slow heterogeneous charge-transfer reactions. *Anal. Chem.* **52,** 2392–2396.

Dayton M. A., Ewing A. G., and Wightman R. M. (1981) Evaluation of amphetamine-induced in vivo electrochemical response. *Eur. J. Pharmacol.* **75,** 141–144.

Dayton M. A., Ewing A. G., and Wightman R. M. (1983) Diffusion processes measured at microvoltammetric electrodes in brain tissue. *J. Electroanal. Chem.* **146,** 189–200.

De Montigny C., Blier P., and Chaput Y. (1984) Electrophysiologically-identified serotonin autoreceptors in the rat CNS. *Neuropharmacology* **23,** 1511–1520.

Dompert W. U., Glaser T., and Traber J. (1985) [^{3}H]-TVX Q 7821: identification of 5-HT_1 binding sites as a target for a novel putative anxiolytic. *Naunyn Schmiedebergs Arch. Pharmacol.* **328,** 467–470.

Doods H. N., Kalkman H. O., Dejonge A., Thooler M. J. M. C., Wilffert B., Timmermans P. B. M. W. M., and Vanzwieten P. (1985) Differential selectivities of RU24969 and 8-OH-DPAT for the purported 5-HT_{1A} and 5-HT_{1B} binding sites, correlation between 5-HT_{1A} affinity and hypotensive activity. *Eur. J. Pharmacol.* **112,** 363–370.

Doxey J C., Roach A. G., and Smith C. F. C. (1983) Studies on RX 781094: a selective, potent and specific antagonist of α_2-adrenoceptors. *Br. J. Pharmacol.* **78,** 489–505.

Echizen H. and Freed C. R. (1983) *In vivo* electrochemical detection of extraneuronal 5-hydroxyindole acetic acid and norepinephrine in the dorsal raphe nucleus of urethane-anaesthetised rats. *Brain Res.* **277,** 55–62.

Echizen H. and Freed, C. R. (1984) Measurement of serotonin turnover rate in rat dorsal raphe nucleus by *in vivo* electrochemistry. *J. Neurochem.* **42,** 1483–1486.

Engel G., Gothert M., Muller-Schweinitzer E., Schlicker E., Sistonen L., and Stadler P. A. (1983) Evidence for common pharmacological properties of [^{3}H]-5-hydroxytryptamine autoreceptors in *CNS* and inhibitory presynaptic 5-hydroxytryptamine receptors on sympathetic nerves. *Naunyn Schmiedebergs Arch. Pharmacol.* **324,** 116–124.

Ennis C., (1986) The 5-HT autoreceptor in the rat hypothalamus resembles a $5HT_{1B}$ receptor. *Br. J. Pharmacol.* **88,** 370P.

Euvard C. and Boissier J. R. (1980) Biochemical assessment of the central 5-HT agonist activity of RU-24969 (a piperidinyl indole). *Eur. J. Pharmacol.* **63,** 65–72.

Ewing A. G. and Wightman R. M. (1984) Monitoring the stimulated release of dopamine with *in vivo* voltammetry. II: Clearance of released dopamine from extracellular fluid. *J. Neurochem.* **43,** 570–577.

Ewing A. G., Bigelow, J. C., and Wightman R. M. (1983) Direct *in vivo* monitoring of dopamine released from two striatal compartments in the rat. *Science* **221,** 169–171.

Ewing A. G., Wightman R. M., and Dayton M. A. (1982) *In vivo* voltammetry with electrodes that discriminate between dopamine and ascorbate. *Brain Res.* **249,** 361–370.

Falat L. and Cheng H.-Y. (1982) Voltammetric differentiation of ascorbic acid and dopamine at an electrochemically treated graphite-epoxy electrode. *Anal. Chem.* **54,** 2108–2111.

Faradji H., Cespuglio A., and Jouvet M. (1983) Voltammetric measurements of 5-hydroxyindole compounds in the suprachiasmatic nucleus: circadian fluctuations. *Brain Res.* **279,** 111–119.

Ford A. P. D. W. and Marsden C. A. (1986) Influence of anaesthetics on rat striatal dopamine metabolism *in vivo. Brain Res.* **379,** 162–166.

Ford A. P. D. W. and Marsden C. A. (1987) Selective mesolimbic neurochemical responsiveness to neurotensin microinjection into rat mesencephalic dopamine cell bodies. *Br. J. Pharmacol.* **90,** 237P.

Fozard J. R. and McDermott I. (1985) The cardiovascular response to 8-hydroxy-2-(di-*n*-propylamino) tetralin (8-OH-DPAT) in the rat. *Br. J. Pharmacol.* **84,** 69P.

Fuxe R., Agnati L., and Everit B. (1975) Effects of metergoline on central monoamine neurons. Evidence for a selective blockade of central 5-HT receptors. *Neurosci. Lett.* **1,** 283–290.

Gazzara R. A. and Howard-Butcher S. (1986) A developmental study of amphetamine-induced dopamine release in rats using *in vivo* voltammetry. *Ann. NY Acad. Sci.* **473,** 527–529.

Gerhardt G. A., Oke A.F., Nagy G., and Adams R. N. (1984) Nafion-coated electrodes with high selectivity for CNS electrochemistry. *Brain Res.* **290,** 390–395.

Gonon F. G. and Buda M. J. (1985) Regulation of dopamine release by impulse flow and by autoreceptors as studied by *in vivo* voltammetry in the rat striatum. *Neuroscience* **14,** 765–774.

Gonon F., Navarre F., and Buda M. (1984a) *In vivo* monitoring of dopamine release in the rat brain with differential normal pulse voltammetry. *Anal. Chem.* **56,** 573–575.

Gonon F., Buda M., and Pujol J. F. (1984b) Treated carbon fibre electrodes for measuring catechols and ascorbic acid, in *Measuring Neurotransmitter Release In Vivo* (Marsden C. A., ed.), Wiley, Chichester, pp. 153–171.

Gonon F., Buda M., Cespuglio R., Jouvet M, and Pujol J. F. (1980) *In vivo* electrochemical detection of catechols in the neostriatum of anaesthetised rats: dopamine or DOPAC? *Nature* **286,** 902–904.

Gonon F., Fombarlet, C. M., Buda M. J., and Pujol J. F. (1981) Electrochemical treatment of pyrolytic fibre electrodes. *Anal. Chem.* **53,** 1386–1389.

Gothert M. (1980) Serotonin receptor-mediated modulation of Ca^{2+}-dependent 5-hydroxytryptamine release from neurons of the rat brain cortex. *Naunyn-Schmiedebergs Arch. Pharmacol.* **314,** 223–228.

Groves P. M., Wilson C. J., Young S. J., and Rebec G. V. (1975) Self-inhibition by dopaminergic neurones. An alternative to the "neuronal feedback loop" hypothesis for the mode of action of certain psychotropic drugs. *Science* **190,** 522–529.

Hjorth S., Carlsson A., Lindberg P., Sanchez D., Wikstrom H., Arvidssen L.-E., Hacksell U., and Nilsson J. L. G. (1982) 8-Hydroxy-2(di-*n*-propylamino) tetralin, (8-OH-DPAT) a potent and selective simplified ergot congener with central 5HT-receptor stimulating activity. *J. Neural. Transm.* **55,** 169–188.

Jolicoeur F. B., Rioux F., and St.-Pierre S., (1985) Neurotensin, in *Handbook of Neurochemistry*, vol. 8 (Lajtha A., ed.), Plenum, New York, pp. 93–114.

Joseph M. H. and Kennett G. A. (1981) *In vivo* voltammetry in the rat hippocampus as an index of drug effects on extraneuronal 5HT. *Neuropharmacology* **20,** 1361–1364.

Joseph M. H. and Kennett G. A. (1983) Stress-induced release of 5HT in the hippocampus and its dependence on increased tryptophan availability: an *in vivo* electrochemical study. *Brain Res.* **270,** 251–257.

Kalivas P. W., Nemeroff C. B., and Prange A. J. (1984) Neurotensin microinjection into the nucleus accumbens antagonises dopamine-induced increase in locomotion and rearing. *Neuroscience* **11,** 919–930.

Kennett G. A. and Joseph M. H. (1982) Does *in vivo* voltammetry in the hippocampus measure 5HT release? *Brain Res.* **236,** 305–316.

Kissinger P. T., Refshauge C. J., Dreiling C. J., and Adams R. N. (1973a) An electrochemical detector for liquid chromatography with picogram sensitivity. *Anal. Lett.* **6,** 465–477.

Kissinger P. T., Hart J. B., and Adams R. N. (1973b) Voltammetry in brain tissue: a new neuropharmacological measurement. *Brain Res.* **55,** 209–213.

Kruk Z. L., Armstrong-James M., and Millar J. (1980) Measurements of the concentration of 5-hydroxytryptamine ejected during iontophoresis using multibarrel carbon fibre microelectrodes. *Life Sci.* **27,** 2093–2098.

Kuhr W. G. and Wightman R. M. (1986) Real-time measurement of dopamine release in rat brain. *Brain Res.* **381,** 168–71.

Kuhr W. G., Ewing, A. G., Caudill W. L., and Wightman R. M. (1984) Monitoring the stimulated release of dopamine with *in vivo* voltammetry. I: Characterisation of the response observed in the caudate nucleus of the rat. *J. Neurochem.* **43,** 560–569.

Lane R. F. and Blaha C. D. (1986) Electrochemistry *In Vivo:* Application to CNS pharmacology. *Ann. NY Acad. Sci.* **473,** 50–69.

Lane R. F. and Hubbard A. T. (1976) Differential double pulse voltammetry at chemically modified platinum electrodes for *in vivo* determination catecholamines. *Anal. Chem.* **48,** 1287–1293.

Lane R. F., Hubbard A. T., Fukunaga K., and Blanchard R. J. (1976) Brain catecholamines: detection *in vivo* by means of differential pulse voltammetry at surface modified platinum electrodes. *Brain Res.* **114,** 346–352.

Lane R. F., Hubbard A. T., and Blaha C. D. (1979) Application of semidifferential electroanalysis to studies of neurotransmitters in the central nervous system. *J. Electroanal. Chem.* **95,** 117–122.

Louilot A., Buda M., Gonon F., Simon H., LeMoal M., and Pujol J. F. (1985) Effect of haloperidol and sulpiride on dopamine metabolism in nucleus accumbens and olfactory tubercle: a study by *in vivo* voltammetry. *Neuroscience* **14,** 775–782.

McCarthy P. S., Walker R. J., Tajima H., Kitagawa, K., and Woodruff, G. N. (1979) The action of Neurotensin on neurones in the nucleus accumbens and cerebellum of the rat. *Gen. Pharmacol.* **10,** 331–333.

McCreery R. L., Dreiling R., and Adams R. N. (1974a) Voltammetry in brain tissue: the fate of injected 6-hydroxydopamine. *Brain Res.* **73,** 15–21.

McCreery R. L., Dreiling R., and Adams R. N. (1974b) Voltammetry in brain tissue: quantitative studies of drug interactions. *Brain Res.* **73,** 23–33.

Maidment N. T. and Marsden C. A. (1985) *In vivo* voltammetric and behavioural evidence for somatodendritic autoreceptor control of mesolimbic dopamine neurones. *Brain Res.* **338,** 312–325.

Maidment N. T. and Marsden C. A. (1987a) Acute administration of dozapine, thioridazine and metaclopramide increases extracellular DOPAC and decreases extracellular 5HIAA, measured in the nucleus accumbens and striatum of the rat using *in vivo* voltammetry. *Neuropharmacology* **26,** 187–193.

Maidment N. T. and Marsden C. A. (1987b) Repeated atypical neuroleptic administration: effects on central dopamine metabolism monitored by *in vivo* voltammetry. *Eur. J. Pharmacol.* **136,** 141–149.

Marcenac F. and Gonon F. G. (1985) Fast *in vivo* monitoring of dopamine release in the rat brain with differential pulse amperometry. *Anal. Chem.* **57,** 1778–1779.

Marsden C. A. (1985) *In vivo* monitoring of pharmacological and physiological changes in endogenous serotonin release and metabolism. In Neuropharmacology of serotonin (Green A. R., ed.), Oxford University Press, Oxford, pp. 218–252.

Marsden C. A. and Martin K. F. (1985a) RU 24969 decreases 5HT release in the SCN by acting on 5HT receptors in the SCN but not the dorsal raphe. *Br. J. Pharmacol.* **85,** 219P.

Marsden C. A. and Martin K. F. (1985b) *In vivo* voltammetric evidence that the 5HT autoreceptor is not of the $5HT_{1A}$ sub-type. *Br. J. Pharmacol.* **86,** 445P.

Marsden C. A. and Martin K. F. (1986) Involvement of $5HT_{1A}$ and α_2 receptors in the decreased 5-hydroxytryptamine release and metabolism in rat suprachiasmatic nucleus after intravenous 8-hydroxy-2-(*n*-dipropylamino) tetralin. *Br. J. Pharmacol.* **89,** 277–286.

Marsden C. A., Conti J., Strope E., Curzon G., and Adams R. N. (1979) Monitoring 5-hydroxytryptamine release in the brain of the freely moving unanaesthetised rat using *in vivo* voltammetry. *Brain Res.* **171,** 85–99.

Marsden C. A., Martin K. F., Routledge C., Brazell M. P., and Maidment N. T. (1986) Application of intracerebral dialysis and *in vivo* voltammetry to pharmacological and physiological studies of amine neurotransmitters. *Ann. NY Acad. Sci.* **473,** 106–125.

Marsden C. A., Joseph M. H., Kruk Z. L., Maidment N. T., O'Neill R. D., Schenk J. O. and Stamford J. A. (1988) *In Vivo* voltammetry—present electrodes and methods. *Neuroscience* **25,** 389–400.

Marsden C. D., Tarsy D., and Baldessarini R. J. (1975) Spontaneous and drug-induced movement disorders in psychotic patients, in *Psychiatric Aspects of Neurological Disease* (Benson D. F. and Blumer, D., eds.), Grune & Stratton, New York, p. 219.

Martin K. F. and Marsden C. A. (1986a) *In vivo* voltammetry in the suprachiasmatic nucleus of the rat: Effects of RU 24969, methiothepin and ketanserin. *Eur. J. Pharmacol.* **121,** 135–140.

Martin K. F. and Marsden C. A. (1986b) Pharmacological manipulation of the serotonergic input to the SCN—an insight into the control of circadian rhythms. *Ann. NY Acad. Sci.* **473,** 542–545.

Martin K. F., Mason R., and McDougall H. (1986) Electrophysiological evidence that isapirone (TVX Q 7821) is a partial agonist at $5HT_{1A}$ receptors in the rat hippocampus. *Br. J. Pharmacol.* **89,** 729P.

Middlemiss D. N. (1984) 8-Hydroxy-2-(di-*n*-propylamino) tetraline is devoid of activity at the 5-hydroxytryptamine autoreceptor in rat brain. Implications for the proposed link between the autoreceptor and the [^{3}H]5-HT recognition site. *Naunyn-Schmiedebergs Arch. Pharmacol.* **327,** 18–22.

Middlemiss D. N. (1985) The putative 5HT*1* receptor agonist, RU 24969, inhibits the efflux of 5-hydroxytryptamine from rat frontal cortex slices by stimulation of the 5HT autoreceptor. *J. Pharm. Pharmacol.* **37,** 434–437.

Middlemiss D. N. and Fozard J. R. (1983) 8-Hydroxy-2-(di-*n*-propylamino) tetralin discriminates between subtypes of the 5HT recognition site. *Eur. J. Pharmacol.* **90,** 151–153.

Millar J., Armstrong-James M., and Kruk Z. L. (1981) Polarographic assay of iontophoretically applied dopamine and low noise unit recording using a multibarrel carbon fibre microelectrode. *Brain Res.* **205,** 419–424.

Millar J., Stamford J. A., Kruk Z. L., and Wightman R. M. (1985) Electrochemical, pharmacological, and electrophysiological evidence of rapid dopamine release and removal in the rat caudate nucleus following electrical stimulation of the median forebrain bundle. *Eur. J. Pharmacol.* **109,** 341–348.

Moore R.Y. (1983) Organisation and function of a central nervous system circadian oscillator: the suprachiasmatic hypothalamic nuclei. *Fed. Proc.* **42,** 2783–2789.

Moret C. (1985) Pharmacology of the serotonin autoreceptor, in *Neuropharmacology of Serotonin* (Green A. R., ed.), Oxford University Press, Oxford, pp. 21–49.

Nagy G., Rice M. E., and Adams R. N. (1982) A new type of enzyme electrode: the ascorbic acid eliminator electrode. *Life Sci.* **31,** 2611–2616.

Nagy G., Moghaddam B., Oke A., and Adams R. N. (1985) Simultaneous monitoring of voltammetric and ion-selective electrodes in mammalian brain. *Neurosci. Lett.* **55,** 119–124.

Nemeroff C. B. and Cain S. T. (1985) Neurotensin-dopamine interactions in the CNS. *Trends Pharmacol. Sci.* **6,** 201–205.

Nieoullon A., Torni C., and Ganouni S. (1986) Contribution to the study of nigrostriatal dopaminergic neuron activity using electrochemical detection of dopamine release in the striatum of freely moving animals. *Ann. NY Acad. Sci.* **473,** 126–139.

O'Neill R. D. and Fillenz M. (1985) Detection of homovanillic acid *in vivo* using microcomputer-controlled voltammetry: simultaneous monitoring of rat motor activity and striatal dopamine release. *Neuroscience* **14,** 753–763.

O'Neill R. D., Fillenz M., Albery W. J., and Goddard N. J. (1983) The monitoring of ascorbate and monoamine transmitter metabolites in the striatum of unanaesthetised rats using microprocessor-based voltammetry. *Neuroscience* **9,** 87–94.

O'Neill R. D., Grunewald R. A., Fillenz M. and Albery W. J. (1982) Linear sweep voltammetry with carbon paste electrodes in the rat striatum. *Neuroscience* **7,** 1945–1954.

O'Neill R. D., Fillenz M., Grunewald R. A., Bloomfield M. R., Albery N. J., Jamieson C. M., Williams J. H., and Gray J. A. (1984) Voltammetric carbon paste electrodes monitor uric acid and not 5HIAA at the 5-hydroxyindole potential in the rat brain. *Neurosci. Lett.* **45,** 39–46.

Pedigo N. W., Yamamura H. J. and Nelson D. L. (1981) Discrimination multiple [^{3}H]-5 hydroxytryptamine binding by neuroleptic spiperone in rat brain. *J. Neurochem.* **36,** 220–226.

Peroutka S. J. and Snyder S. H. (1979) Multiple serotonin receptors: Differential binding of [^{3}H]-5 hydroxytryptamine, [^{3}H]-lysergic acid diethylamide and [H]-spiroperidol. *Mol. Pharmacol.* **16,** 687–699.

Ponchon J. L., Cespuglio R., Gonon F., Jouvet M., and Pujol J. F. (1979) Normal pulser polarography with carbon fibre electrodes for *in vitro* and *in vivo* determination of catecholamines. *Anal. Chem.* **51,** 1483–1486.

Reynolds L. S., Seymour P. A. and Heym J. (1986) Inhibition of the behavioural effects of 8-OH-DPAT by the novel anxiolytics buspirone, gepirone and isapirone. *Soc. Neurosci. Abstr.* **12,** 481.

Saller C. F. and Salama A. I. (1985) Alterations in dopamine metabolism after chronic administration of haloperidol. Possible role of increased autoreceptor sensitivity. *Neuropharmacology* **24,** 123–129.

Scatton B. (1977) Differential regional development of tolerance to increase in dopamine turnover upon repeated neuroleptic administration. *Eur. J. Pharmacol.* **46,** 363–369.

Schenk J. O. and Adams R. N. (1984) Chronoamperometric measurements in the central nervous system, in *Measurement of Neurotransmitter Release IN VIVO* (Marsden C. A., ed.), Wiley, Chichester, pp. 193–208.

Schenk J. O., Miller E., Rice M. E., and Adams R. N. (1983) Chronoamperometry in brain slices: quantitative evaluation of *in vivo* electrochemistry. *Brain Res.* **277,** 1–8.

Schnellman R. G., Waters S. J., and Nelson D. L. (1984) [^{3}H]5-Hydroxytryptamine binding sites: species and tissue variation. *J. Neurochem.* **42,** 65–70.

Schuumann T., Davies M. A., Dompert W. U., and Traber J. (1984) TVX Q 7821: a new nonbenzodiazepine putative anxiolytic. *Proc. 9th Int. Congr. Pharmacol.* London, 1463P.

Sharp T. Maidment N. T., Brazell M. P., Zetterstrom T., Ungerstedt U., Bennett G. W., and Marsden C. A. (1984) Changes in monoamine metabolites measured by simultaneous *in vivo* differential pulse voltammetry and intracerebral dialysis. *Neuroscience* **12,** 1213–1221.

Sills M. A., Wolfe B. B., and Frazer A. (1984) Determination of selective and nonselective compounds for the $5HT_{1A}$ and $5HT_{1B}$ receptor subtypes in rat frontal cortex. *J. Pharmacol. Exp. Ther.* **231,** 480–487.

Skirboll L. R., Grace A. A., and Bunney B. S. (1979) Dopamine auto- and postsynaptic receptors: electrophysiological evidence for differential sensitivity to dopamine agonists. *Science* **206,** 80–82.

Stamford J. A., Kruk Z. L. and Millar J. (1986) Sub-second striatal dopamine release measured by *in vivo* voltammetry. *Brain Res.* **381,** 351–355.

Stamford J. A., Kruk Z. L., Millar J., and Wightman R. M. (1984) Striatal dopamine uptake in the rat: *in vivo* analysis by fast cyclic voltammetry. *Neurosci. Lett.* **51,** 133–138.

Ungerstedt U. (1984) Measurement of neurotransmitter release by intracranial dialysis, in *Measurement of Neurotransmitter Release In Vivo* (Marsden C. A., ed.), Wiley, Chichester, pp. 81–105.

Wehr T. A., Sack D., Rosenthal N., Duncan W., and Gillin J. C. (1983) Circadian rhythm disturbances in manic-depressive illness. *Fed. Proc.* **42,** 2809–2814.

White F. J. and Wang R. Y. (1983) Differential effects of classical and atypical antipsychotic drugs on A9 and A10 dopamine neurones. *Science* **221,** 1054–1057.

Widerlov E. and Breese G. K. (1982) *Neurotensin, a Brain and Gastrointestinal Peptide* (Nemeroff C. B. and Prange A. J., eds.), New York Academy of Sciences, New York.

Wightman R. M. (1981) Microvoltammetric electrodes. *Anal. Chem.* **53,** 1125–1130A.

Wightman R. M., Strope E., Plotsky P. M., and Adams R. N. (1976) Monitoring of transmitter metabolites by voltammetry in cerebrospinal fluid following neuronal pathway stimulation. *Nature* **262,** 145–146.

Wirz-Justice A., Groos, G. A., and Wehr T. A. (1982) The neuropharmacology of circadian timekeeping in mammals, in *Vertebrate Circadian Systems, Structure and Physiology,* (Aschoff J., Daan S., and Groos G. A., eds.), Springer-Verlag, Berlin, pp. 183–193.

Zetterstrom T., Sharp T., Marsden C. A., and Ungerstedt U. (1983) *In Vivo* measurement of dopamine and its metabolites by intracerebral dialysis: changes after D-amphetamine. *J. Neurochem.* **41,** 1769–1773.

Zetterstrom T., Sharp T., and Ungerstedt U. (1986) Effect of dopamine D-1 and D-2 receptor selective drugs on dopamine release and metabolism in rat striatum *in vivo. Naunyn Schmiedebergs Arch. Pharmacol.* **334,** 117–124.

From: ***Neuromethods, Vol. 14: Neurophysiological Techniques: Basic Methods and Concepts*** **Edited by: A. A. Boulton, G. B. Baker, and C. H. Vanderwolf Copyright © 1990 The Humana Press Inc., Clifton, NJ**

Index